Principles of Cell Biology

Principles of Cell Biology

Oliver Taylor

STATES
ACADEMIC PRESS
www.statesacademicpress.com

Published by States Academic Press,
109 South 5th Street,
Brooklyn, NY 11249, USA
www.statesacademicpress.com

Principles of Cell Biology
Oliver Taylor

International Standard Book Number: 978-1-63989-435-2 (Hardback)

Cataloging-in-Publication Data

Principles of cell biology / Oliver Taylor.
p. cm.
Includes bibliographical references and index.
ISBN 978-1-63989-435-2
1. Cytology. 2. Cells. 3. Biology. I. Taylor, Oliver.
QH581.2 .P75 2022
571.6--dc23

TABLE OF CONTENTS

Preface VII

Chapter 1 **Cell Biology: An Introduction** 1
- Cell Diversity 2
- Cell Metabolism 4

Chapter 2 **The Basic Structure of a Cell** 9
- Cell Architecture 9
- Cell Organelles 29
- Cell Membrane 35
- Cell Wall 105

Chapter 3 **Primary Components of a Cell** 108
- Cytoplasm 108
- Structure and Function of Cytoplasm 116
- Cell Nucleus 117
- Properties of Cytoplasmic Matrix 137
- Mitochondrion 142

Chapter 4 **Concepts of Cell Biology** 163
- Cell Disruption 163
- Cell Theory 166
- Cell Culture 174
- Autophagy 190
- Mitophagy 196

Chapter 5 **Cell Cycle** 201
- Mitosis 210
- Cell-cycle Analysis 220
- Cell Division 223
- Cellular Differentiation 225

Permissions

Index

PREFACE

Cell biology is a sub-discipline of biology which attempts to study the structure and functions of cells. They are classified into prokaryotic and eukaryotic cells. Prokaryotic cells include bacteria and archaea. These organisms lack an enclosed cell nucleus and reproduce using binary fission. Eukaryotic cells can be unicellular or multicellular in nature. They include fungi, plant, protozoa and animals. The discipline of cell biology can be further divided into cell communication, cell metabolism, cell cycle, cell composition and biochemistry. Cell culture, cell fractionation and microscopy are some of the prominent techniques used for studying the cells. Cell biology often overlaps with various other fields such as molecular biology, genetics and medical microbiology. The book aims to shed light on some of the unexplored aspects of cell biology. It elucidates the concepts and innovative models around prospective developments with respect to cell biology. This book will serve as a valuable source of reference for those interested in this field.

A detailed account of the significant topics covered in this book is provided below:

Chapter 1- Cell Biology is the study of cells, which are the basic units of life. The study of their structure and functions is important to understand the various tissues and organs that are composed of cells, which in turn are vital to understand life. This chapter serves as an introduction to cell biology.

Chapter 2- Cells are the smallest units of life and are made up of various parts. A cell consists of three primary parts: a membrane, a cytoplasm and a nucleus. Biomolecules and small fiber-like structures called organelles are also present within the cytoplasm. This chapter discusses in detail the types of cells, their architecture, structure, membrane and their composition.

Chapter 3- A cell is mainly composed of a membrane, cytoplasm, nucleus and the mitochondrion suspended in the cytoplasm. The cytoplasm contains all the organelles of the cell and forms the non-nuclear content. Contained within the cytoplasm is the nucleus, separated from the cytoplasm by an enclosing membrane. This chapter covers these topics in a clear and concise manner.

Chapter 4- The pursuit of cell biology would be impossible without its various concepts and theories. Cell Theory is one such concept which states that all living organisms are made up of cells. Cell disruption and cell culture are two crucial developments in cell biology made possible by the scientific research based on this theory. These topics along with autophagy and mitophagy are discussed in this chapter.

Chapter 5- The events through which a cell grows and divides into two new cells is known as the cell cycle. It is a series of ordered steps during which the cell grows, replicates its DNA and finally separates to form two new cells through division. The various phases of the cell cycle are explained in a detailed manner in this chapter.

It gives me an immense pleasure to thank our entire team for their efforts. Finally in the end, I would like to thank my family and colleagues who have been a great source of inspiration and support.

Oliver Taylor

Chapter 1

Cell Biology: An Introduction

Cell Biology is the study of cells, which are the basic units of life. The study of their structure and functions is important to understand the various tissues and organs that are composed of cells, which in turn are vital to understand life. This chapter serves as an introduction to cell biology.

Cell biology is the study of cells and how they function, from the subcellular processes which keep them functioning, to the way that cells interact with other cells. Whilst molecular biology concentrates largely on the molecules of life (largely the nucleic acids and proteins), cell biology concerns itself with how these molecules are used by the cell to survive, reproduce and carry out normal cell functions.

In biomedical research, cell biology is used to find out more about how cells normally work, and how disturbances in this normal function can result in disease. An understanding of these processes can lead to therapies which work by targeting the abnormal function.

Common Cell Biology Techniques

The following list covers some of the more commonly used cell biology techniques – it is by no means exhaustive:

- Cell/Tissue Culture: In the same way that bacteria and other simple organisms can be grown in the laboratory outside their normal environment, cells and tissues from more complicated organisms can be cultured as well. The techniques are slightly different, and the culture media are more complex to reflect the complex internal environment inside the host from which the cells are derived, however cell and tissue culture is a powerful tool which provides an almost limitless supply of test material for researchers to use without resorting to using whole organisms. In addition, the controlled condition in cell and tissue culture allows researchers to carry out experiments with a lower number of variables which may affect the outcome of the test. Cell culture may use cells removed directly from an organism (primary culture), or it may use lines of cultured cancer cells. The benefit of the latter approach is that cancer cells continue to divide, while primary cultures cease dividing after a number of cycles.

- Microscopy: The basic tool of cell biology is microscopy. A recent advance in imaging technology has allowed an unprecedented amount of information to be gleaned from microscopic analysis. Types of microscopic techniques which are used include:

    - Brightfield: Traditional microscopy, where cells are illuminated by visible light. Brightfield microscopy gives a general picture of cell function, although that information is not very detailed or specific. As animal cells lack cell walls, brightfield microscopy may use special techniques such as phase contrast to show cellular structures in more detail. Brightfield microscopy allows imaging of live or fixed (dead) cells and tissues).

  - Electron Microscopy: Uses a focused beam of electrons instead of light. Electron microscopy permits a much higher magnification of specimens than light microscopy and is useful in obtaining detailed information about sub-cellular structures. Electron microscopy requires extensive processing and so can only be performed on fixed specimens. Transmission electron microscopy provides a cross section of a specimen, while scanning electron microscopy gives a three-dimensional image of the surface of a specimen.

  - Fluorescence Microscopy: Uses fluorescent materials to indicate structures in a specimen. Fluorescence occurs when light of one wavelength "excites" a material and causes it to emit light of a different wavelength. Most fluorescent materials give off visible light after excitation by ultraviolet light. Structures may be naturally fluorescent (autofluorescence) or they may be labeled with a compound which is fluorescent (e.g. DAPI is a dye which binds onto DNA. The DNA and nuclei of cells stained with DAPI emit a blue light under ultraviolet light).

  - Immunofluorescence: Antibodies are proteins made by the immune system which bind onto specific parts of proteins. Antibodies can be raised against any protein in the cell. If these antibodies are attached to a fluorescent tag, the tag will only show up where that antibody attached (ie. where the target protein is found in the cell). Immunofluorescence allows very specific targeting of cellular structures.

  - RNA Interference: RNA interference uses short sequences of RNA which are complementary to the mRNA which carries to instructions to translate proteins from the DNA to the ribosomes. The interfering RNA binds to the target sequence, preventing it from being translated. As a result, careful selection of interfering RNA can be used to silence a particular gene. This allows researchers to study what role a protein plays in a cell, by observing what happens when that protein is absent.

- Timelapse Microscopy: Many cellular processes (eg. mitosis) occur over a period of time which is not practical for direct observation. Imaging cells over a period of time (eg. a photograph is taken every 20 minutes for 24 hours) allows us to combine these images in a "movie" which compresses a long time period into a shorter one.

Cell Diversity

Cells show great diversity in form and functions. Because of this, it was not easy to realize that all living organisms are made up of units sharing a common basic structure. Every unit is a cell. The other major issue for the discovering of the cell was the very small size they usually show.

Cell Size

Cell size is measured in micrometers (µm). One micrometer, or micron, is one thousandth of a millimeter (10^{-3} millimeters), and one millionth of a meter (10^{-6} meters). A typical eukaryote cell is between 10 and 30 µm in size. This is true for the cells of a worm and for those of an elephant, but there are many more cells in the elephant. To be aware of how small the cells are, imagine a 1.70 meters tall person which is stretched to match the height of the Everest, which is about 8500 meters. The stretched giant cells of that person would measure only 1.3 centimeters, i.e., smaller than an euro cent coin (then, it would be a giant formed by a huge amount of euro cent coins).

However, there are eukaryote cells having uncommon dimensions. They can be very small, like sperm cells, whose head may be smaller than 4 µm in diameter, while others like the eggs of some birds and reptiles can be larger than 10 centimeters (thousands of microns) in their larger axis, but we should measure only the yolk, since the egg white is not part of the cell. An extreme example is the egg of ostriches. Some cells may have cytoplasmic extensions as large as several meters in length, such as the brain neurons of giraffes that innervate the more caudal part of the spinal cord. Smaller than eukaryote cells are prokaryote cells, which typically are around 1 to 2 µm in diameter, being Mycoplasma the smallest with 0.5 µm.

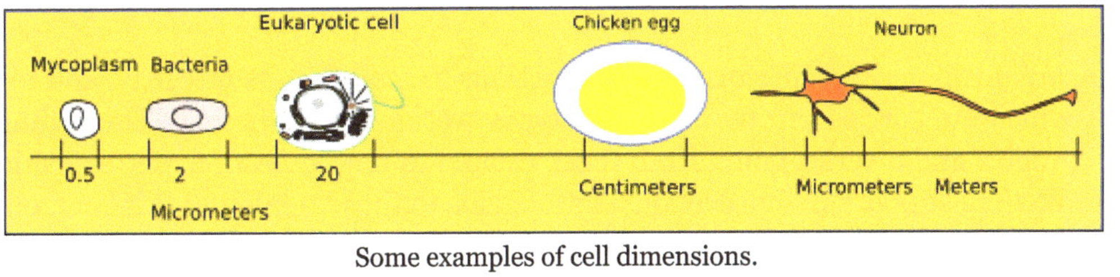

Some examples of cell dimensions.

Number

Most living organisms are unicellular, i.e., a single cell. Prokaryotes (bacteria and archaea) are the most abundant unicellular organisms. Unicellular eukaryote species are abundant too. Organisms that can be observed without microscopes are mostly multicellular, i.e., they are made up of many cells. Multicellular organisms are animals, plants, fungi and some algae. In general, larger multicellular organisms contain higher number of cells since they have a similar average cell size. Estimates of the total number of cells of an organism with similar size to humans may range from 10^{13} (1 followed by 13 zeros) to 10^{14} (1 followed by 14 zeros). To be aware of these numbers, it is estimated that the total number of cells in the human brain is about 86×10^9 neurons and of a mouse brain is about 15×10^9. The most abundant cells of the human body are red blood cells and glial/neuron cells of the nervous system.

Morphology

Cell morphology is typically sketched as rounded, but this is probably the most uncommon shape (except for a few types of cells). Cell morphology in animal tissues is diverse, enormously diverse! It can range from rounded to star-like, from multi-lobed to filiform. Plant cells also show a wide diversity of forms, which is determined by the cell wall, being cuboidal and columnar shapes the most common shapes.

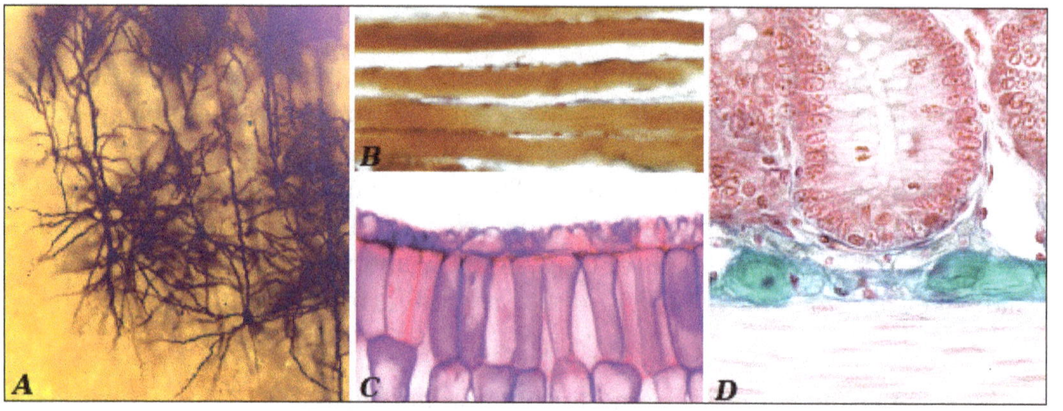

Figure shows cell shapes. A) Neurons of the cerebral cortex. B) Skeletal muscle cells in longitudinal view. C) Cells of a leaf. Different morphologies can be observed in the parenchyma, lower part, with large and elongated cells, and in the epidermis at the upper part, with small and irregular cells located. D) Different cell types of small intestine. The purple upper cells are epithelial cells, the pale elongated cells at the bottom are smooth muscle cells, and the greenish are connective tissue cells.

Function

Every living organism needs to perform many functions to maintain its integrity, and to grow and proliferate, which are carried out by many cell types working coordinately. These functions are extremely complex and diverse, going from those related to food digestion, detoxification, movement, reproduction, support, defense against pathogens, to those related to thinking, emotions or consciousness. All these functions are carried out by specialized cells, such as those of the gastrointestinal epithelium, liver, muscle, germ cells, bone, lymphocytes and neurons, respectively. Cells need a particular molecular framework, mainly based on proteins, to carry out their functions. In an organism, some functions may be carried out by only one cell type, but the cooperation of several cell types acting in a coordinated way is commonly needed.

Cell Metabolism

A cell's daily operations are accomplished through the biochemical reactions that take place within the cell. Reactions are turned on and off or sped up and slowed down according to the cell's immediate needs and overall functions. At any given time, the numerous pathways involved in building up and breaking down cellular components must be monitored and balanced in a coordinated fashion. To achieve this goal, cells organize reactions into various enzyme-powered pathways.

Enzymes

Enzymes are protein catalysts that speed biochemical reactions by facilitating the molecular rearrangements that support cell function. Recall that chemical reactions convert substrates into products, often by attaching chemical groups to or breaking off chemical groups from the substrates. For example, in the final step of glycolysis, an enzyme called pyruvate kinase transfers a phosphate group from one substrate (phosphoenolpyruvate) to another substrate (ADP), thereby generating pyruvate and ATP as products.

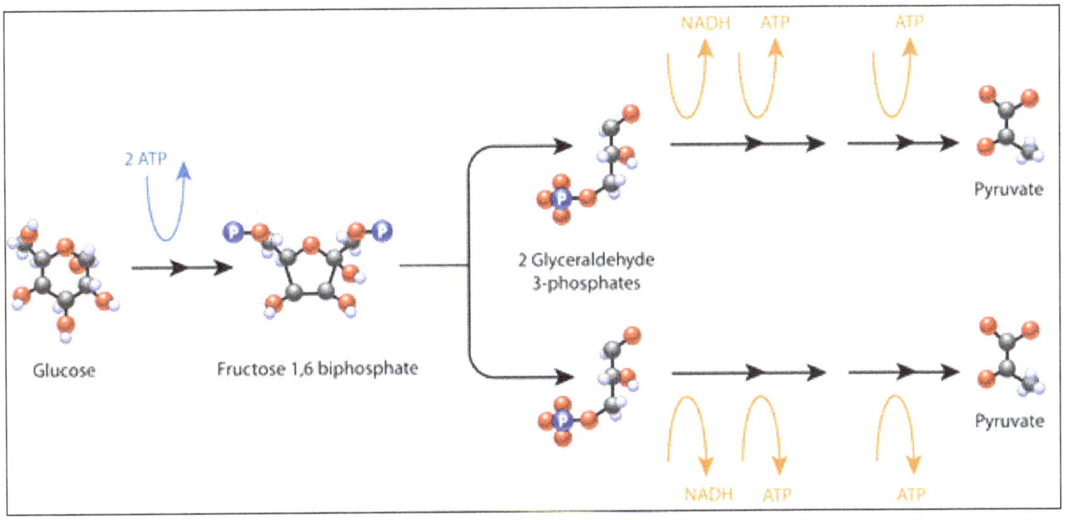

Glycolysis: Energy is used to convert glucose to a 6 carbon form. Thereafter, energy is generated to create two molecules of pyruvate.

Enzymes are flexible proteins that change shape when they bind with substrate molecules. In fact, this binding and shape changing ability is how enzymes manage to increase reaction rates. In many cases, enzymes function by bringing two substrates into close proximity and orienting them for easier electron transfer. Shape or conformational changes can also act as an on/off switch. For example, when inhibitor molecules bind to a site on an enzyme distinct from the substrate site, they can make the enzyme assume an inactive conformation, thereby preventing it from catalyzing a reaction. Conversely, the binding of activator molecules can make an enzyme assume an active conformation, essentially turning it on.

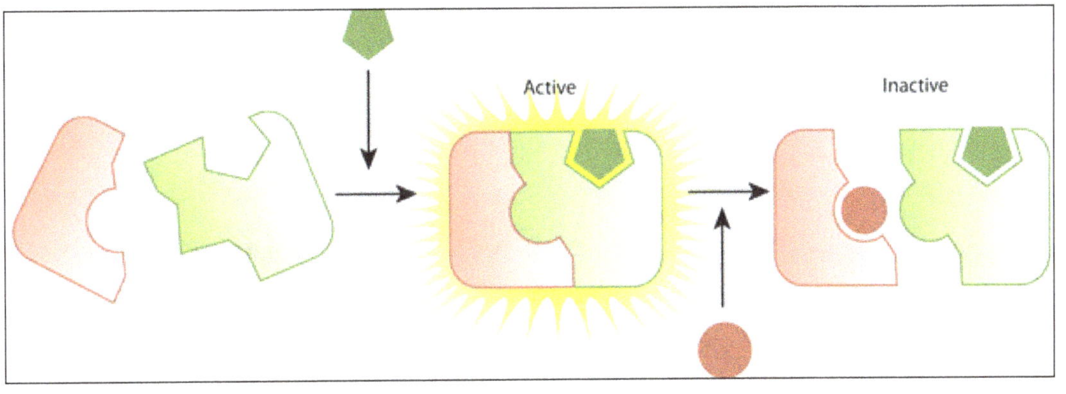

Figure shows activation and inactivation of enzyme reaction. Enzymes are proteins that can change shape and therefore become active or inactive. An activator molecule (green pentagon) can bind to an enzyme (light green puzzle shape) and change its overall shape. Note the transformation of the triangular point on the green enzyme into a rounded shape. This transformation enables the enzyme to better bind with its substrate (light pink puzzle piece). In contrast, an inhibitor molecule (pink circle) can prevent the interaction of an enzyme with its substrate and render it inactive.

Metabolic Pathways

Many of the molecular transformations that occur within cells require multiple steps to accomplish. Recall, for instance, that cells split one glucose molecule into two pyruvate molecules by way of a ten-step process called glycolysis. This coordinated series of chemical reactions is an example of a metabolic pathway in which the product of one reaction becomes the substrate for the next reaction. Consequently, the intermediate products of a metabolic pathway may be short-lived.

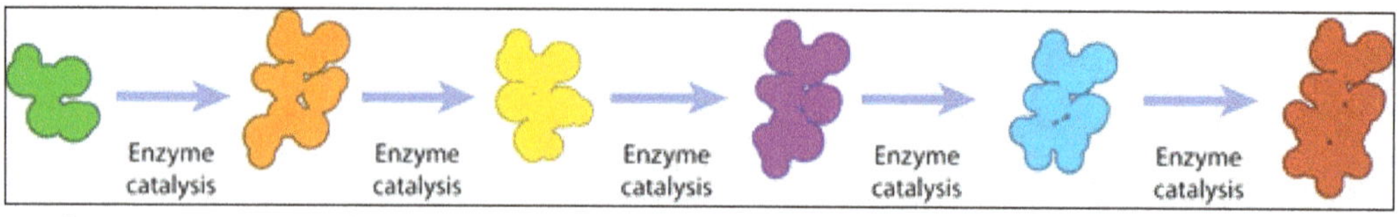

Reaction pathway: Enzymes can be involved at every step in a reaction pathway. At each step, the molecule is transformed into another form, due to the presence of a specific enzyme. Such a reaciton pathway can create a new molecule (biosynthesis), or it can break down a molecule (degradation).

Sometimes, the enzymes involved in a particular metabolic pathway are physically connected, allowing the products of one reaction to be efficiently channeled to the next enzyme in the pathway. For example, pyruvate dehydrogenase is a complex of three different enzymes that catalyze the path from pyruvate (the end product of glycolysis) to acetyl CoA (the first substrate in the citric acid cycle). Within this complex, intermediate products are passed directly from one enzyme to the next.

Cells are expert recyclers. They disassemble large molecules into simpler building blocks and then use those building blocks to create the new components they require. The breaking down of complex organic molecules occurs via catabolic pathways and usually involves the release of energy. Through catabolic pathways, polymers such as proteins, nucleic acids, and polysaccharides are reduced to their constituent parts: amino acids, nucleotides, and sugars, respectively. In contrast, the synthesis of new macromolecules occurs via anabolic pathways that require energy input.

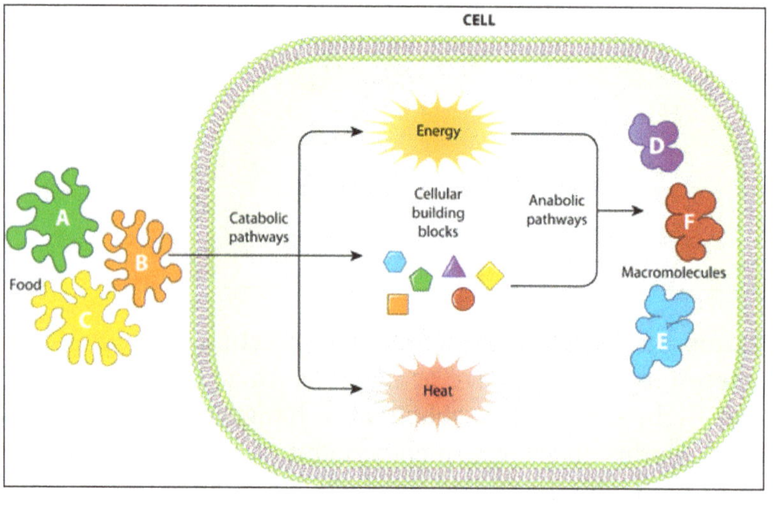

Cells must balance their catabolic and anabolic pathways in order to control their levels of critical metabolites — those molecules created by enzymatic activity — and ensure that sufficient energy is available. For example, if supplies of glucose start to wane, as might happen in the case of starvation, cells will synthesize glucose from other materials or start sending fatty acids into the citric acid cycle to generate ATP. Conversely, in times of plenty, excess glucose is converted into storage forms, such as glycogen, starches, and fats.

Figure show catabolic and anabolic pathways in cell metabolism. Catabolic pathways involve the breakdown of nutrient molecules (Food: A, B, C) into usable forms (building blocks). In this process, energy is either stored in energy molecules for later use, or released as heat. Anabolic pathways then build new molecules out of the products of catabolism, and these pathways typically use energy. The new molecules built via anabolic pathways (macromolecules) are useful for building cell structures and maintaining the cell.

Not only do cells need to balance catabolic and anabolic pathways, but they must also monitor the needs and surpluses of all their different metabolic pathways. In order to bolster a particular pathway, cells can increase the amount of a necessary (rate-limiting) enzyme or use activators to convert that enzyme into an active conformation. Conversely, to slow down or halt a pathway, cells can decrease the amount of an enzyme or use inhibitors to make the enzyme inactive.

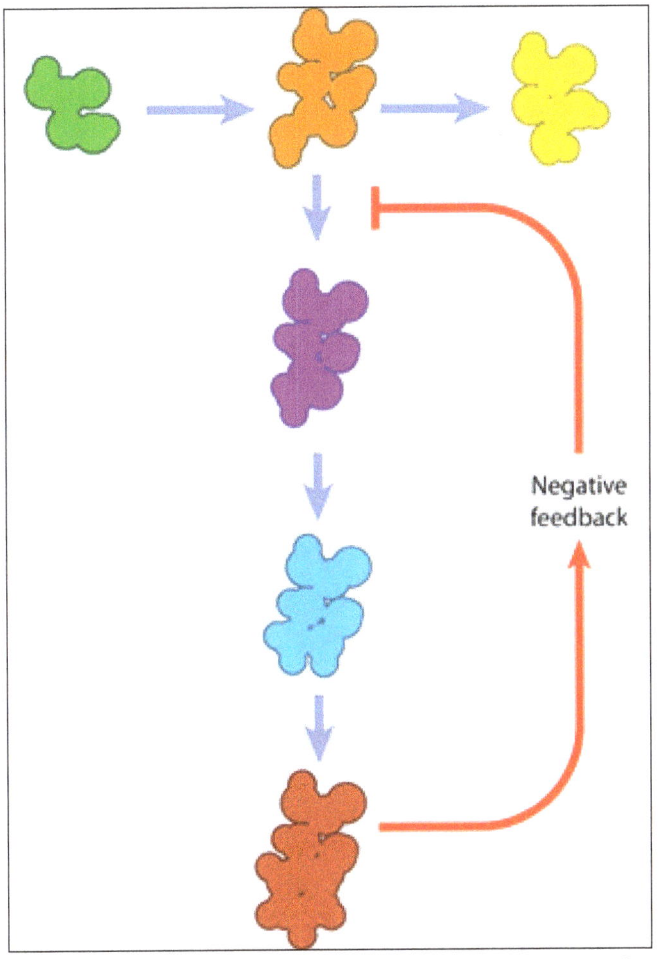

Feedback inhibition: When there is enough product at the end of a reaction pathway (red macromolecule), it can inhibit its own synthesis by interacting with enzymes in the synthesis pathway (red arrow).

Such up- and down-regulation of metabolic pathways is often a response to changes in concentrations of key metabolites in the cell. For example, a cell may take stock of its levels of intermediate metabolites and tune the glycolytic pathway and the synthesis of glucose accordingly. In some instances, the products of a metabolic pathway actually serve as inhibitors of their own synthesis, in a process known as feedback inhibition. For example, the first intermediate in glycolysis, glucose-6-phosphate, inhibits the very enzyme that produces it, hexokinase.

References

- Introduction-cell-biology, cell-and-molecular-biology-experiences, community-and-alumni: uq.edu.au, Retrieved 25, May 2020
- Cell-metabolism-14026182: nature.com, Retrieved 16, March 2020

Chapter 2

The Basic Structure of a Cell

Cells are the smallest units of life and are made up of various parts. A cell consists of three primary parts: a membrane, a cytoplasm and a nucleus. Biomolecules and small fiber-like structures called organelles are also present within the cytoplasm. This chapter discusses in detail the types of cells, their architecture, structure, membrane and their composition.

Cell Architecture

Cell architecture it is important to realize that for cells, as for buildings, the underpinning for the external shape is provided by a complex internal superstructure. And for cells, this cytoskeletal underpinning must be highly dynamic to effect the changes in morphology and organization associated with division, growth and differentiation.

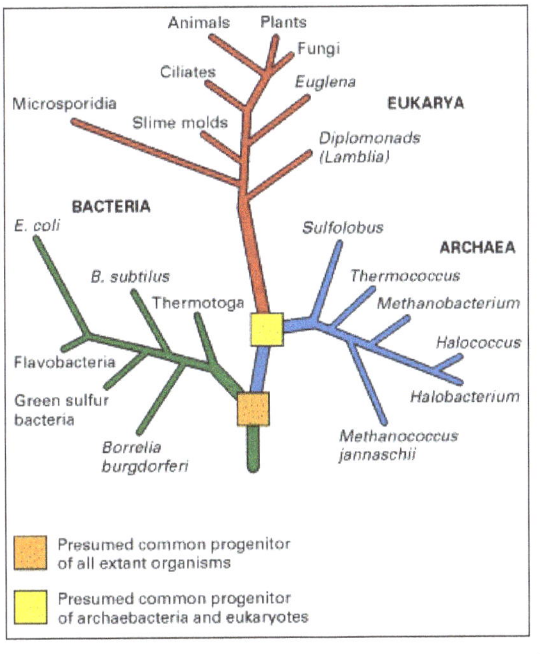

The three kingdoms of organisms are related through common sequences of their ribosomal RNAs. Their lineage depicts a view of how all life on earth, from simple bacteria to complex mammals, evolved from a common, single-celled progenitor.

Although generalizations in biology usually lack the theoretical underpinnings found in physics, there are very clear commonalities among living systems that give biology a unity. One is the style of cellular construction. The biological universe consists of two types of cells – prokaryotic cells, which lack a defined nucleus and have a simplified internal organization, and eukaryotic cells, which have a more complicated internal structure including a defined, membrane-limited nucleus.

Detailed analysis of the DNA from a variety of prokaryotic organisms in recent years has revealed two distinct types: bacteria (often called "true" bacteria or eubacteria) and archaea (also called archaebacteria or archaeans). The archaea are in some respects more similar to eukaryotic organisms than to the true bacteria.

Based on the assumption that organisms with more similar genes evolved from a common progenitor more recently than those with more dissimilar genes, researchers have developed the lineage tree shown in figure. According to this tree, the archaea and eukarya (eukaryotes) are thought to have diverged from the bacteria before they diverged from each other. Despite the differences in the organization of prokaryotic and eukaryotic cells, all cells share certain structural features and carry out many complicated processes in basically the same way.

Cells are Surrounded by Water-Impermeable Membranes

A cell, because it is a limited space, must have an outer border. The construction of that border represents one of the most fundamental considerations in biological organization. The outer shell of cells, like any shell, is built to keep the interior contents from leaking out into the surrounding environment. The chemical processes of cellular life generally take place in a watery solution, and the intracellular constituents of cells are largely molecules that are easily dissolved in water. Similarly, the environment around cells is a watery one, the blood and other bodily fluids being solutions in water. Cells then, in order to maintain their integrity, need to be surrounded by an environment through which water cannot flow. A membrane composed of fatty molecules serves this purpose.

We all know from common experience that "oil and water don't mix." That maxim is all one needs to appreciate how a cell is constructed. When oil is poured on water, the oil spreads into a thin film; that film is analogous to the film of fat that surrounds cells, called the plasma membrane. Biological membranes differ from a pure oil film in that the molecules that make the membrane have both oily and watery portions; they have long fatty chains, but they also have a head group that is water-soluble by virtue of being electrically charged. Thus membranes are formed because these bipartite molecules, called phospholipids, spontaneously orient themselves to form a double layer, or bilayer, having a fatty interior with external surfaces bonded to the surrounding water by the charged head groups. The membrane is given rigidity by interspersion of cholesterol, a molecule we have come to hate because of its association with heart disease, but one that is required to build the outer membrane of all our cells. Hence from an understanding of the contrasting properties of watery solutions and oily layers, an understanding of cellular construction emerges.

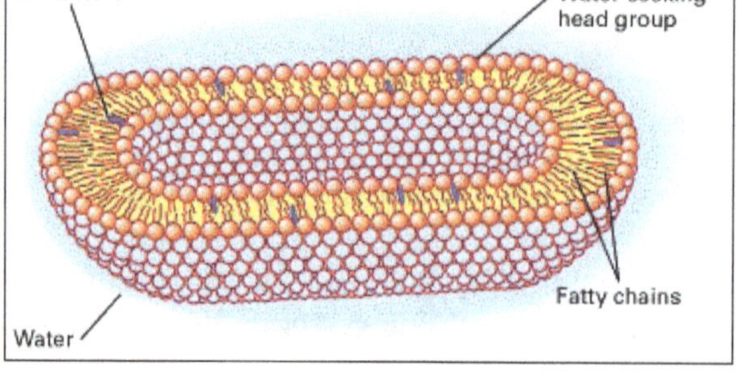

Figure shows the watery interior of cells is surrounded by the plasma membrane, a two-layered shell of phospholipids. Cholesterol molecules provide some rigidity to the fatty layer. The phospholipid molecules are oriented with their fatty chains facing inward and their water-seeking head groups (red spheres) facing outward. Thus both sides of the membrane are lined by head groups, mainly charged phosphates, adjacent to the watery spaces inside and outside the cell. In actuality, the interior space is much larger relative to the volume of lipid. All biological membranes have the same basic phospholipid bilayer structure.

In spite of the rigidity provided by cholesterol, membranes composed of fat are not very strong, so numerous mechanisms for strengthening the borders of cells have evolved. In plants the plasma membrane is surrounded by a rigid cell wall. Although most animal cells lack a cell wall, proteins attached to their exterior surfaces provide some stability; the linking of cells together through these proteins helps maintain the integrity of tissues. Tissues and organs are often covered by strong networks of proteins and other molecules that strengthen and protect them, and also wall off the various compartments of the body. Single-celled organisms, like bacteria, have special outer coats to protect them.

Membranes Serve Functions other than Segregation

Although membranes are valuable as a way to segregate the watery interior of the cell from its environment, or to segregate intracellular events from one another, they have other important functions, including energy storage. Because membranes separate watery compartments from one another, if an ion or a molecule dissolved in water is moved through a membrane into a new cellular compartment, it will not be able to diffuse freely out of the compartment into which it was moved. It takes energy to move the molecule, but once moved, the molecule stores that energy by virtue of its entrapment. Formally, this storage of energy is just like the storage of energy in a battery. Therefore, membranes not only delineate compartments, but also serve as active participants in the cell's dynamism.

The functions of many proteins depend on their mode of association with membranes. For instance, the passage of water-soluble molecules through membranes is carried out by protein transporters that are embedded in the membrane. Also, cells send information to one another by releasing signaling molecules. The outer membranes of cells have proteins, known appropriately as receptors, that bind the circulating signaling molecules. These signaling molecules allow the individual activities of the many cells in the body to be coordinated. The receipt of a signaling molecule by a receptor causes the transient organization of particular types of intracellular proteins, called signal-transduction proteins, into an activated complex at the interior face of the cell's outer membrane, from which it directs alterations of events in the cell's cytoplasm or nucleus.

Prokaryotes Comprise a Single Membrane-Limited Compartment

All prokaryotes are single-celled organisms, or protists. The bacterial lineage includes Escherichia coli, found in animal intestines and a favorite experimental organism, and the photosynthetic organisms formerly known as blue-green algae but better known today as cyanobacteria. Many members of the archaeal lineage grow in unusual, often extreme, environments. For instance, the halophiles require high concentrations of salt to survive, and the thermoacidophiles grow in hot (80 °C) sulfur springs, where a pH of less than 2 is common. Other archaeans, called methanogens, live in oxygen-free milieus and generate methane (CH_4) by the reduction of carbon dioxide.

In general, prokaryotes consist of a single closed compartment containing the cytosol and bounded by the plasma membrane. Although bacterial cells do not have a defined nucleus, the genetic material, DNA, is condensed into the central region of the cell. In addition, most ribosomes — the cell's protein-synthesizing particles — are found in the DNA-free region of the cell. Some bacteria also have an invagination of the cell membrane, called a mesosome, which is associated with synthesis of DNA and secretion of proteins. Thus bacterial cells are not completely devoid of internal organization.

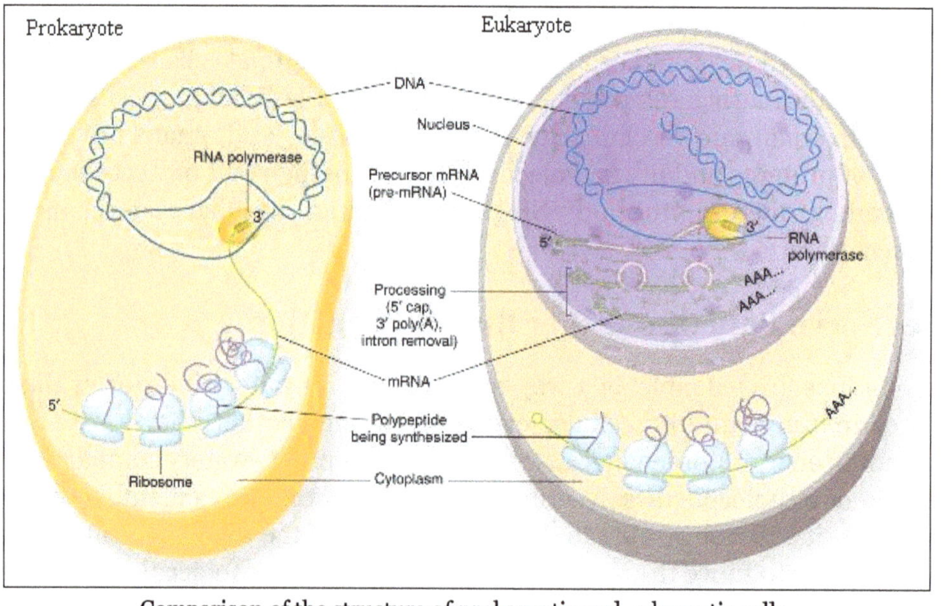

Comparison of the structure of prokaryotic and eukaryotic cells.

Bacterial cells possess a cell wall, which lies adjacent to the external side of the plasma membrane. The cell wall is composed of layers of peptidoglycan, a complex of proteins and oligosaccharides; it helps protect the cell and maintain its shape. Some bacteria (e.g., E. coli) have a thin cell wall and an unusual outer membrane separated from the cell wall by the periplasmic space. Such bacteria are not stained by the Gram technique and thus are classified as gram-negative. Other bacteria (e.g., Bacillus polymyxa) that have a thicker cell wall and no outer membrane take the Gram stain and thus are classified as gram-positive.

Eukaryotic Cells Contain Many Organelles and a Complex Cytoskeleton

Eukaryotes comprise all members of the plant and animal kingdoms, including the unicellular fungi (e.g., yeasts, mushrooms, molds) and protozoans. Eukaryotic cells, like prokaryotic cells, are surrounded by a plasma membrane. However, unlike prokaryotic cells, most eukaryotic cells also contain extensive internal membranes that enclose specific compartments, the organelles, and separate them from the rest of the cytoplasm, the region of the cell lying outside the nucleus.

Most organelles are surrounded by a single phospholipid membrane, but several, including the nucleus, are enclosed by two membranes. Each type of organelle plays a unique role in the growth and metabolism of the cell, and each contains a collection of specific enzymes that catalyze requisite chemical reactions. The membranes defining these subcellular compartments control their internal ionic composition so that it commonly differs from that of the cytosol (the portion of the cytoplasm outside the organelles) and among the various organelles.

The largest organelle in a eukaryotic cell is generally the nucleus, which houses most of the cellular DNA. In addition to the nucleus, several other organelles are present in nearly all eukaryotic cells: the mitochondria, in which much of the cell's energy metabolism is carried out; the rough and smooth endoplasmic reticula, a network of membranes in which glycoproteins and lipids are synthesized; Golgi vesicles, which direct membrane constituents to appropriate places in the cell; and peroxisomes, in which fatty acids and amino acids are degraded. Animal cells, but not plant cells, contain lysosomes, which degrade worn-out cell constituents and foreign materials taken in by the cell. Chloroplasts, where photosynthesis occurs, are found only in certain leaf cells of plants and some single-celled organisms. Both plant cells and some single-celled eukaryotes contain one or more vacuoles, large, fluid-filled organelles in which nutrients and waste compounds are stored and some degradative reactions occur.

The cytosol of eukaryotic cells contains an array of fibrous proteins collectively called the cytoskeleton. Three classes of fibers compose the cytoskeleton: microtubules (20 nm in diameter), built of polymers of the protein tubulin; microfilaments (7 nm in diameter), built of the protein actin; and intermediate filaments (10 nm in diameter), built of one or more rod-shaped protein subunits. The cytoskeleton gives the cell strength and rigidity, thereby helping to maintain cell shape. Cytoskeletal fibers also control movement of structures within the cell; for example, some cytoskeletal fibers connect to organelles or provide tracks along which organelles move.

The rigid cell wall, composed of cellulose and other polymers, that surrounds plant cells contributes to their strength and rigidity. Fungi are also surrounded by a cell wall, but its composition differs from that of bacterial or plant cell walls.

Cellular DNA Is Packaged within Chromosomes

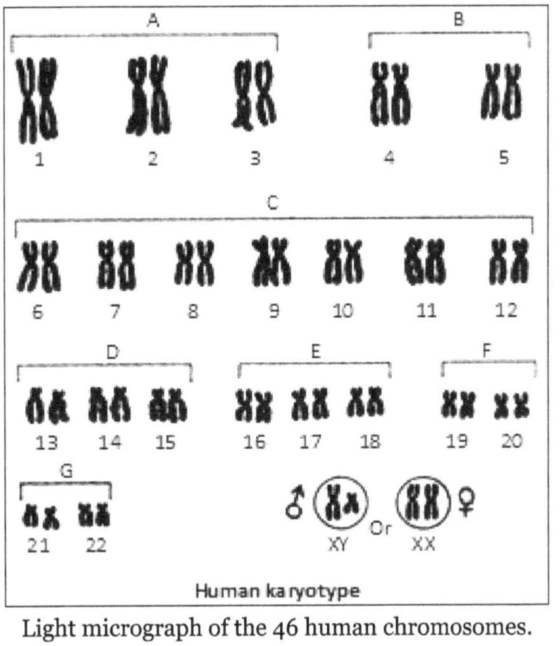

Light micrograph of the 46 human chromosomes.

The DNA in the nuclei of eukaryotic cells is distributed among 1 to more than 50 long linear structures called chromosomes. The number and size of the chromosomes are the same in all cells of an organism, but vary among different types of organisms. Each chromosome comprises a single DNA molecule associated with numerous proteins, and the total DNA in the chromosomes of an organism is referred to as its genome. Chromosomes, which stain intensely with basic dyes, are visible in the light microscope only during cell division when the DNA becomes tightly compacted.

In all prokaryotic cells, most of or all the genetic information resides in a single circular DNA molecule, about a millimeter in length; this molecule lies, folded back on itself many times, in the central region of the cell. Although the large genomic DNA molecule in prokaryotes is associated with proteins and often is referred to as a chromosome, the arrangement of DNA within a bacterial chromosome differs greatly from that within the chromosomes of eukaryotic cells.

The concept that genes are like "beads" strung on a long "string," the chromosome, was proposed early in the 1900s based on genetic work with the fruit fly Drosophila. The early Drosophila workers could position, or map, the genes responsible for various mutant traits on a chromosome, even though they did not yet know that genes were segments of DNA or that the function of a gene was due to a protein whose sequence was encoded by that gene.

Design Principles of Cellular Architecture

Cells are highly complex structures. This complexity affects biological functions at many size scales. But how does this complexity arise? What mechanisms pattern the architecture of the cell and where does the information to specify a distinct architecture for different cell types come from?

The amount of information that is needed to direct the action of the cellular assembly pathways could be minimal because cellular morphogenesis is a self-organizing process. Subject to simple inputs that can bias how the system develops, self-organizing processes can generate complex, patterned responses. In principle, therefore, the cell architecture could be specified by a small number of distinct genes, which would serve to modulate the inherent self-organizing processes that lead to the physical shape of cells and organelles. Understanding the origins of cell organization thus requires us to answer the question of how the self-organizing physical and biochemical properties of cellular constituents can convert simple temporal changes in gene-expression levels into complex changes in physical, three-dimensional structures.

In all cases, we will attempt to identify design principles by which a simple input containing minimal information can produce a complex output in terms of morphological changes. In engineering terminology a design principle is a simple rule that, when followed in the design of a machine, ensures or at least increases the likelihood of proper assembly and function. We note that our use of the term does not imply the existence of a 'designer', but given the current interest in synthetic biology we suggest that the design principles used in evolution could now be used to reprogramme cells in a rational way. For example, we argue that a common strategy used by cells to regulate organelle size is to limit either the assembly or disassembly rate of an organelle by a process that is inherently dependent on organelle size. Other specific principles that we discuss are the control of organelle number by the balance of organelle formation and partitioning; the establishment of cellular polarity by self-organization that is biased by external inputs; the control of organelle position by direct motion on polarized cytoskeletal arrays; and the control of organelle shape by membrane-binding proteins that drive local curvature changes.

How Cells Measure Size and Length

Length and size are fundamental parameters for specifying geometry. Organelle size must be appropriately scaled as a function of cell size and physiological demands. Mitochondrion size in yeast, for example, is proportional to cell volume. Organelle size is tightly controlled: for example, the coefficient of variation in the lengths of eukaryotic cilia and flagella is less than 10%. But how are the sizes of subcellular structures determined? Genetic screens have identified genes that are involved in size control for various organelles, including cilia and flagella, peroxisomes and lipid droplets. However, in most cases the mechanistic basis for size alterations in these mutants remains unknown.

Measuring Length using Molecular Rulers

One way to control the size of linear structures is for the cell to measure the length of the structure with a molecular ruler protein which has a physical length that equals the desired length of the overall structure. For example, the length of the λ-phage tail is dictated by the size of the gene H product. Deletions in gene H produce shorter tails and insertions produce longer tails. One end of the gene H product is attached to the capsid and the other tracks the growing tail, protecting it from a growth-terminating factor until the tail finally exceeds the length of the gene H product and growth terminates. Rulers have been proposed for length control of bacterial flagellar hooks, bacterial type III secretion needles and sarcomere actin filaments in muscle.

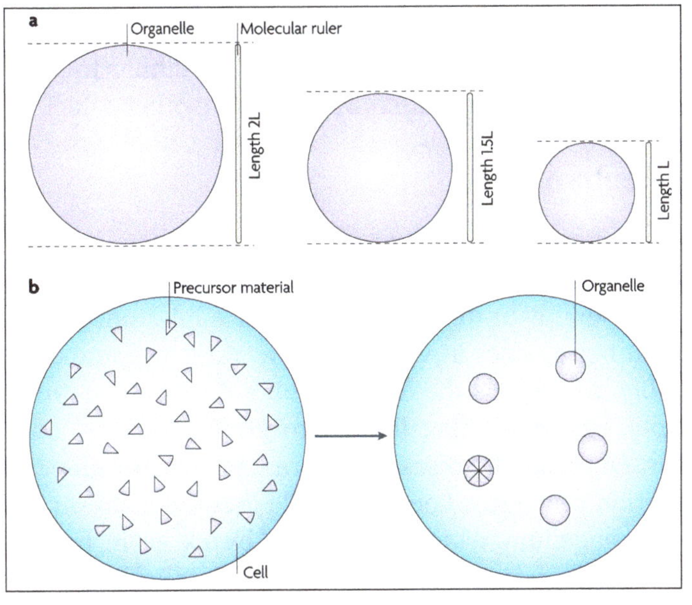

Figure shows design principles of organelle size control. a Organelle size can be controlled by molecular rulers. The physical length of a molecular ruler protein determines the size of a cellular organelle. b. Organelle size can be controlled by quantal synthesis whereby the precise amount of precursor material is synthesized by the cell in order to build the correct number of organelles of the correct size.

Size control by quantal synthesis. Size could be determined by synthesizing precisely enough material to build a structure of the correct size, a model that has been termed quantal synthesis13. Precursor protein levels affect the length of cilia/flagella in the unicellular green alga Chlamydomonas reinhardtii and are under tight regulation during flagellum regeneration. In sea urchin cells, the lengths of cilia correlate with the concentration of the protein tektin, which is produced by the cells. However, new flagella can grow to roughly half of their normal length when protein synthesis is inhibited in C. reinhardtii, which shows that cells retain a pool of unused precursor protein. Moreover, a simple model in which the amount of precursor directly determines flagellum length would predict a geometric dependence of flagellum length on number. For example, if flagella double in number, then the average length of each flagellum should halve. However, mutants with varying numbers of flagella show a significantly weaker dependence of flagellar length on number. Therefore, it seems that the levels of precursor protein might not be the only length-controlling mechanism at work in flagella.

Another example of an organelle whose size is apparently regulated by the synthesis of components is the endoplasmic reticulum (ER). Forced upregulation of lipid-biosynthesis enzymes leads to an increase in the size of the ER. In this case there seems to be a feedback pathway that controls lipid synthesis as a function of ER size: the unfolded protein response (UPR), a signaling pathway that monitors proper ER function, and controls the transcription of many genes, including those that encode lipid-biosynthesis enzymes. If the amount of lipid controls the size of the ER, and the UPR activates lipid synthesis, then it is predicted that constitutive activation of the UPR pathway should cause an increase in the size of the ER membrane. This prediction has been experimentally verified.

One prediction of the quantal synthesis model is that multiple organelles with common limiting components will compete with each other, such that one organelle can only grow if the others shrink. This has been demonstrated experimentally for flagella in C. reinhardtii, in which the severing and regeneration of one flagellum in a biflagellate cell induces the other intact flagellum to shorten.

Dynamic Balance Mechanisms

Most cellular structures undergo continuous turnover, and hence size maintenance requires the rates of assembly and disassembly to precisely equal each other only when the correct structure size has been reached. If assembly or disassembly rates are inherently size-dependent then there will only be a limited number of sizes at which the two rates become equal. If there is only one size at which the assembly rate equals the disassembly rate then this will be the steady-state organelle size. For example, eukaryotic flagellar microtubules undergo continuous assembly and disassembly at their tips. Of these, the disassembly rate is length-independent and the assembly rate is a decreasing function of length. Longer flagella assemble slower than shorter flagella because transport within the flagellum, which is mediated by a kinesin-based system called intraflagellar transport, becomes less efficient. This limits the availability of precursor protein to support tip assembly, thus reducing the assembly rate with increasing length. As the rate of assembly is a continuously decreasing function of length (whereas the rate of disassembly is length-independent), there is only one length at which the two rates balance, and this sets the steady-state flagellar length.

A similar dynamic balance mechanism might control the length of actin-based structures, such as microvilli or the stereocilia of the mouse inner ear. Actin filaments in these structures constantly treadmill back towards the cell body, with disassembly at the base balanced by continuous assembly at the tip. In turn, this depends on the diffusion of free actin to the tip. The requirement for actin to diffuse to the tip provides an inherent dependence of the assembly rate on the length of the filament. Diffusion-induced length dependence of the assembly rate has also been suggested for the control of bacterial flagella length.

Yeast kinesin8 binds along the length of microtubules, moves processively to their tips and then catalyses disassembly. Longer microtubules recruit more kinesin8, which results in a length-dependent disassembly process that, when combined with a constant rate of assembly, yields a unique steady-state microtubule length. Unlike flagella and microvilli, it seems to be the disassembly of the microtubules, rather than the assembly, that is regulated as a function of length. In complex cytoskeletal assemblies, such as the mitotic spindle, length control might involve multiple opposing length-dependent processes in a more complicated dynamic balance.

The same types of dynamic balance mechanisms that are used to control length could also be used to control the volume or surface area of the organelle. If an organelle such as the Golgi apparatus exchanges vesicles with the ER, vesicle fusion on the organelle occurs at a rate that is dictated by the rate of vesicle budding from the ER, rather than by the organelle size. By contrast, vesicle budding from the organelle could occur at a rate that is proportional to the surface area of the organelle (assuming a constant density of budding sites). The rates of membrane addition and membrane removal would only balance each other for a single surface area value, thereby dictating the surface area of the organelle. If the organelle became larger than this the budding rate would increase, leading to a reduction in surface area: if the organelle became too small the budding rate would decrease relative to the fusion rate and hence it would enlarge to its steady-state surface area.

How Cells Control Organelle Number

A typical eukaryotic cell has one nucleus, two centrioles, one ER, multiple mitochondria, scores of endosomes and many ribosomes. How is the number of organelles controlled?

Number Control as a Dynamic Balance

The number of organelles reflects the balance between two competing processes: the increase in number due to de novo synthesis or fission of pre-existing organelles; and the decrease in number due to fusion with other organelles, degradation processes such as autophagy, or the partitioning of organelles into daughter cells during cell division. To control the number of organelles, one or more of these processes must be regulated. It has been proposed, for example, that the synthesis of chloroplasts is regulated as a function of the number of chloroplasts in algal cells. Cells that inherit too few chloroplasts will make more, whereas cells that inherit too many will stop chloroplast synthesis. A similar scheme has been experimentally demonstrated for centrioles. Therefore, the control of organelle number shares a conceptually similar strategy to the dynamic balance mechanisms of organelle size control.

Figure show the dynamic balance mechanism to control size and number. a. Dynamic balance to control organelle size. The size of the organelle feeds back onto either the assembly or disassembly rate such that one of these is inherently size-dependent. Thus, the assembly rate and disassembly rate are equal to each other only at the correct size of the organelle. The organelle size changes if the regulation of either the assembly or the disassembly is altered. b. Dynamic balance to control organelle number. The mechanisms that regulate either the increase or decrease of organelle number are themselves a function of the organelle such that the correct number of organelles is maintained. The number of organelles changes if the underlying regulation of either the increase (thick arrow) or the decrease (thin arrow) in organelle number is altered.

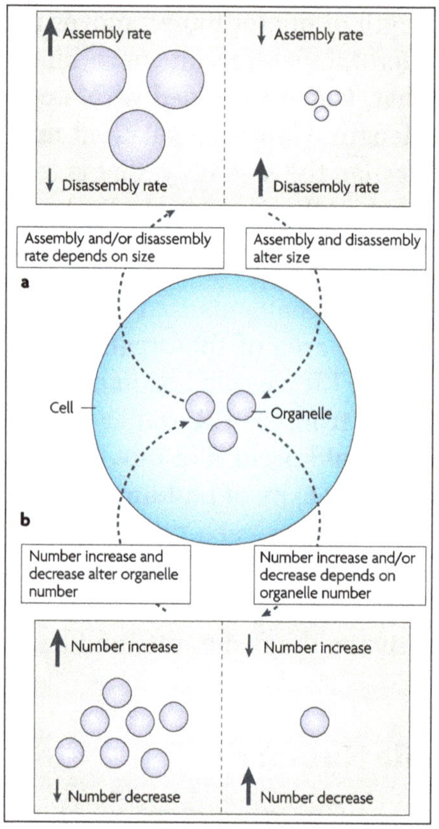

Controlling the Increase in Organelle Number

The fact that organelle production can be regulated by the existing number of organelles suggests that cells can count. For example, an absence of centrioles activates a de novo assembly pathway, whereas an excess of centrioles (> 2) inactivates further centriole assembly. Peroxisomes undergo highly regulated proliferation, the rate of which is adjusted in response to unknown intracellular cues that control steady-state peroxisome number. Apparently, cells can detect whether the organelle number has drifted outside of a preset range. One possible mechanism by which cells count organelle number would be for the organelles to generate a diffusible molecular signal at a constant rate per organelle. This signalling molecule would degrade with a constant half-life. As a result, the steady-state concentration of the signal would be proportional to the number of organelles. The concentration could then be used to regulate the assembly of new organelles. One such mechanism seems to function in the control of plasmid replication, in which plasmids encode diffusible trans-acting autoinhibitors of plasmid replication at a constant rate per plasmid. This type of mechanism relies on measuring the ratio of organelle number to cell volume and could thus be tested by manipulating cell size.

Reducing Organelle Number

The number (N) of organelles can be reduced by their partitioning between daughter cells. If N organelles distribute randomly between two daughter cells, the number of copies that a given daughter receives will vary randomly according to binomial statistics. For each organelle that segregates at random, it is as though one flips a coin to determine which daughter cell receives that organelle. If one flips a coin a few times, the number of flips that come up heads can vary substantially.

Similarly, if organelles partition at random the actual number that a given daughter inherits can fluctuate substantially. One way to dampen such fluctuations is to fragment the organelle into a large number of small pieces so that the effective value of N becomes very large. In this case most daughter cells will inherit a number that is close to N/2. Alternatively, the mitotic spindle can be used to segregate organelles nonrandomly between the two daughters, as with centrioles. For many organelles, it is still not entirely clear how they are partitioned and whether this partitioning is truly a random process or not. For example, endosomes and lysosomes associate with spindle poles but still distribute according to binomial statistics. Golgi segregate using the mitotic spindle poles in some organisms, whereas in others they fragment and scatter randomly between the daughters. It should be noted that partitioning by the spindle will not necessarily reduce fluctuations in organelle numbers if individual organelles associate with the spindle poles at random. Regardless, partitioning is certain to reduce organelle number because the initial number, N, must somehow be divided among the two daughter cells.

Organelle number might also be reduced by the active destruction of existing organelles when the number exceeds a threshold. Such a process has been documented for peroxisomes and involves an autophagy-like mechanism. In ciliates, the number of cilia in various cortical structures is proportional to the cell size. When the cell volume is experimentally reduced in the absence of cell division, the number of cilia is reduced so as to maintain a cilia number that is in constant proportion to cell volume. These experiments suggest that active degradation pathways, such as autophagy, can sense the relative concentration of organelles per unit volume, rather than a strict absolute number. We speculate that active organelle degradation might be triggered by the accumulation of diffusible factors produced by individual organelles, a model that is formally equivalent to the inhibitor-dilution model for control of organelle production that is discussed above. However, the diffusible molecule would trigger degradation rather than inhibit biosynthesis.

Spatial Organization and Polarity

A key aspect of cell geometry is the fact that both cell shape and intracellular organization are polarized. Cell polarity is an asymmetry of shape or of protein distribution, for example, between the front and the back in motile cells or between the top and the bottom in epithelial cells. A non-polarized cell becomes polarized through the loss of symmetry, occurring as a result of an extrinsic or intrinsic localized cue or occurring spontaneously as a result of underlying self-organizational mechanisms that are inherent in the cell or system. Polarization and directed motility do not require gene transcription or even the nucleus (as demonstrated by experiments using anucleate cytoplasts).

Spontaneous polarization has been observed within cells, including the formation of a polarized distribution of the signalling protein Cdc42 at the plasma membrane in Saccharomyces cerevisiae, the polarization of neutrophils in a uniform bath of chemoattractant, and the initiation of motility in stationary keratocytes. In the case of keratocytes, spontaneous myosindependent actin contractility at the presumptive rear of the cell leads to the increased inward flow of filamentous (F)actin and the retraction of the rear. These changes in Factin organization and dynamics propagate globally and result in the decrease of Factin flow at the cell rear and the induction of actin-polymerization-based protrusion at the front of the cell, thus leading to persistent keratocyte motility. The common design principle in establishing cell polarity seems to be the existence of intrinsic selforganizing polarity systems, in which the cell exhibits fluctuating asymmetry in the localization of signaling proteins or cytoskeletal dynamics. The cell then responds to a stochastic amplification in these fluctuations by initiating a positive-feedback loop, which involves both signalling networks and cytoskeletal structures that locally amplify the initial symmetry-breaking event and further regulate other mechanochemical polarizing processes throughout the cell. Positive feedback alone would lead to constant uniform activation of polarizing processes. Therefore, polarization to a discrete site also requires long-range negative interactions to suppress the spread of polarization. The combination of both local positive feedback and the long-range inhibition of polarizing processes allows spontaneous symmetry breaking and self-organization. Such systems are then biased by upstream spatial inputs in order to drive polarization in a desired direction.

Organelle Positioning in Response to Polarity

Cell polarity directs the positioning of intracellular organelles. This positioning is often crucial for their function, for example, in determining the orientation of stereocilia bundles in response to planar cell polarity cues in the cochlea or the orientation of the secretory apparatus towards the targets of natural killer immune cells. In migrating cells, many organelles, including mitochondria and the Golgi apparatus, are positioned relative to the axis of polarization. A particularly dramatic example of polarized organelle positioning occurs during the development of neurons, in which the positions of the Golgi apparatus and the centrosomes strongly correlate with sites of neurite outgrowth.

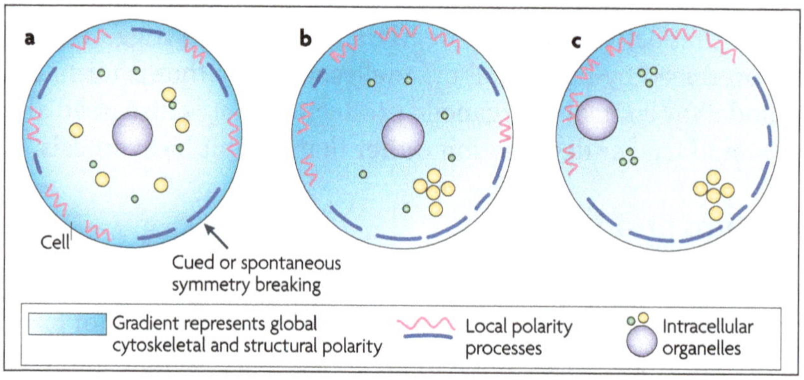

Figure shows design principles of polarity and organelle positioning. a. Local polarity processes (including cytoskeletal dynamics and protein localization) undergo stochastic fluctuating asymmetry. The cell responds to either a stochastic or cued amplification in these fluctuations by locally amplifying this symmetry-breaking event. Meanwhile, organelles are randomly positioned within the cell. b. Local polarity processes induce global cytoskeletal or structural polarity (shown by the gradient). Long-range suppression of these polarity processes prevents them from being activated uniformly throughout the cell. Together, the local amplification and long-range suppression of polarity processes amplify the overall polarity of the cell. Some organelles (yellow) are positioned by this global cell polarity. c. The positioning of some organelles (yellow) further amplifies the global cytoskeletal or structural polarity (the gradient steepens) and controls the positioning of other organelles (green and grey) in the cell.

Mechanistically, it is not clear how organelles become repositioned in response to polarity cues. Organelles might move by random diffusion until they collide with a target docking site, although it is unclear whether the diffusion of organelles in the crowded cytoplasm is fast enough to allow such a diffuse-to-capture mechanism to operate. Perhaps for this reason, organelle positioning often involves the cytoskeleton, which can allow for more rapid, directed transport of organelles by motor proteins or specific subcellular localization through the regulation of cytoskeletal distribution and dynamics. For example, cortical actin flow drives the polarized redistribution of the cortical ER and myoplasm (mitochondrially enriched cytoplasm) in Ascidian oocytes and establishes the nucleus–centrosome axis in migrating cells during wound healing.

Feedback from Organelle Positioning

One common theme in polarized cells is the non-random distribution and dynamics of the microtubule array. Work has recently focused on the plus-end microtubuletipbinding proteins and motors that track with microtubule plus ends and interact with other signalling or motor proteins that are localized to the cell cortex. These proteins allow the selective stabilization of microtubule dynamics and the proper localization of the microtubule cytoskeleton within the cell. For example, the microtubule array is polarized and microtubules are selectively stabilized towards the leading edge in motile cells.

The position of the microtubule array can in turn affect the position of organelles, such as the Golgi apparatus in mammalian cells. The Golgi is located adjacent to the centrosome and its position is dependent on both the microtubules and the motor protein dynein. The cisternal compartments of the Golgi have a polarized organization, and their orientation depends on the cell type and on both dynein and kinesin molecular motors moving on polarized microtubules65. For example, in motile cells the trans-Golgi network (TGN), from which vesicles that are destined for the plasma membrane bud, is orientated towards the outside of the cell, whereas the cis-Golgi is orientated inward. The TGN has been shown to harbour microtubule-nucleation ability and Golgi-attached microtubules exhibit increased stability. It was recently further shown that the TGN protein GCC185 recruits CLASPs (CLIP-associated proteins), which stabilize small microtubule seeds that are nucleated at the TGN69. This leads to a population of stabilized microtubules emanating from the TGN, preferentially towards the leading edge, which might be required for cell motility. Thus, the positioning of the Golgi in response to cell polarity might itself contribute further to processes that promote cell polarity. This sort of organelle position-dependent positive-feedback loop could potentially lead to the spontaneous formation of a polarized intracellular axis, although to our knowledge this has not been tested.

Organelles can also feedback on the positioning of other organelles. In C. reinhardtii, which exhibits stereotyped positioning of its intracellular organelles, the mother centriole directs the positioning of the daughter centriole, the nucleus and possibly the contractile vacuole. This suggests that perhaps the mother centriole acts more generally to coordinate the overall geometry of the many subcellular structures in this organism.

After the initial polarization, some organelles are repositioned in the cell in response to the now orientated cytoskeleton. These organelles can in turn redirect the position of the cytoskeleton and further enhance the global cell polarity, and can cause the reposition of other organelles through an organelle positiondependent feedback loop. Because the underlying polarity system is self-organizing, it does not require high information content in its inputs, but because it is coupled to the cytoskeleton and ultimately to organelle positioning, the polarity system can lead to complicated responses to simple inputs.

Dynamic Shape Specification

Organelles take on a striking variety of shapes that range from interconnected tubular networks of the mitochondria and peripheral ER to carefully localized stacks of membranes in the Golgi cisterna, and to the large continuous membrane sheets in the nuclear envelope and regions of the ER. How are these distinct shapes achieved?

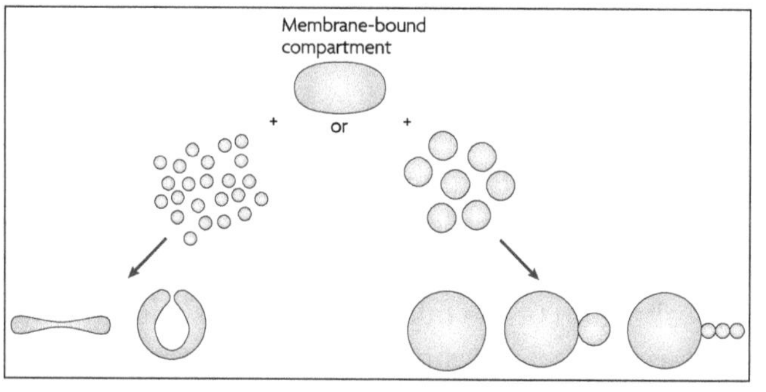

Figure shows the regulation of organelle shape by changing the vesicle size. Depending on the number and size of the vesicles that fuse with a membrane-bound compartment, the resultant compartment can display significant shape differences. Small vesicles would increase the surface area of the compartment more than the volume, whereas large vesicles would increase the volume over the surface area. While the possible resulting shapes of the compartment are extreme, this sort of control mechanism could also create many more subtle variations in compartment shape and function.

Shaping Membrane Curvature

The shape of an organelle can be completely defined by specifying the local curvature of the surrounding membrane at each point. The membrane curvature depends on the surfaceareato-volume ratio and the preferred intrinsic curvature of the membrane. Indeed, changes in the surface-area-to-volume ratio and the difference in area between the two lipid monolayers can lead to a wide range of dramatic transitions in shape, such as to spheres, flattened discs, cup-shapes and even complex star-like shapes. These shapes can be explained by theoretical models that minimize the energetics of membrane deformation. Different lipids preferentially stabilize different degrees of membrane curvature and can profoundly affect the organelle shape. Local membrane curvature can drive the segregation of lipids, which can then influence further curvature changes, suggesting a potential for complicated feedback loops between local curvature and lipid composition. Phase-separated vesicles and tubules can even spontaneously bud owing to the induced local membrane curvature of segregated domains.

Membrane-associated proteins can also affect membrane curvature, as is believed to be the case for the BAR-domain-containing proteins. An extreme example is that when purified matrix protein of Newcastle disease virus is added to membranes, they deform and bud off vesicles of similar sizes to virus particles in infected cells.

Determining the Tubular Morphology of ER

The ER consists of a membrane-bound compartment that is continuous with the nuclear envelope.

There are two types of ER: the rough ER, which is usually composed of membrane sheets, and the smooth ER, which is usually composed of a tubular network that extends to the cell periphery in mammalian cells or is attached to the periphery in yeast. Membrane tubes can be formed in vitro by the activity of motor proteins, which pull the lipid membrane into tubes as they move along microtubules. These membrane tubes resemble tubular ER networks. However, membrane tubes can also be formed from purified ER-membrane fractions and cytosol, in the absence of microtubules, in a process that is dependent on cytosol-regulated vesicle–membrane fusion. This suggests that a microtubule-independent regulation of ER networks exists. The movement of the ER along and the direct attachment of the ER to microtubules, however, is required for its peripheral localization in mammalian cells.

Insights into how the tubular morphology compared with the sheet-like morphology of the ER is regulated has come from the identification of the reticulon proteins in yeast and mammalian cells, which localize to the tubular ER and are excluded from the sheets. Altering the amounts of these proteins in cells induces transitions between a more tubular ER and a more sheet-like ER. Reticulons contain two long hydrophobic domains that are thought to induce membrane tubulation by constraining local membrane curvature.

Dynamic Morphology of Mitochondria

Depending on the cell type, mitochondrial shapes range widely from small individual organelles, to large interconnected networks. Mitochondrial morphology involves a combination of the underlying tubular structure, the constant tubule fusion and fission dynamics which regulate the degree of mitochondrial fragmentation, and the tubule movement along or the attachment to the cytoskeleton and cell periphery. In mammalian cells mitochondria form tubular networks that spread throughout the cell along the microtubule cytoskeleton, whereas in budding yeast they form an interconnected tubular network that is localized to the inner cell surface which forms a cage-like structure around the cell. The cell regulates the overall degree of mitochondrial fragmentation in response to the cell cycle or apoptotic signals. The mitochondrial tubule fusion and fission dynamics constantly remodel the connections within the mitochondrial network, and therefore perhaps provide a mechanism for the cell to remodel the mitochondrial network, for example, from a more simplified to a more complex and ramified network, as observed in yeast cells in response to changes in growth condition.

The wealth of imaging and genetic tools available to study mitochondrial morphology makes this organelle an ideal model for determining the minimum requirements that are needed to establish a complex organelle morphology from simple algorithmic rules. It is also an ideal model for understanding the basic design principles that underlie the precise regulation of mitochondrial morphology in response to the functional needs of a cell. These sorts of studies, however, require rigorous methods to quantify the dynamic morphology of mitochondria to compare them with modelled structures.

Membrane-bound organelle morphology results from basic physical effects (including surface-to-volume-driven curvature distributions and curvature-induced lipid partitioning) combined with protein-induced curvature changes and the attachment to and motor-driven movement along cellular structures. Thermodynamically, energy input into the system by the cytoskeleton, by the production and turnover of membrane-associated proteins and by the generation of specific lipid composition changes allows the membranous systems to attain more complicated structures than a simple lipid-bound vesicle could attain at thermodynamic equilibrium. Further precise control of the shape-altering dynamics of these structures imparts a flexibility to their morphologies that allows them, in a controlled manner, to regulate the cellular functions accomplished by these complex structures.

Propagation of Structure

Because cells are dynamic systems, their organization must be continuously regulated. It seems most likely that the genome exerts its control over cellular structure not by explicitly dictating where each part of the cell has to go and how it is to be shaped, but rather by providing small course corrections that steer the evolving organization of the cell in particular directions. The steering of dynamic systems by small perturbations is an important principle of control-system theory and has been applied to control biological systems, such as heart rhythms.

The dynamic nature of the cell structure also means that the influence of a given pattern of gene expression on the cell organization will be a strong function of the current state of the cell — the history of a structure of a cell directly affects its future dynamic organization. It has been observed that cells can re-form their original shape after the removal of chemical inhibitors that disrupt cell shape. Cells thus have a persistent state which is gradually modified by dynamic processes.

If cells can restore their geometric state following transient perturbations, one might wonder how much of cellular structure persists following a round of division: cell division is, in essence, a geometrical perturbation. This question can be potentially addressed by comparing the structural similarity between sister cells. In fact, several studies have suggested that sister cells are similar and might even be mirror images of each other. However, these studies have been based on visual observation and the statistical strength of the similarity between sisters is yet to be rigorously demonstrated. The degree to which sister cells are mirror images of each other is therefore an open question, whose answer will have strong implications for the importance of pre-existing structure in the dynamic organization of the cell.

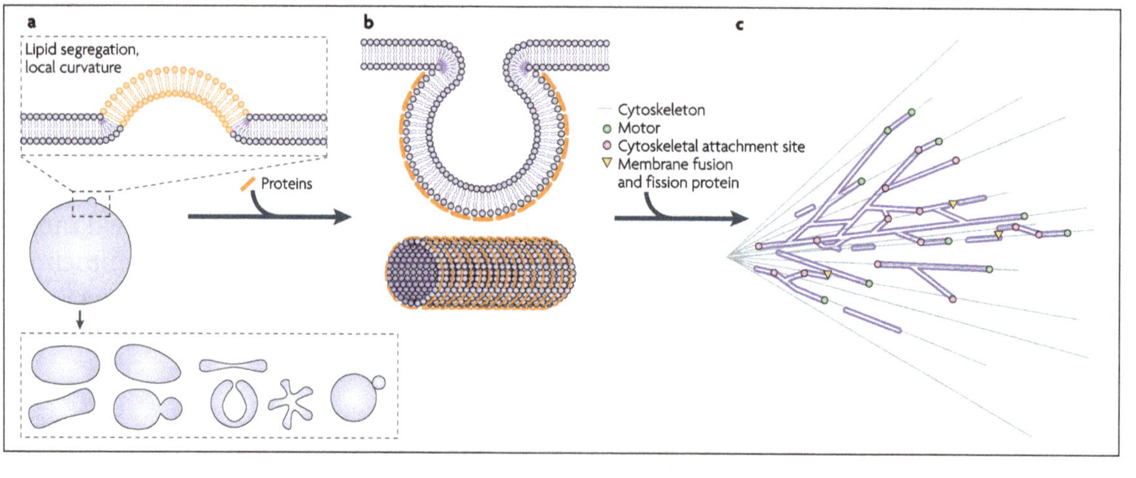

Figure shows Design principles of dynamic shape specification. a. Possible changes in membrane curvature and vesicle shape based on lipid size and composition. The segregation of lipid domains occurs spontaneously within a lipid bilayer and can affect localized curvature (top). Changing the surface-area-to-volume ratio and the relative surface areas of each monolayer within the bilayer can induce a wide range of vesicle shapes simply by minimizing the energetics of membrane curvature (bottom). b. The addition of membrane-binding proteins or integral membrane proteins can induce or stabilize localized membrane curvature, including the extreme curvatures that are seen during the budding of smaller vesicles from the membrane (top) and during the formation of membrane tubes (bottom). c. The cytoskeleton provides a framework along which membrane-protein structures can be moved and attached. Regulated fusion and fission dynamics can constantly remodel the resultant complex membrane networks. The overall network morphology depends on the localization of the cytoskeletal attachment sites and the localization of the fusion and fission proteins within the network.

Table: Design principles for cellular architecture.

Parameter	Design principle
Size	Dynamic balance of assembly and disassembly.
Number	Balance of organelle formation and partitioning.
Polarity	Self-organization biased by external inputs.
Position	Directed motion on polarized cytoskeletal arrays.
Shape	Energetics of protein binding drive curvature change.

In cells with sufficiently complex structures, it has been possible to test for the inheritance of spatial organization by experimentally rearranging cellular components and testing whether the rearranged geometry persists. The famous experiments of Beisson and Sonneborn on inverted ciliary rows in Paramecium aurelia clearly show that artificially rearranged structures can be propagated through division. This phenomenon has been extensively investigated with a series of elegant experiments in various types of ciliates, which have shown that certain aspects of a rearranged organization can be inherited and other rearrangements can self-correct. An example is provided by the oral apparatus, a spiral-shaped array of cilia. The oral apparatus is a polarized structure with a distinct anterior and posterior end, and is also chiral, showing a pronounced left–right asymmetry. When the oral apparatus is repositioned to different places in the cell, it retains its normal left–right asymmetry but switches its anterior–posterior asymmetry in response to the new cellular context. This suggests that the structure relies strictly on local self-propagating cues for left–right position but responds to global cellular cues for anterior–posterior positioning. This has been interpreted as reflecting a varying level of input of local information (which can be rearranged and thus propagated) for some aspects of structure compared with global cellular polarity information (which cannot be rearranged by local manipulations and therefore cannot be propagated). If genomic tools could be used to probe the underlying events of these rearrangements, they might aid the understanding of the origins of cellular geometry and its propagation.

Cells are considered the basic units of life in part because they come in discrete and easily recognizable packages. That's because all cells are surrounded by a structure called the cell membrane — which, much like the walls of a house, serves as a clear boundary between the cell's internal and external environments. The cell membrane is sometimes also referred to as the plasma membrane.

Cell membranes are based on a framework of fat-based molecules called phospholipids, which physically prevent water-loving, or hydrophilic, substances from entering or escaping the cell. These membranes are also studded with proteins that serve various functions. Some of these proteins act as gatekeepers, determining what substances can and cannot cross the membrane. Others function as markers, identifying the cell as part of the same organism or as foreign. Still others work like fasteners, binding cells together so they can function as a unit. Yet other membrane proteins serve as communicators, sending and receiving signals from neighboring cells and the environment — whether friendly or alarming.

Within this membrane, a cell's interior environment is water based. Called cytoplasm, this liquid environment is packed full of cellular machinery and structural elements. In fact, the concentrations of proteins inside a cell far outnumber those on the outside — whether the outside is ocean water (as in the case of a single-celled alga) or blood serum (as in the case of a red blood cell). Although cell membranes form natural barriers in watery environments, a cell must nonetheless expend quite a bit of energy to maintain the high concentrations of intracellular constituents necessary for its survival. Indeed, cells may use as much as 30 percent of their energy just to maintain the composition of their cytoplasm.

Other Components of Cells

A cell's cytoplasm is home to numerous functional and structural elements. These elements exist in the form of molecules and organelles — picture them as the tools, appliances, and inner rooms of the cell. Major classes of intracellular organic molecules include nucleic acids, proteins, carbohydrates, and lipids, all of which are essential to the cell's functions.

Nucleic acids are the molecules that contain and help express a cell's genetic code. There are two major classes of nucleic acids: deoxyribonucleic acid (DNA) and ribonucleic acid (RNA). DNA is the molecule that contains all of the information required to build and maintain the cell; RNA has several roles associated with expression of the information stored in DNA. Of course, nucleic acids alone aren't responsible for the preservation and expression of genetic material: Cells also use proteins to help replicate the genome and accomplish the profound structural changes that underlie cell division.

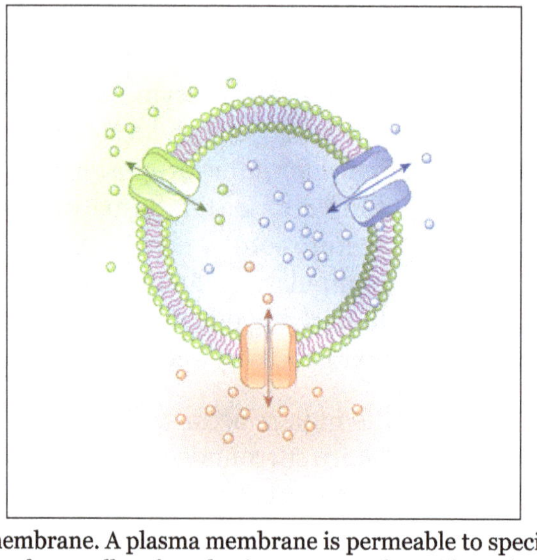

Transport proteins in the cell membrane. A plasma membrane is permeable to specific molecules that a cell needs. Transport proteins in the cell membrane allow for selective passage of specific molecules from the external environment. Each transport protein is specific to a certian molecule (indicated by matching colors).

Proteins are a second type of intracellular organic molecule. These substances are made from chains of smaller molecules called amino acids, and they serve a variety of functions in the cell, both catalytic and structural. For example, proteins called enzymes convert cellular molecules (whether proteins, carbohydrates, lipids, or nucleic acids) into other forms that might help a cell meet its energy needs, build support structures, or pump out wastes.

Carbohydrates, the starches and sugars in cells, are another important type of organic molecule. Simple carbohydrates are used for the cell's immediate energy demands, whereas complex carbohydrates serve as intracellular energy stores. Complex carbohydrates are also found on a cell's surface, where they play a crucial role in cell recognition.

Finally, lipids or fat molecules are components of cell membranes — both the plasma membrane and various intracellular membranes. They are also involved in energy storage, as well as relaying signals within cells and from the bloodstream to a cell's interior.

Some cells also feature orderly arrangements of molecules called organelles. Similar to the rooms in a house, these structures are partitioned off from the rest of a cell's interior by their own intracellular membrane. Organelles contain highly technical equipment required for specific jobs within the cell. One example is the mitochondrion — commonly known as the cell's "power plant" — which is the organelle that holds and maintains the machinery involved in energy-producing chemical reactions.

Figure shows the relative scale of biological molecules and structures. Cells can vary between 1 micrometer (μm) and hundreds of micrometers in diameter. Within a cell, a DNA double helix is approximately 10 nanometers (nm) wide, whereas the cellular organelle called a nucleus that encloses this DNA can be approximately 1000 times bigger (about 10 μm). See how cells compare along a relative scale axis with other molecules, tissues, and biological structures (blue arrow at bottom). Note that a micrometer (μm) is also known as a micron.

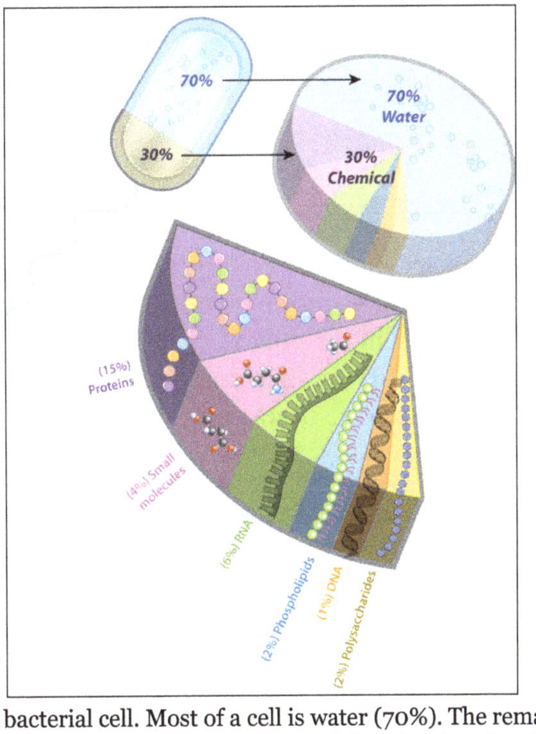

The composition of a bacterial cell. Most of a cell is water (70%). The remaining 30% contains varying proportions of structural and functional molecules.

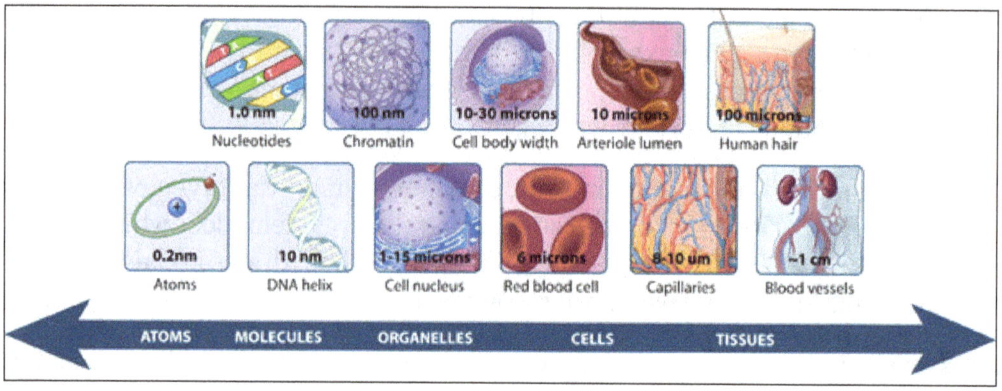

Different Categories of Cells

Rather than grouping cells by their size or shape, scientists typically categorize them by how their genetic material is packaged. If the DNA within a cell is not separated from the cytoplasm, then that cell is a prokaryote. All known prokaryotes, such as bacteria and archaea, are single cells. In contrast, if the DNA is partitioned off in its own membrane-bound room called the nucleus, then that cell is a eukaryote. Some eukaryotes, like amoebae, are free-living, single-celled entities. Other eukaryotic cells are part of multicellular organisms. For instance, all plants and animals are made of eukaryotic cells — sometimes even trillions of them.

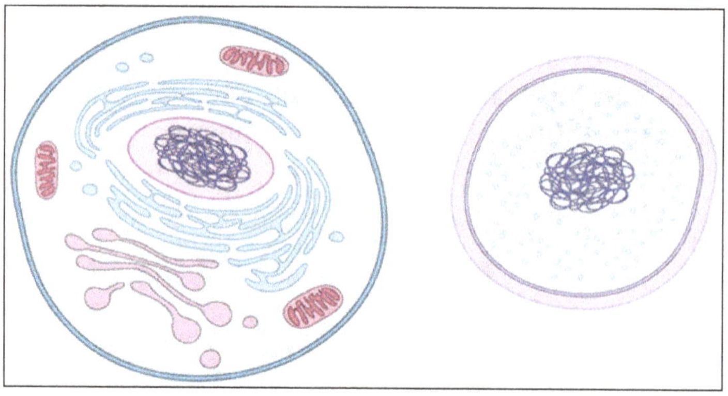

Figure shows comparing basic eukaryotic and prokaryotic differences. A eukaryotic cell (left) has membrane-enclosed DNA, which forms a structure called the nucleus (located at center of the eukaryotic cell; note the purple DNA enclosed in the pink nucleus). A typical eukaryotic cell also has additional membrane-bound organelles of varying shapes and sizes. In contrast, a prokaryotic cell (right) does not have membrane-bound DNA and also lacks other membrane-bound organelles as well.

Researchers hypothesize that all organisms on Earth today originated from a single cell that existed some 3.5 to 3.8 billion years ago. This original cell was likely little more than a sac of small organic molecules and RNA-like material that had both informational and catalytic functions. Over time, the more stable DNA molecule evolved to take over the information storage function, whereas proteins, with a greater variety of structures than nucleic acids, took over the catalytic functions.

The absence or presence of a nucleus — and indeed, of all membrane-bound organelles — is important enough to be a defining feature by which cells are categorized as either prokaryotes or eukaryotes. Scientists believe that the appearance of self-contained nuclei and other organelles represents a major advance in the evolution of cells. But where did these structures come from? More than one billion years ago, some cells "ate" by engulfing objects that floated in the liquid environment in which they existed. Then, according to some theories of cellular evolution, one of the early eukaryotic cells engulfed a prokaryote, and together the two cells formed a symbiotic relationship. In particular, the engulfed cell began to function as an organelle within the larger eukaryotic cell that consumed it. Both chloroplasts and mitochondria, which exist in modern eukaryotic cells and still retain their own genomes, are thought to have arisen in this manner.

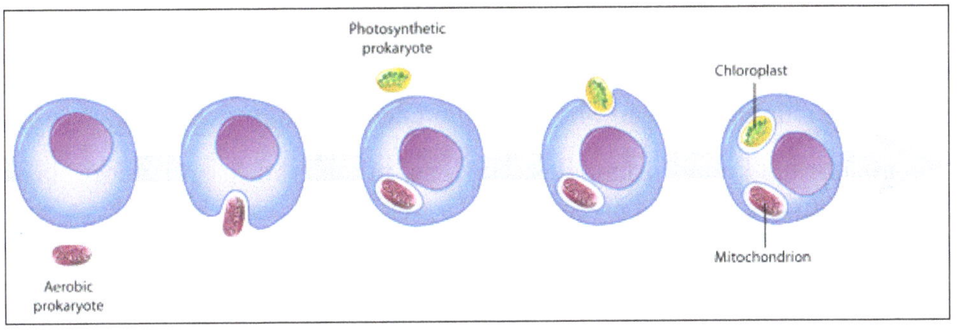

Figure shows the origin of mitochondria and chloroplasts. Mitochondria and chloroplasts likely evolved from engulfed prokaryotes that once lived as independent organisms. At some point, a eukaryotic cell engulfed an aerobic prokaryote, which then formed an endosymbiotic relationship with the host eukaryote, gradually developing into a mitochondrion. Eukaryotic cells containing mitochondria then engulfed photosynthetic prokaryotes, which evolved to become specialized chloroplast organelles.

Of course, prokaryotic cells have continued to evolve as well. Different species of bacteria and archaea have adapted to specific environments, and these prokaryotes not only survive but thrive without having their genetic material in its own compartment. For example, certain bacterial species that live in thermal vents along the ocean floor can withstand higher temperatures than any other organisms on Earth.

Cell Organelles

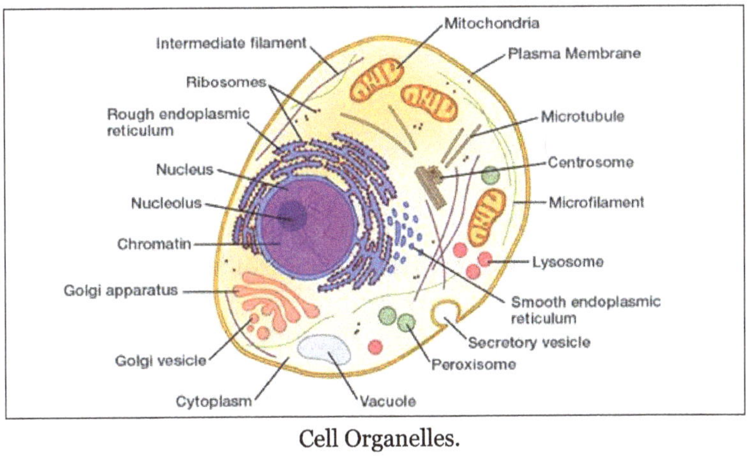

Cell Organelles.

The cellular components are called cell organelles. These cell organelles include both membrane and non-membrane bound organelles, present within the cells and are distinct in their structures and functions. They coordinate and function efficiently for the normal functioning of the cell. A few of them function by providing shape and support, whereas some are involved in the locomotion and reproduction of a cell. There are various organelles present within the cell and are classified into three categories based on the presence or absence of membrane.

- Organelles without membrane: The Cell wall, Ribosomes, and Cytoskeleton are non-membrane-bound cell organelles. They are present both in prokaryotic cell and the eukaryotic cell.
- Single membrane-bound organelles: Vacuole, Lysosome, Golgi Apparatus, Endoplasmic Reticulum are single membrane-bound organelles present only in a eukaryotic cell.
- Double membrane-bound organelles: Nucleus, mitochondria and chloroplast are double membrane-bound organelles present only in a eukaryotic cell.

List of Cell Organelles and their Functions

Plasma Membrane

The plasma membrane is also termed as a Cell Membrane or Cytoplasmic Membrane. It is a selectively permeable membrane of the cell, which is composed of a lipid bilayer and proteins.

The plasma membrane is present both in plant and animal cell, which functions as the selectively permeable membrane, by permitting the entry of selective materials in and out of the cell according to the requirement. In an animal cell, the cell membrane functions by providing shape and protects the inner contents of the cell. Based on the structure of the plasma membrane, it is regarded as the fluid mosaic model. According to the fluid mosaic model, the plasma membranes are subcellular structures, made of a lipid bilayer in which the protein molecules are embedded.

Cytoplasm

The cytoplasm is present both in plant and animal cells. They are jelly-like substances, found between the cell membrane and nucleus. They are mainly composed of water, organic and inorganic compounds. The cytoplasm is one of the essential components of the cell, where all the cell organelles are embedded. These cell organelles contain enzymes, mainly responsible for controlling all metabolic activity taking place within the cell and are the site for most of the chemical reactions within a cell.

Nucleus

The nucleus is a double-membraned organelle found in all eukaryotic cells. It is the largest organelle, which functions as the control centre of the cellular activities and is the storehouse of the cell's DNA. By structure, the nucleus is dark, round, surrounded by a nuclear membrane. It is a porous membrane (like cell membrane) and forms a wall between cytoplasm and nucleus. Within the nucleus, there are tiny spherical bodies called nucleolus. It also carries another essential structure called chromosomes.

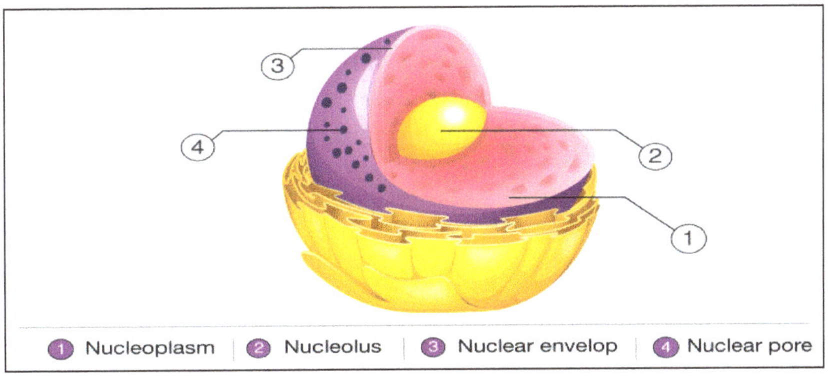

Chromosomes are thin and thread-like structures which carry another important structure called a gene. Genes are a hereditary unit in organisms i.e., it helps in the inheritance of traits from one generation (parents) to another (offspring). Hence, the nucleus controls the characters and functions of cells in our body. The primary function of the nucleus is to monitor cellular activities including metabolism and growth by making use of DNA's genetic information. Nucleoli in the nucleus are responsible for the synthesis of protein and RNA.

Endoplasmic Reticulum

The Endoplasmic Reticulum is a network of membranous canals filled with fluid. They are the transport system of the cell, involved in transporting materials throughout the cell. There are two different types of Endoplasmic Reticulum:

- Rough Endoplasmic Reticulum: They are composed of cisternae, tubules, and vesicles, which are found throughout the cell and are involved with protein manufacture.
- Smooth Endoplasmic Reticulum: They are the storage organelle, associated with the production of lipids, steroids, and also responsible for detoxifying the cell.

Mitochondria

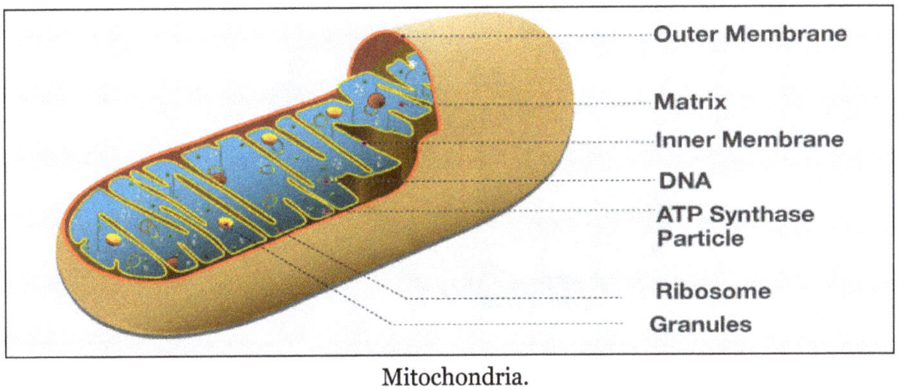

Mitochondria.

Mitochondria are called the powerhouses of the cell as they produce energy-rich molecules for the cell. The mitochondrial genome is inherited maternally in several organisms. It is a double membrane-bound, sausage-shaped organelle, found in almost all eukaryotic cells.

The double membranes divide its lumen into two distinct aqueous compartments. The inner compartment is called 'matrix' which is folded into cristae whereas the outer membrane forms a continuous boundary with the cytoplasm. They usually vary in their size and are found either round or oval in shape. Mitochondria are the sites of aerobic respiration in the cell, produces energy in the form of ATP and helps in the transformation of the molecules.

For instance, glucose is converted into adenosine triphosphate – ATP. Mitochondria have their own circular DNA, RNA molecules, ribosomes (the 70s), and a few other molecules that help in protein synthesis.

Plastids

Plastids are large, membrane-bound organelles which contain pigments. Based on the type of pigments, plastids are of three types:

- Chloroplasts: Chloroplasts are double membrane-bound organelles, which usually vary in their shape – from a disc shape to spherical, discoid, oval and ribbon. They are present in mesophyll cells of leaves, which store chloroplasts and other carotenoid pigments. These pigments are responsible for trapping light energy for photosynthesis. The inner membrane encloses a space called the stroma. Flattened disc-like chlorophyll-containing structures known as thylakoids are arranged in a stacked manner like a pile of coins. Each pile is called as granum (plural: grana) and the thylakoids of different grana are connected by flat membranous tubules known as stromal lamella. Just like the mitochondrial matrix, the stroma of chloroplast also contains a double-stranded circular DNA, 70S ribosomes, and enzymes which required for the synthesis of carbohydrates and proteins.
- Chromoplasts: The chromoplasts include fat-soluble, carotenoid pigments like xanthophylls, carotene, etc. which provide the plants with their characteristic color – yellow, orange, red, etc.
- Leucoplasts: Leucoplasts are colorless plastids which store nutrients. Amyloplasts store carbohydrates (like starch in potatoes), aleuroplasts store proteins, and elaioplasts store oils and fats.

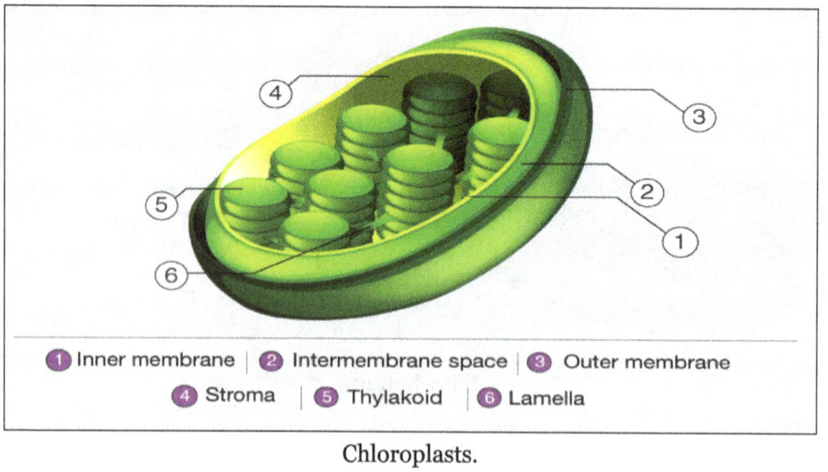

Chloroplasts.

Ribosomes

Ribosomes are nonmembrane-bound and important cytoplasmic organelles found in close association with the endoplasmic reticulum. Ribosomes are found in the form of tiny particles in a large number of cells and are mainly composed of 2/3rd of RNA and 1/3rd of protein. They are named as the 70s (found in prokaryotes) or 80s (found in eukaryotes) The letter S refers to the density and the size, known as Svedberg's Unit. Both 70S and 80S ribosomes are composed of two sub-units. Ribosomes are either encompassed within the endoplasmic reticulum or are freely traced in the cell's cytoplasm. Ribosomal RNA and Ribosomal proteins are the two components that together constitute ribosomes. The primary function of the ribosomes includes protein synthesis in all living cells that ensure the survival of the cell.

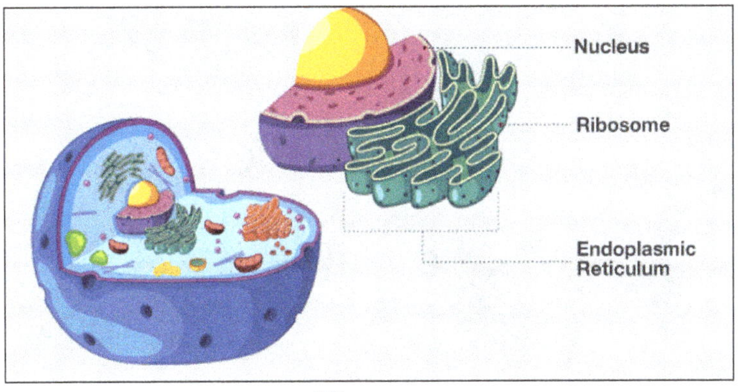

Ribosomes.

Golgi Apparatus

Golgi Apparatus also termed as Golgi Complex. It is a membrane-bound organelle, which is mainly composed of a series of flattened, stacked pouches called cisternae. This cell organelle is primarily responsible for transporting, modifying, and packaging proteins and lipid to targeted destinations. Golgi Apparatus is found within the cytoplasm of a cell and are present in both plant and animal cells.

Microbodies

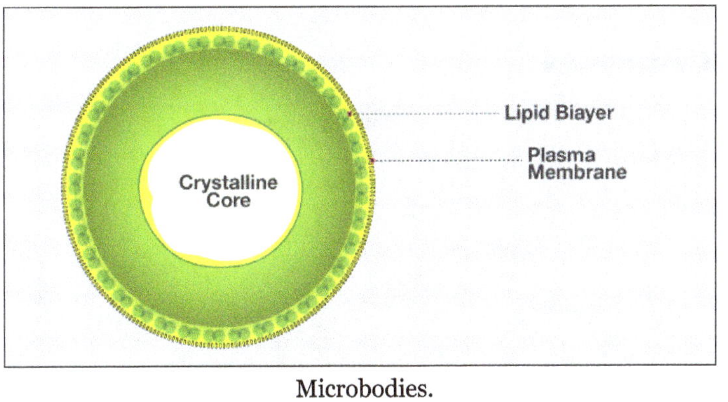

Microbodies.

Microbodies are membrane-bound, minute, vesicular organelles, found in both plant and animal cell. They contain various enzymes and proteins and can be visualized only under the electron microscope.

Cytoskeleton

It is a continuous network of filamentous proteinaceous structures that run throughout the cytoplasm, from the nucleus to the plasma membrane. It is found in all living cells, notably in the eukaryotes. The cytoskeleton matrix is composed of different types of proteins that can divide rapidly or disassemble depending on the requirement of the cells. The primary functions include providing the shape and mechanical resistance to the cell against deformation, the contractile nature of the filaments helps in motility and during cytokinesis.

Cilia and Flagella

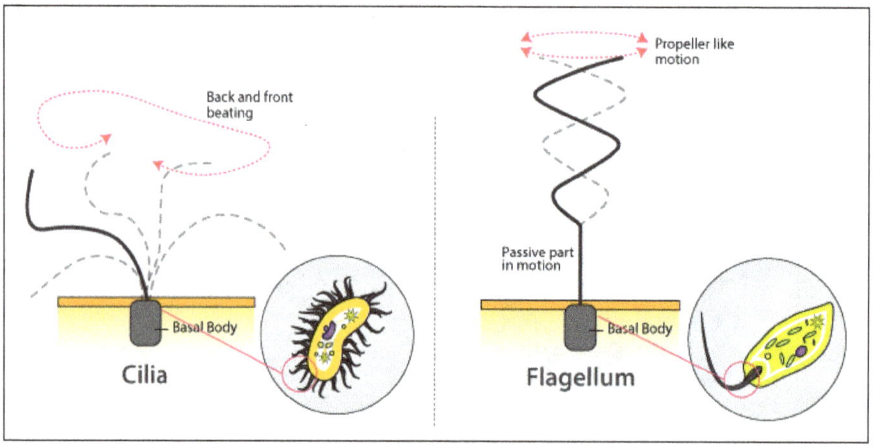

Cilia and Flagella.

Cilia are hair-like projections, small structures, present outside the cell wall and work like oars to either move the cell or the extracellular fluid. Flagella are slightly bigger and are responsible for the cell movements. The eukaryotic flagellum structurally differs from its prokaryotic counterpart. The core of the cilium and flagellum is called a axoneme, which contains nine pairs of gradually arranged peripheral microtubules and a set of central microtubules running parallel to the axis. The central tubules are interconnected by a bridge and are embedded by a central sheath. One of the peripheral microtubular pairs is also interconnected to the central sheath by a radial spoke. Hence there is a total of 9 radial spokes. The cilia and flagella emerge from centriole-like structures called basal bodies.

Centrosome and Centrioles

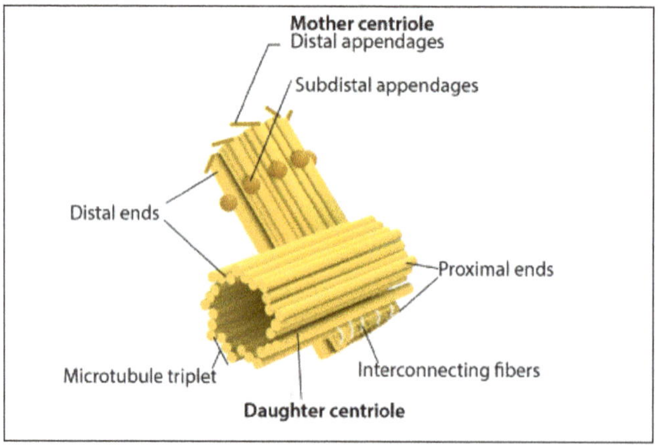

Centrosome and Centrioles.

The centrosome organelle is made up of two mutually perpendicular structures known as centrioles. Each centriole is composed of 9 equally spaced peripheral fibrils of tubulin protein, and the fibril is a set of interlinked triplets. The core part of the centriole is known as a hub and is proteinaceous. The hub connects the peripheral fibrils via radial spoke, which is made up of proteins. The centrioles from the basal bodies of the cilia and flagella give rise to spindle fibres during cell division.

Vacuoles

Vacuoles are mostly defined as storage bubbles of irregular shapes which are found in cells. They are fluid-filled organelles enclosed by a membrane. The vacuole stores the food or a variety of nutrients that a cell might need to survive. In addition to this, it also stores waste products. The waste products are eventually thrown out by vacuoles. Thus, the rest of the cell is protected from contamination. The animal and plant cell have different size and number of vacuoles. Compared to the animals, plant cell have larger vacuoles.

Cell Membrane

Their roots in synaptic dysfunction. world by a thin, fragile structure called the plasma membrane that is only 5 to 10 nm wide.

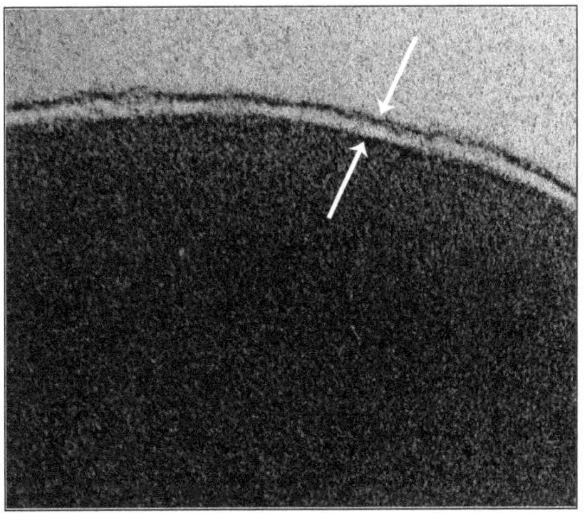

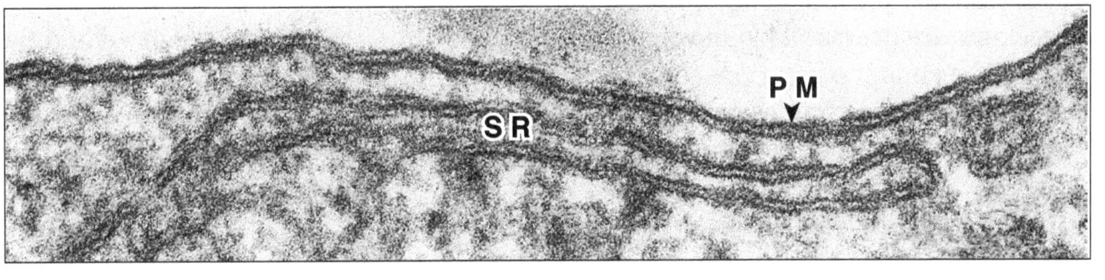

Figure shows the trilaminar appearance of membranes. (a) Electron micrograph showing the three-layered (trilaminar) structure of the plasma membrane of an erythrocyte after staining the tissue with the heavy metal osmium. Osmium binds preferentially to the polar head groups of the lipid bilayer, producing the trilaminar pattern. The arrows denote the inner and outer edges of the membrane. (b) The outer edge of a differentiated muscle cell grown in culture showing the similar trilaminar structure of both the plasma membrane (PM) and the membrane of the sarcoplasmic reticulum (SR), a calcium-storing compartment of the cytoplasm.

Because it is so thin, no hint of the plasma membrane is detected when a section of a cell is examined under a light microscope. In fact, it wasn't until the late 1950s that techniques for preparing and staining tissue had progressed to the point where the plasma membrane could be resolved in the electron microscope. These early electron micrographs, such as those taken by J. D. Robertson of Duke University, portrayed the plasma membrane as a three-layered structure, consisting of darkly staining inner and outer layers and lightly staining middle layer. All membranes that were examined closely—whether they were plasma, nuclear, or cytoplasmic membranes, or taken from plants, animals, or microorganisms—showed this same ultrastructure. In addition to providing a visual image of this critically important cellular structure, these electron micrographs touched off a vigorous debate as to the molecular composition of the various layers of a membrane, an argument that went to the very heart of the subject of membrane structure and function. As we will see shortly, cell membranes contain a lipid bilayer, and the two dark-staining layers in the electron micrographs of figure correspond primarily to the inner and outer polar surfaces of the bilayer. We will return to the structure of membranes below, but first we will survey some of the major functions of membranes in the life of a cell.

An Overview of Membrane Functions

Compartmentalization

Membranes are continuous, unbroken sheets and, as such, inevitably enclose compartments. The plasma membrane encloses the contents of the entire cell, whereas the nuclear and cytoplasmic membranes enclose diverse intracellular spaces. The various membrane-bounded compartments of a cell possess markedly different contents. Membrane compartmentalization allows specialized activities to proceed without external interference and enables cellular activities to be regulated independently of one another.

Scaffold for Biochemical Activities

Membranes not only enclose compartments but are also a distinct compartment themselves. As long as reactants are present in solution, their relative positions cannot be stabilized and their interactions are dependent on random collisions. Because of their construction, membranes provide the cell with an extensive framework or scaffolding within which components can be ordered for effective interaction.

Providing a Selectively Permeable Barrier

Membranes prevent the unrestricted exchange of molecules from one side to the other. At the same time, membranes provide the means of communication between the compartments they separate. The plasma membrane, which encircles a cell, can be compared to a moat around a castle: both serve as a general barrier, yet both have gated "bridges" that promote the movement of select elements into and out of the enclosed living space.

Transporting Solutes

The plasma membrane contains the machinery for physically transporting substances from one side of the membrane to another, often from a region where the solute is present at low concentration into a region where that solute is present at much higher concentration. The membrane's transport machinery allows a cell to accumulate substances, such as sugars and amino acids, that are necessary to fuel its metabolism and build its macromolecules. The plasma membrane is also able to transport specific ions, thereby establishing ionic gradients across itself. This capability is especially critical for nerve and muscle cells.

Responding to External Stimuli

The plasma membrane plays a critical role in the response of a cell to external stimuli, a process known as signal transduction. Membranes possess receptors that combine with specific molecules (ligands) or respond to other types of stimuli such as light or mechanical tension. Different types of cells have membranes with different receptors and are, therefore, capable of recognizing and responding to different environmental stimuli. The interaction of a plasma membrane receptor with an external stimulus may cause the membrane to generate a signal that stimulates or inhibits internal activities. For example, signals generated at the plasma membrane may tell a cell to manufacture more glycogen, to prepare for cell division, to move toward a higher concentration of a particular compound, to release calcium from internal stores, or possibly to commit suicide.

Intercellular Interaction

Situated at the outer edge of every living cell, the plasma membrane of multicellular organisms mediates the interactions between a cell and its neighbors. The plasma membrane allows cells to recognize and signal one another, to adhere when appropriate, and to exchange materials and information. Proteins within the plasma membrane may also facilitate the interaction between extracellular materials and the intracellular cytoskeleton.

Energy Transduction

Membranes are intimately involved in the processes by which one type of energy is converted to another type (energy transduction). The most fundamental energy transduction occurs during photosynthesis when energy in sunlight is absorbed by membrane-bound pigments, converted into chemical energy, and stored in carbohydrates. Membranes are also involved in the transfer of chemical energy from carbohydrates and fats to ATP. In eukaryotes, the machinery for these energy conversions is contained within membranes of chloroplasts and mitochondria.

Figure show a summary of membrane functions in a plant cell. (1) An example of membrane compartmentalization in which hydrolytic enzymes (acid hydrolases) are sequestered within the membrane bounded vacuole. (2) An example of the role of cytoplasmic membranes as a site of enzyme localization. The fixation of CO_2 by the plant cell is catalyzed by an enzyme that is associated with the outer surface of the thylakoid membranes of the chloroplasts. (3) An example of the role of membranes as a selectively permeable barrier. Water molecules are able to penetrate rapidly through the plasma membrane, causing the plant cell to fill out the available space and exert pressure against its cell wall. (4) An example of solute transport. Hydrogen ions, which are produced by various metabolic processes in the cytoplasm, are pumped out of plant cells into the extracellular space by a transport protein located in the plasma membrane. (5) An example of the involvement of a membrane in the transfer of information from one side to another (signal transduction). In this case, a hormone (e.g., abscisic acid) binds to the outer surface of the plasma membrane and triggers the release of a chemical message (such as IP_3) into the cytoplasm. In this case, IP_3 causes release of Ca^{2+} ions from a cytoplasmic warehouse. (6) An example of the role of membranes in cell–cell communication. Openings between adjoining plant cells, called plasmodesmata, allow materials to move directly from the cytoplasm of one cell into its neighbors. (7) An example of the role of membranes in energy transduction. The conversion of ADP to ATP occurs in close association with the inner membrane of the mitochondrion.

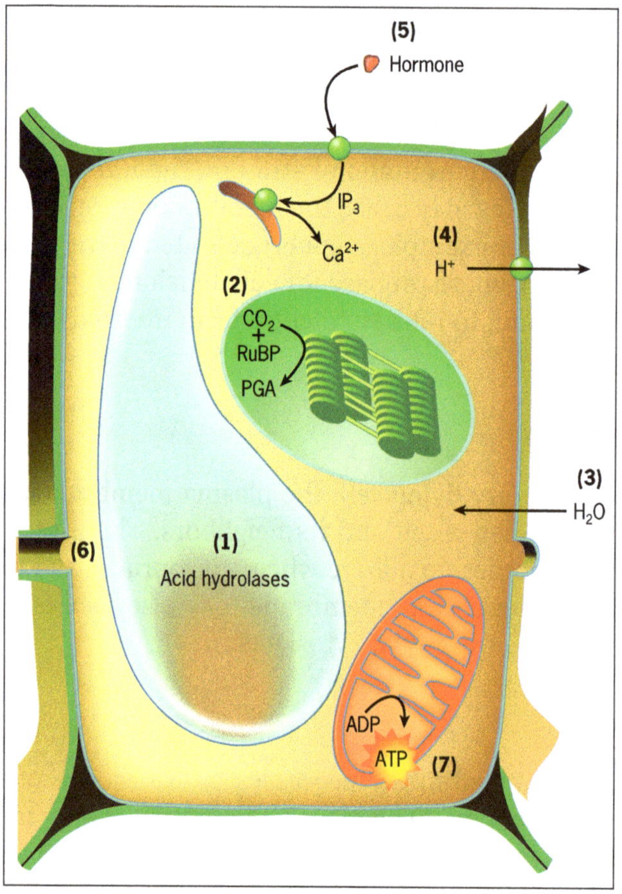

The first insights into the chemical nature of the outer boundary layer of a cell were obtained by Ernst Overton of the University of Zürich during the 1890s. Overton knew that nonpolar solutes dissolved more readily in nonpolar solvents than in polar solvents, and that polar solutes had the opposite solubility. Overton reasoned that a substance entering a cell from the medium would first have to dissolve in the outer boundary layer of that cell. To test the permeability of the outer boundary layer, Overton placed plant root hairs into hundreds of different solutions containing a diverse array of solutes. He discovered that the more lipid-soluble the solute, the more rapidly it would enter the root hair cells. He concluded that the dissolving power of the outer boundary layer of the cell matched that of a fatty oil.

The first proposal that cellular membranes might contain a lipid bilayer was made in 1925 by two Dutch scientists, E. Gorter and F. Grendel. These researchers extracted the lipid from human red blood cells and measured the amount of surface area the lipid would cover when spread over the surface of water. Since mature mammalian red blood cells lack both nuclei and cytoplasmic organelles, the plasma membrane is the only lipid-containing structure in the cell. Consequently, all of the lipids extracted from the cells can be assumed to have resided in the cells' plasma membranes. The ratio of the surface area of water covered by the extracted lipid to the surface area calculated for the red blood cells from which the lipid was extracted varied between 1.8 to 1 and 2.2 to 1. Gorter and Grendel speculated that the actual ratio was 2:1 and concluded that the plasma membrane contained a bimolecular layer of lipids, that is, a lipid bilayer. They also suggested that the polar groups of each molecular layer (or leaflet) were directed outward toward the aqueous environment, as shown in figure b, c. This would be the thermodynamically favored arrangement, because the polar head groups of the lipids could interact with surrounding water molecules, just as the hydrophobic fatty acyl chains would be protected from contact with the aqueous environment. Thus, the polar head groups would face the cytoplasm on one edge and the blood plasma on the other. Even though Gorter and Grendel made several experimental errors (which fortuitously canceled one another out), they still arrived at the correct conclusion that membranes contain a lipid bilayer.

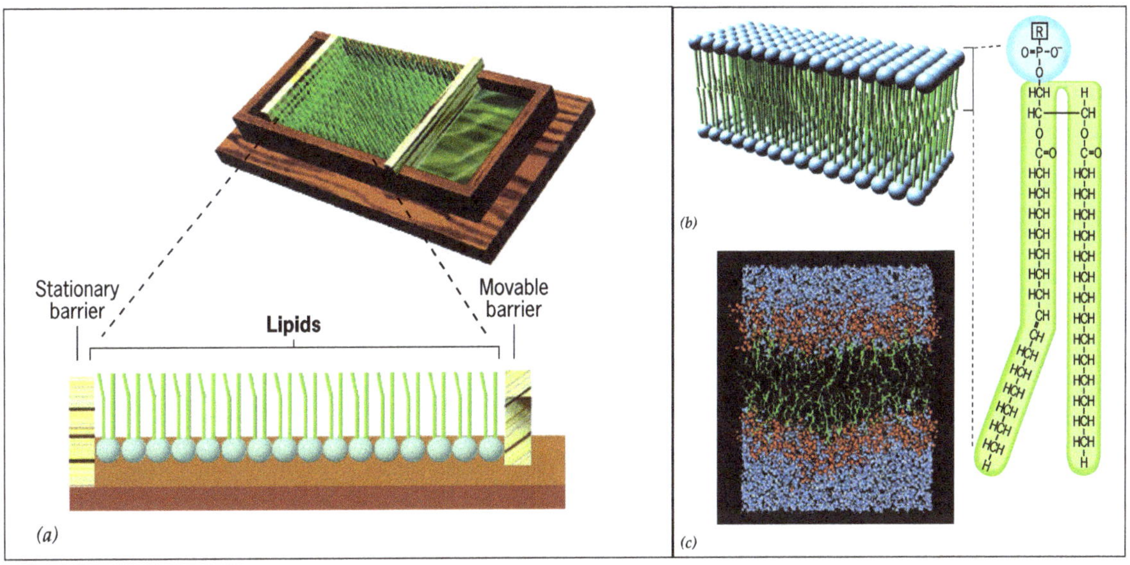

Figure shows the plasma membrane contains a lipid bilayer. (a) Calculating the surface area of a lipid preparation. When a sample of phospholipids is dissolved in an organic solvent, such as hexane, and spread over an aqueous surface, the phospholipid molecules form a layer over the water that is a single molecule thick: a monomolecular layer. The molecules in the layer are oriented with their hydrophilic groups bonded to the surface of the water and their hydrophobic chains directed into the air. To estimate the surface area the lipids would cover if they were part of a membrane, the lipid molecules can be compressed into the smallest possible area by means of movable barriers. Using this type of apparatus, which is called a Langmuir trough after its inventor, Gorter and Grendel concluded that red blood cells contained enough lipid to form a layer over their surface that was two molecules thick: a bilayer. (b) As Gorter and Grendel first proposed, the core of a membrane contains a bimolecular layer of phospholipids oriented with their water-soluble head groups facing the outer surfaces and their hydrophobic fatty acid tails facing the interior. (c) Simulation of a fully hydrated lipid bilayer composed of the phospholipid phosphatidylcholine. Phospholipid head groups are orange, water molecules are blue and white, fatty acid chains are green.

In the 1920s and 1930s, cell physiologists obtained evidence that there must be more to the structure of membranes than simply a lipid bilayer. It was found, for example, that lipid solubility was not the sole determining factor as to whether or not a substance could penetrate the plasma membrane. Similarly, the surface tensions of membranes were calculated to be much lower than those of pure lipid structures. This decrease in surface tension could be explained by the presence of protein in the membrane. In 1935, Hugh Davson and James Danielli proposed that the plasma membrane was composed of a lipid bilayer that was lined on both its inner and outer surface by a layer of globular proteins. They revised their model in the early 1950s to account for the selective permeability of the membranes they had studied. In the revised version, Davson and Danielli suggested that, in addition to the outer and inner protein layers, the lipid bilayer was also penetrated by protein-lined pores, which could provide conduits for polar solutes and ions to enter and exit the cell.

In the fluid-mosaic model, which has served as the "central dogma" of membrane biology for more than three decades, the lipid bilayer remains the core of the membrane, but attention is focused on the physical state of the lipid. The bilayer of a fluid-mosaic membrane is present in a fluid state, and individual lipid molecules can move laterally within the plane of the membrane.

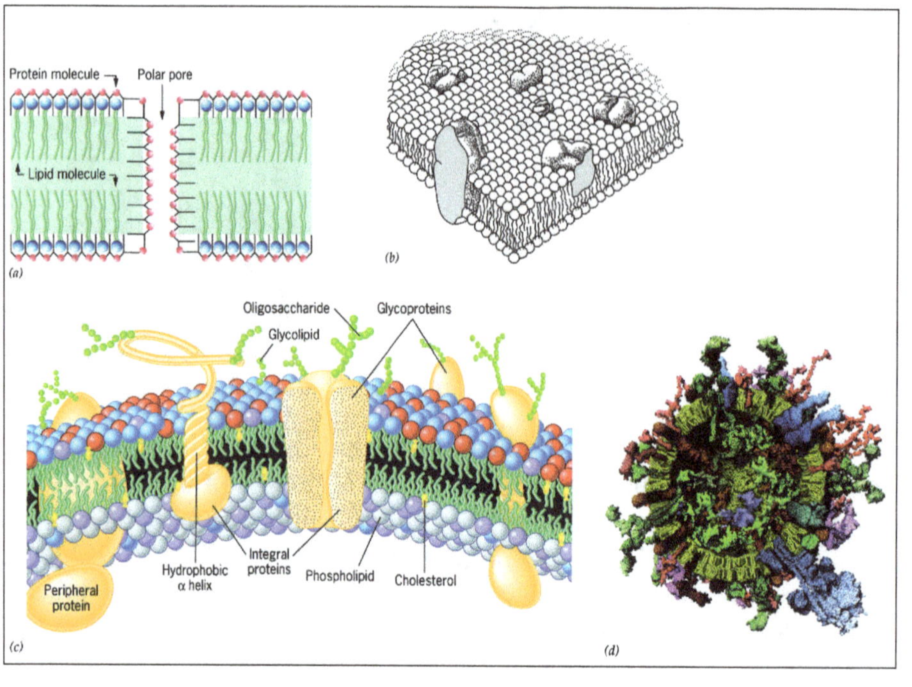

Figure shows a brief history of the structure of the plasma membrane. (a) A revised 1954 version of the Davson-Danielli model showing the lipid bilayer, which is lined on both surfaces by a monomolecular layer of proteins that extends through the membrane to form protein-lined pores. (b) The fluid-mosaic model of membrane structure as initially proposed by Singer and Nicolson in

1972. Unlike previous models, the proteins penetrate the lipid bilayer. Although the original model shown here depicted a protein that was only partially embedded in the bilayer, lipid-penetrating proteins that have been studied span the entire bilayer. (c) A current representation of the plasma membrane showing the same basic organization as that proposed by Singer and Nicolson. The external surface of most membrane proteins, as well as a small percentage of the phospholipids, contain short chains of sugars, making them glycoproteins and glycolipids. Those portions of the polypeptide chains that extend through the lipid bilayer typically occur as a helices composed of hydrophobic amino acids. The two leaflets of the bilayer contain different types of lipids as indicated by the differently colored head groups. (d) Molecular model of the membrane of a synaptic vesicle constructed using known structures of the various proteins along with information on their relative numbers obtained from the analysis of purified synaptic vesicles. The high protein density of the membrane is apparent. Most of the proteins in this membrane are required for the interaction of the vesicle with the plasma membrane. The large blue protein at the lower right pumps H^+ ions into the vesicle.

The structure and arrangement of membrane proteins in the fluid-mosaic model differ from those of previous models in that they occur as a "mosaic" of discontinuous particles that penetrate the lipid sheet. Most importantly, the fluid-mosaic model presents cellular membranes as dynamic structures in which the components are mobile and capable of coming together to engage in various types of transient or semipermanent interactions. In the following sections, we will examine some of the evidence used to formulate and support this dynamic portrait of membrane structure and look at some of the recent data that bring the model up to date.

The Chemical Composition of Membranes

Membranes are lipid–protein assemblies in which the components are held together in a thin sheet by noncovalent bonds. As noted above, the core of the membrane consists of a sheet of lipids arranged in a bimolecular layer. The lipid bilayer serves primarily as a structural backbone of the membrane and provides the barrier that prevents random movements of water-soluble materials into and out of the cell. The proteins of the membrane, on the other hand, carry out most of the specific functions summarized in figure. Each type of differentiated cell contains a unique complement of membrane proteins, which contributes to the specialized activities of that cell type.

The ratio of lipid to protein in a membrane varies, depending on the type of cellular membrane (plasma vs. endoplasmic reticulum vs. Golgi), the type of organism (bacterium vs. plant vs. animal), and the type of cell (cartilage vs. muscle vs. liver). For example, the inner mitochondrial membrane has a very high ratio of protein/lipid in comparison to the red blood cell plasma membrane, which is high in comparison to the membranes of the myelin sheath that form a multilayered wrapping around a nerve cell. To a large degree, these differences can be correlated with the basic functions of these membranes. The inner mitochondrial membrane contains the protein carriers of the electron-transport chain, and relative to other membranes, lipid is diminished. In contrast, the myelin sheath acts primarily as electrical insulation for the nerve cell it encloses, a function that is best carried out by a thick lipid layer of high electrical resistance with a minimal content of protein. Membranes also contain carbohydrates, which are attached to the lipids and proteins as indicated in figure.

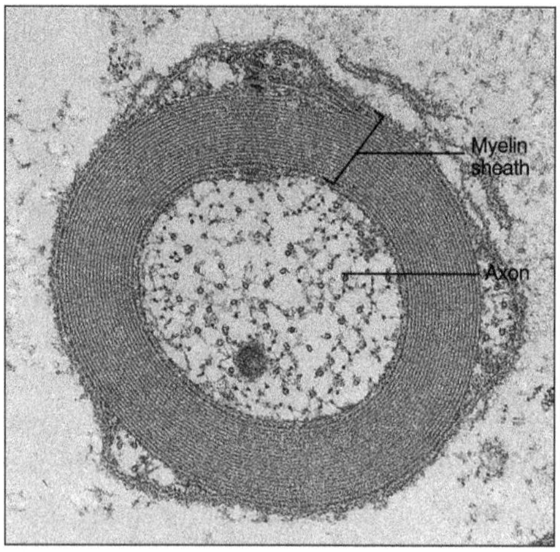

Figure shows the myelin sheath. Electron micrograph of a nerve cell axon surrounded by a myelin sheath consisting of concentric layers of plasma membrane that have an extremely low protein/lipid ratio. The myelin sheath insulates the nerve cell from the surrounding environment, which increases the velocity at which impulses can travel along the axon. The perfect spacing between the layers is maintained by interlocking protein molecules (called P_0) that project from each membrane.

Membrane Lipids

Membranes contain a wide diversity of lipids, all of which are amphipathic; that is, they contain both hydrophilic and hydrophobic regions. There are three main types of membrane lipids: phosphoglycerides, sphingolipids, and cholesterol.

Phosphoglycerides

Most membrane lipids contain a phosphate group, which makes them phospholipids. Because most membrane phospholipids are built on a glycerol backbone, they are called phosphoglycerides. Unlike triglycerides, which have three fatty acids and are not amphipathic, membrane glycerides are diglycerides—only two of the hydroxyl groups of the glycerol are esterified to fatty acids; the third is esterified to a hydrophilic phosphate group. Without any additional substitutions beyond the phosphate and the two fatty acyl chains, the molecule is called phosphatidic acid, which is virtually absent in most membranes. Instead, membrane phosphoglycerides have an additional group linked to the phosphate, most commonly either choline (forming phosphatidylcholine, PC), ethanolamine (forming phosphatidylethanolamine, PE), serine (forming phosphatidylserine, PS), or inositol (forming phosphatidylinositol, PI). Each of these groups is small and hydrophilic and, together with the negatively charged phosphate to which it is attached, forms a highly water-soluble domain at one end of the molecule, called the head group. At physiologic pH, the head groups of PS and PI have an overall negative charge, whereas those of PC and PE are neutral. In contrast, the fatty acyl chains are hydrophobic, unbranched hydrocarbons approximately 16 to 22 carbons in length. A membrane fatty acid may be fully saturated (i.e., lack double bonds), monounsaturated (i.e., possess one double bond), or polyunsaturated (i.e., possess more than one double bond). Phosphoglycerides often contain one unsaturated and one saturated fatty acyl chain. Recent interest has focused on the apparent health benefits of two highly unsaturated fatty acids (EPA and DHA) found at high concentration in fish oil. EPA and

DHA contain five and six double bonds, respectively, and are incorporated primarily into PE and PC molecules of certain membranes, most notably in the brain and retina. EPA and DHA are described as omega-3 fatty acids because their last double bond is situated 3 carbons from the omega (CH_3) end of the fatty acyl chain. With fatty acid chains at one end of the molecule and a polar head group at the other end, all of the phosphoglycerides exhibit a distinct amphipathic character.

Sphingolipids

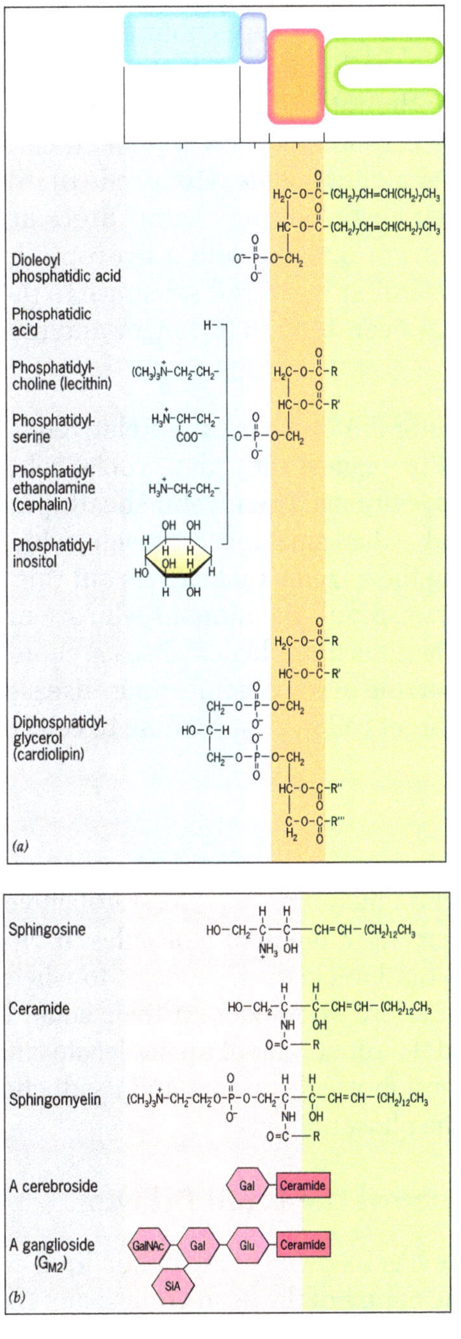

Figure shows the chemical structures of membrane lipids. (a) The structures of phosphoglycerides.

(b) The structures of sphingolipids. Sphingomyelin is a phospholipid; cerebrosides and gangliosides are glycolipids. A third membrane lipid is cholesterol, which is shown in the next figure. (R 5 fatty acyl chain). [The green portion of each lipid, which represents the hydrophobic tail(s) of the molecule, is actually much longer than the hydrophilic head group.]

A less abundant class of membrane lipids, called sphingolipids, are derivatives of sphingosine, an amino alcohol that contains a long hydrocarbon chain. Sphingolipids consist of sphingosine linked to a fatty acid by its amino group. This molecule is a ceramide. The various sphingosine-based lipids have additional groups esterified to the terminal alcohol of the sphingosine moiety. If the substitution is phosphorylcholine, the molecule is sphingomyelin, which is the only phospholipid of the membrane that is not built with a glycerol backbone. If the substitution is a carbohydrate, the molecule is a glycolipid. If the carbohydrate is a simple sugar, the glycolipid is called a cerebroside; if it is a small cluster of sugars that includes sialic acid, the glycolipid is called a ganglioside. Hundreds of different gangliosides have been identified by differences in their carbohydrate chains. Since all sphingolipids have two long, hydrophobic hydrocarbon chains at one end and a hydrophilic region at the other, they are also amphipathic and basically similar in overall structure to the phosphoglycerides. The fatty acyl chains of sphingolipids, however, tend to be longer and more highly saturated than those of phosphoglycerides.

Glycolipids are interesting membrane components. Relatively little is known about them, yet tantalizing hints have emerged to suggest they play crucial roles in cell function. The nervous system is particularly rich in glycolipids. The myelin sheath pictured in figure contains a high content of a particular glycolipid, called galactocerebroside, which is formed when a galactose is added to ceramide. Mice lacking the enzyme that carries out this reaction exhibit severe muscular tremors and eventual paralysis. Similarly, humans who are unable to synthesize a particular ganglioside (G_{M3}) suffer from a serious neurological disease characterized by severe seizures and blindness. Glycolipids also play a role in certain infectious diseases; the toxins that cause cholera and botulism both enter their target cell by first binding to cell surface gangliosides, as does the influenza virus.

Cholesterol

Another lipid component of certain membranes is the sterol cholesterol, which in certain animal cells may constitute up to 50 percent of the lipid molecules in the plasma membrane. Plant cells contain cholesterol-like sterols, but biologists disagree as to whether or not they completely lack cholesterol. Cholesterol molecules are oriented with their small hydrophilic hydroxyl group toward the membrane surface and the remainder of the molecule embedded in the lipid bilayer. The hydrophobic rings of a cholesterol molecule are flat and rigid, and they interfere with the movements of the fatty acid tails of the phospholipids.

The Nature and Importance of the Lipid Bilayer

Each type of cellular membrane has its own characteristic lipid composition, differing from one another in the types of lipids, the nature of the head groups, and the particular species of fatty acyl chain(s). Because of this structural variability, it is estimated that some biological membranes contain hundreds of chemically distinct species of phospholipids, which can be catalogued by mass spectrometry. The biological significance of this remarkable diversity of lipid species remains the subject of interest and speculation.

The percentages of some of the major types of lipids of a variety of membranes are given in table. The lipids of a membrane are more than simple structural elements; they can have important effects on the biological properties of a membrane. Lipid composition can determine the physical state of the membrane and influence the activity of particular membrane proteins. Membrane lipids also provide the precursors for highly active chemical messengers that regulate cellular function.

Various types of measurements indicate that the combined fatty acyl chains of both leaflets of the lipid bilayer span a width of about 30 Å and that each row of head groups (with its adjacent shell of water molecules) adds another 15 Å. Thus, the entire lipid bilayer is only about 60 Å (6 nm) thick. The presence in membranes of this thin film of amphipathic lipid molecules has remarkable consequences for cell structure and function. Because of thermodynamic considerations, the hydrocarbon chains of the lipid bilayer are never exposed to the surrounding aqueous solution. Consequently, membranes are never seen to have a free edge; they are always continuous, unbroken structures. As a result, membranes form extensive interconnected networks within the cell. Because of the flexibility of the lipid bilayer, membranes are deformable and their overall shape can change, as occurs during locomotion or cell division. The lipid bilayer is thought to facilitate the regulated fusion or budding of membranes. For example, the events of secretion, in which cytoplasmic vesicles fuse to the plasma membrane, or of fertilization, where two cells fuse to form a single cell, involve processes in which two separate membranes come together to become one continuous sheet. The importance of the lipid bilayer in maintaining the proper internal composition of a cell, in separating electric charges across the plasma membrane, and in many other cellular activities.

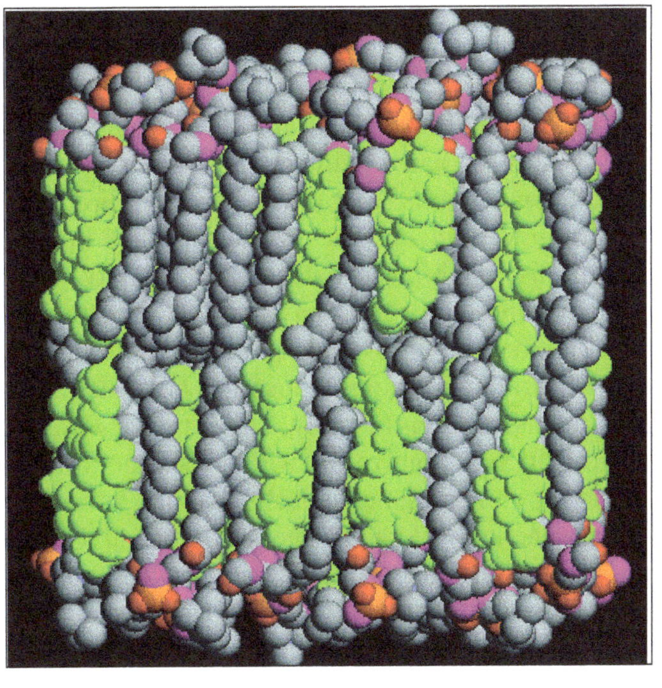

Figure shows the cholesterol molecules (shown in green) of a lipid bilayer are oriented with their small hydrophilic end facing the external surface of the bilayer and the bulk of their structure packed in among the fatty acid tails of the phospholipids. The placement of cholesterol molecules interferes with the flexibility of the lipid hydrocarbon chains, which tends to stiffen the bilayer while maintaining its overall fluidity. Unlike other lipids of the membrane, cholesterol is often rather evenly distributed between the two layers (leaflets).

Table: Lipid Compositions of Some Biological Membranes.

Lipid	Human erythrocyte	Human myelin	Beef heart mitochondria	E. coli
Phosphatidic acid	1.5	0.5	0	0
Phosphatidylcholine	19	10	39	0
Phosphatidylethanolamine	18	20	27	65
Phosphatidylgycerol	0	0	0	18
Phosphatidylserine	8.5	8.5	0.5	0
Cardiolipin	0	0	22.5	12
Sphingomyelin	17.5	8.5	0	0
Glycolipids	10	26	0	0
Cholesterol	25	26	3	0

Another important feature of the lipid bilayer is its ability to self-assemble, which can be demonstrated more easily within a test tube than a living cell. If, for example, a small amount of phosphatidylcholine is dispersed in an aqueous solution, the phospholipid molecules assemble spontaneously to form the walls of fluid-filled spherical vesicles, called liposomes. The walls of these liposomes consist of a continuous lipid bilayer that is organized in the same manner as that of the lipid bilayer of a natural membrane. Liposomes have proven invaluable in membrane research. Membrane proteins can be inserted into liposomes and their function studied in a much simpler environment than that of a natural membrane. Liposomes have also been developed as vehicles to deliver drugs or DNA molecules within the body. The drugs or DNA can be linked to the wall of the liposome or contained at high concentration within its lumen. In these studies, the walls of the liposomes are constructed to contain specific proteins (such as antibodies or hormones) that allow the liposomes to bind selectively to the surfaces of particular target cells where the drug or DNA is intended to go. Most of the early clinical studies with liposomes met with failure because the injected vesicles were rapidly removed by phagocytic cells of the immune system. This obstacle has been overcome with the development of so-called stealth liposomes (e.g., Caelyx) that contain an outer coating of a synthetic polymer that protects the liposomes from immune destruction.

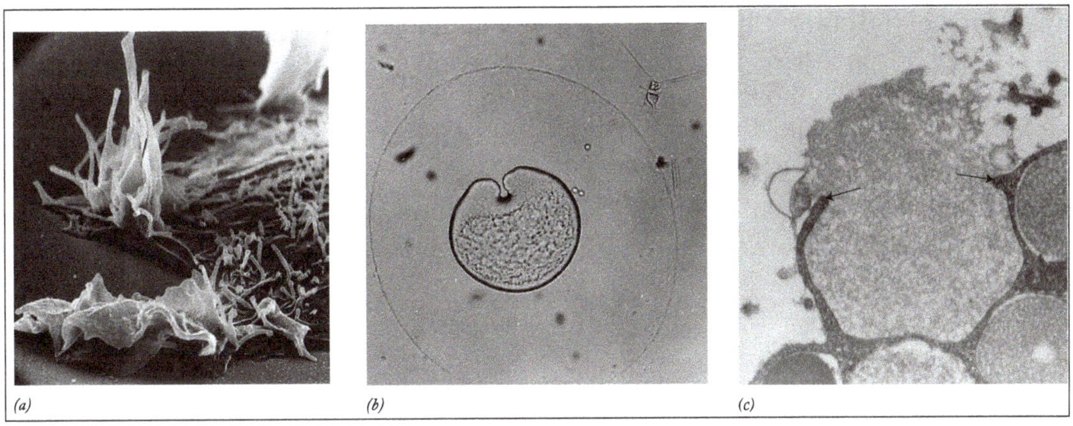

Figure shows the dynamic properties of the plasma membrane. (a) The leading edge of a moving cell often contains sites where the plasma membrane displays undulating ruffles. (b) Division of a cell is accompanied by the deformation of the plasma membrane as it is pulled toward the center of the cell. Unlike most dividing cells, the cleavage furrow of this dividing ctenophore egg begins at one pole and moves unidirectionally through the egg. (c) Membranes are capable of fusing with other membranes. This electron micrograph shows a secretory granule discharging its contents after fusion with the overlying plasma membrane (arrows).

The Asymmetry of Membrane Lipids

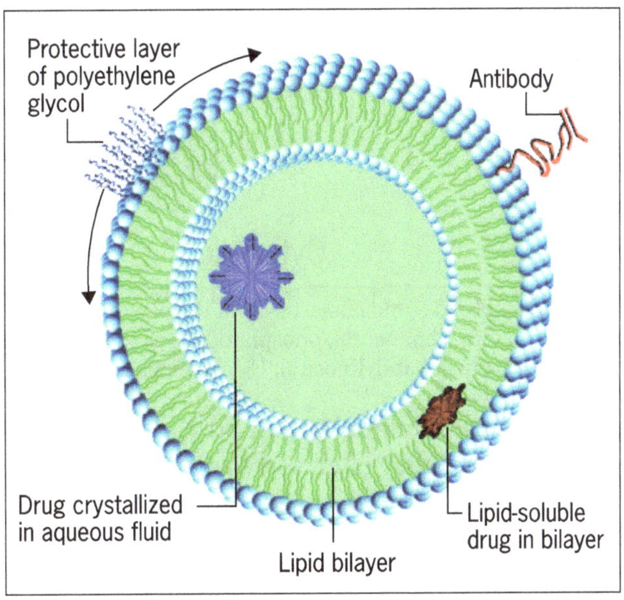

Liposomes: A schematic diagram of a stealth liposome containing a hydrophilic polymer (such as polyethylene glycol) to protect it from destruction by immune cells, antibody molecules that target it to specific body tissues, a water-soluble drug enclosed in the fluidfilled interior chamber, and a lipid-soluble drug in the bilayer.

The lipid bilayer consists of two distinct leaflets that have a distinctly different lipid composition. One line of experiments that has led to this conclusion takes advantage of the fact that lipid-digesting enzymes cannot penetrate the plasma membrane and, consequently, are only able to digest lipids that reside in the outer leaflet of the bilayer. If intact human red blood cells are treated with a lipid-digesting phospholipase, approximately 80 percent of the phosphatidylcholine (PC) of the membrane is hydrolyzed, but only about 20 percent of the membrane's phosphatidylethanolamine (PE) and less than 10 percent of its phosphatidylserine (PS) are attacked. These data indicate that, compared to the inner leaflet, the outer leaflet has a relatively high concentration of PC (and sphingomyelin) and a low concentration of PE and PS. It follows that the lipid bilayer can be thought of as composed of two more-or-less stable, independent monolayers having different physical and chemical properties.

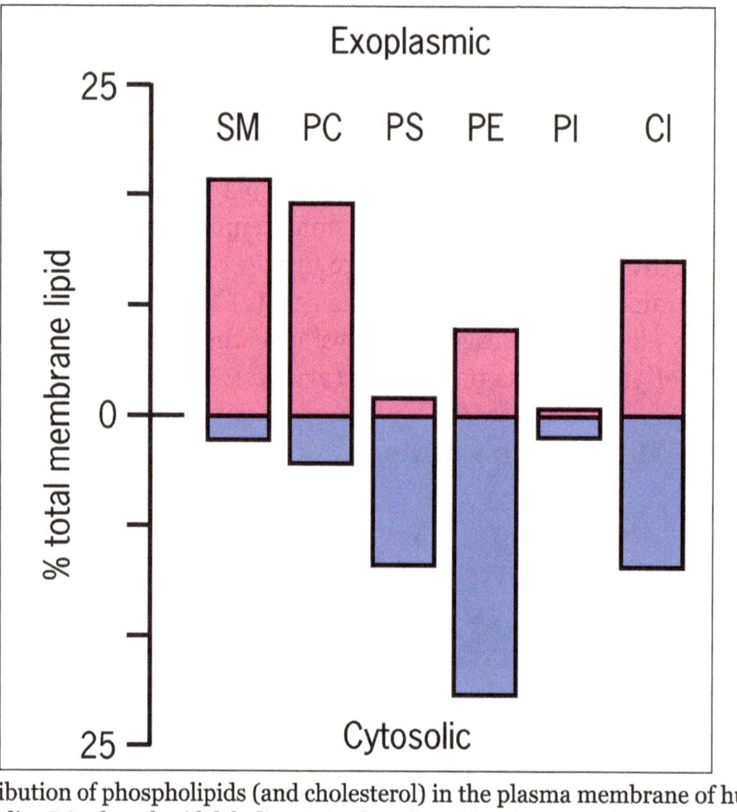

The asymmetric distribution of phospholipids (and cholesterol) in the plasma membrane of human erythrocytes. (SM, sphingomyelin; PC, phosphatidylcholine; PS, phosphatidylserine; PE, phosphatidylethanolamine; PI, phosphatidylinositol; Cl, cholesterol.)

The different classes of lipids in figure exhibit different properties. All the glycolipids of the plasma membrane are in the outer leaflet where they often serve as receptors for extracellular ligands. Phosphatidylethanolamine, which is concentrated in the inner leaflet, tends to promote the curvature of the membrane, which is important in membrane budding and fusion. Phosphatidylserine, which is concentrated in the inner leaflet, has a net negative charge at physiologic pH, which makes it a candidate for binding positively charged lysine and arginine residues, such as those adjacent to the membrane-spanning a helix of glycophorin A in figure. The appearance of PS on the outer surface of aging lymphocytes marks the cells for destruction by macrophages, whereas its appearance on the outer surface of platelets leads to blood coagulation. Phosphatidylinositol (PI), which is concentrated in the inner leaflet, can be phosphorylated at different sites on the inositol ring, which converts the lipid into a phosphoinositide. Phosphoinositides play a key role in the transfer of stimuli from the plasma membrane to the cytoplasm and the recruitment of proteins to the cytosolic face of the plasma membrane.

Membrane Carbohydrates

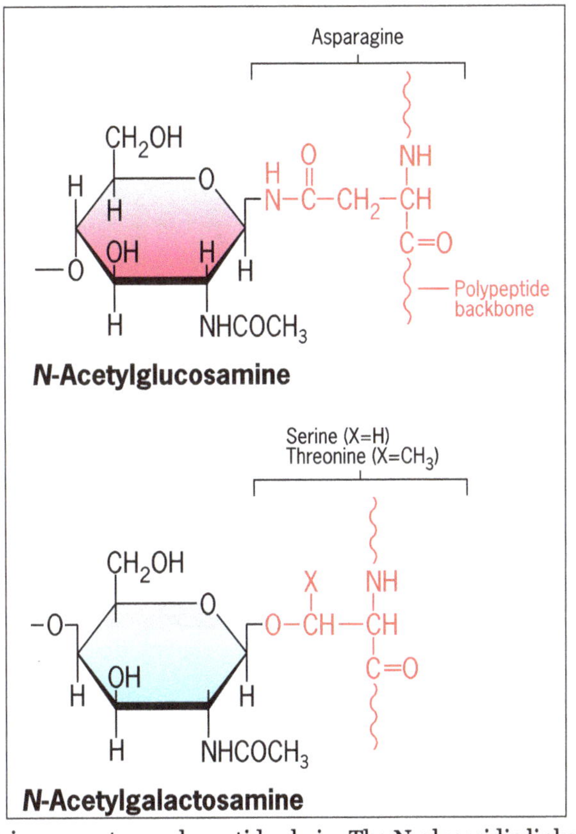

Two types of linkages that join sugars to a polypeptide chain. The N-glycosidic linkage between asparagine and N-acetylglucosamine is more common than the O-glycosidic linkage between serine or threonine and N-acetylgalactosamine.

The plasma membranes of eukaryotic cells also contain carbohydrate. Depending on the species and cell type, the carbohydrate content of the plasma membrane ranges between 2 and 10 percent by weight. More than 90 percent of the membrane's carbohydrate is covalently linked to proteins to form glycoproteins; the remaining carbohydrate is covalently linked to lipids to form glycolipids. As indicated in figure, all of the carbohydrate of the plasma membrane faces outward into the extracellular space.1 The carbohydrate of internal cellular membranes also faces away from the cytosol. The addition of carbohydrate, or glycosylation, is the most complex of these modifications. The carbohydrate of glycoproteins is present as short, branched hydrophilic oligosaccharides, typically having fewer than about 15 sugars per chain. In contrast to most high-molecular-weight carbohydrates (such as glycogen, starch, or cellulose), which are polymers of a single sugar, the oligosaccharides attached to membrane proteins and lipids can display extensive variability in composition and structure. Even the same protein can display different chains of sugars in different cells and tissues. Oligosaccharides may be attached to several different amino acids by two major types of linkages. These carbohydrate projections play an important role in mediating the interactions of a cell with its environment and sorting of membrane proteins to different cellular compartments. The carbohydrates of the glycolipids of the red blood cell plasma membrane determine whether a person's blood type is A, B, AB, or O. A person having blood type A has an enzyme that adds an Nacetylgalactosamine to the end of the chain, whereas a person with type B blood has an enzyme that adds galactose to the chain terminus. These two enzymes are encoded by alternate versions of the same gene, yet they recognize different substrates. People with AB blood type possess both enzymes, whereas people with O blood type lack enzymes capable of attaching either terminal sugar. The function of the ABO blood-group antigens remains a mystery.

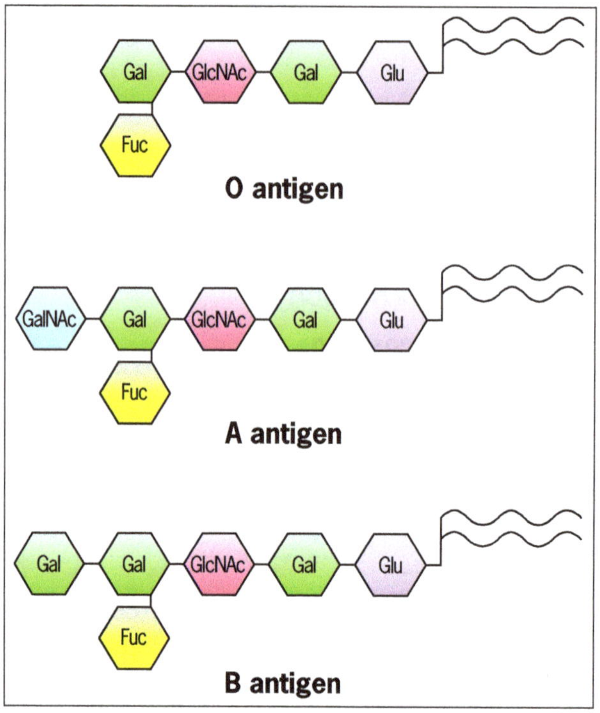

Figure shows blood-group antigens.Whether a person has type A, B, AB, or O blood is determined by a short chain of sugars covalently attached to membrane lipids and proteins of the red blood cell membrane. The oligosaccharides attached to membrane lipids (forming a ganglioside) that produce the A, B, and O blood types are shown here. A person with type AB blood has gangliosides with both the A and B structure. (Gal, galactose; GlcNAc, N-acetylglucosamine; Glu, glucose; Fuc, fucose; GalNAc, N-acetylgalactosamine.)

The Structure and Functions of Membrane Proteins

Depending on the cell type and the particular organelle within that cell, a membrane may contain hundreds of different proteins. Each membrane protein has a defined orientation relative to the cytoplasm, so that the properties of one surface of a membrane are very different from those of the other surface. This asymmetry is referred to as membrane "sidedness." In the plasma membrane, for example, those parts of membrane proteins that interact with other cells or with extracellular substances are exposed to the extracellular space, whereas those parts of membrane proteins that interact with cytoplasmic molecules are exposed to the cytosol. Membrane proteins can be grouped into three distinct classes distinguished by the intimacy of their relationship to the lipid bilayer. These are:

- Integral proteins that penetrate the lipid bilayer. Integral proteins are transmembrane proteins; that is, they pass entirely through the lipid bilayer and thus have domains that protrude from both the extracellular and cytoplasmic sides of the membrane. Some integral proteins have only one membrane-spanning segment, whereas others are multispanning. Genome-sequencing studies suggest that integral proteins constitute 25–30 percent of all encoded proteins and roughly 60 percent of all current drug targets.

- Peripheral proteins that are located entirely outside of the lipid bilayer, on either the cytoplasmic or extracellular side, yet are associated with the surface of the membrane by noncovalent bonds.
- Lipid-anchored proteins that are located outside the lipid bilayer, on either the extracellular or cytoplasmic surface, but are covalently linked to a lipid molecule that is situated within the bilayer.

Integral Membrane Proteins

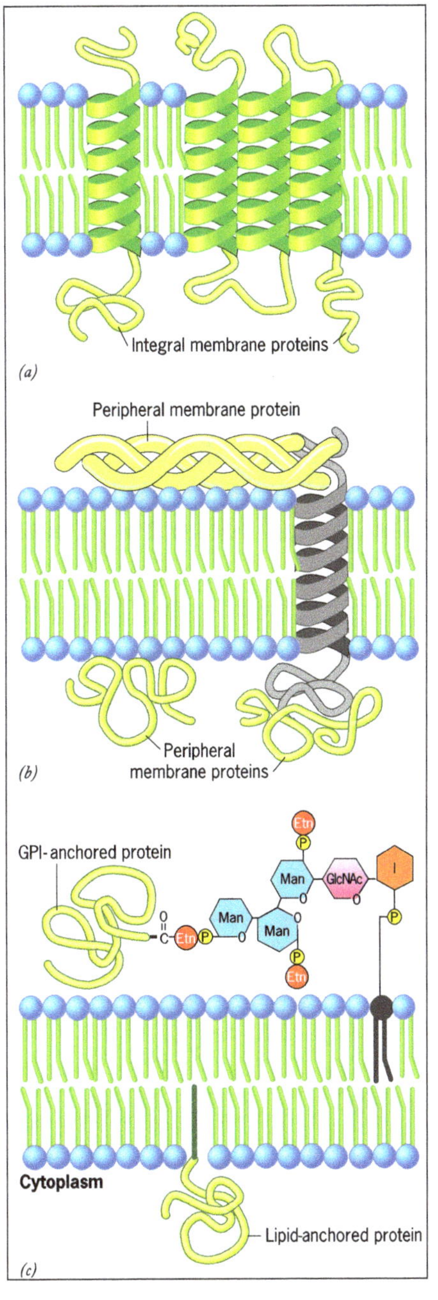

Figure shows three classes of membrane protein. (a) Integral proteins typically contain one or more transmembrane helices. (b) Peripheral proteins are noncovalently bonded to the polar head groups of the lipid bilayer and/or to an integral membrane protein. (c) Lipid-anchored proteins are covalently bonded to a lipid group that resides within the membrane. The lipid can be phosphatidylinositol, a fatty acid, or a prenyl group (a long-chain hydrocarbon built from five-carbon isoprenoid units). I, inositol; GlcNAc, N-acetylglucosamine; Man, mannose; Etn, ethanolamine; GPI, glycosylphosphatidylinositol.

Most integral membrane proteins function in the following capacities: as receptors that bind specific substances at the membrane surface, as channels or transporters involved in the movement of ions and solutes across the membrane, or as agents that transfer electrons during the processes of photosynthesis and respiration. Like the phospholipids of the bilayer, integral membrane proteins are also amphipathic, having both hydrophilic and hydrophobic portions. Those portions of an integral membrane protein that reside within the lipid bilayer—the transmembrane domains—tend to have a hydrophobic character. Amino acid residues in transmembrane domains form van der Waals interactions with the fatty acyl chains of the bilayer, which seals the protein into the lipid "wall" of the membrane. As a result, the permeability barrier of the membrane is preserved, the protein is anchored within the bilayer, and the protein is brought into direct contact with surrounding lipid molecules. According to current consensus, most of the lipid molecules that make contact with a transmembrane domain, such as those shown in figure a, are simply passive bystanders and are rapidly exchanged with other lipid molecules in the bilayer. There is increasing evidence, however, that certain sites on the surface of many membrane proteins do form important functional interactions with specific lipid molecules. An example of this is shown in figure b, where anionic lipid molecules are seen to bind in the crevice at the interfaces between the subunits of a tetrameric KcsA K^+ channel. The channel does not open normally in a bilayer that lacks these specific lipid molecules.

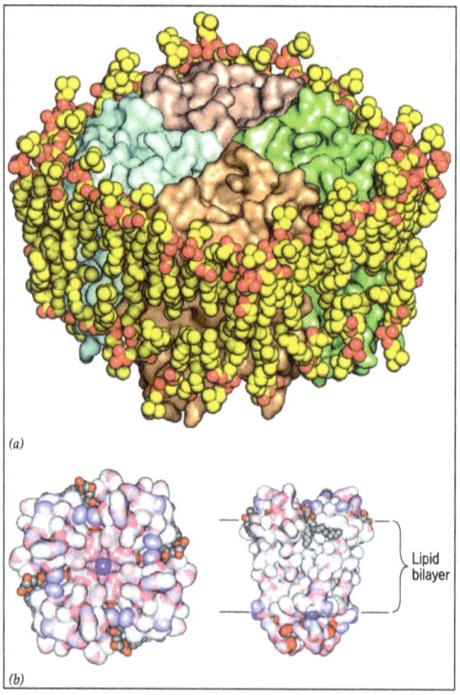

Figure shows the interactions between membrane proteins and lipids. (a) Aquaporin is a membrane protein containing four subunits (colored differently in the illustration) with each subunit containing an aqueous channel. Analysis of the protein's structure revealed the presence of a surrounding layer of bound lipid molecules. In this case, these lipid molecules are not likely to play a role in the function of aquaporin because the protein retains its function as a water channel in bilayers containing nonnative lipids. (b) Two views of another tetrameric membrane protein, in this case the bacterial K^+ channel, KcsA. Anionic phosphatidylglycerol molecules (red/gray) are seen in each crevice between the subunits and are thought to be required for normal channel function. A K^+ ion (purple sphere) is seen in transit through the pore.

Those portions of an integral membrane protein that project into either the cytoplasm or extracellular space tend to be more like the globular proteins. These nonembedded domains tend to have hydrophilic surfaces that interact with water-soluble substances (lowmolecular-weight substrates, hormones, and other proteins) at the edge of the membrane. In some membrane proteins, the transmembrane domains are essentially devoid of water molecules, whereas others allow the aqueous solvent to penetrate deep into the protein's membrane-embedded regions. Several large families of membrane proteins contain an interior channel that provides an aqueous passageway through the lipid bilayer. The linings of these channels typically contain key hydrophilic residues at strategic locations. Integral proteins need not be fixed structures but may be able to move laterally within the membrane.

Distribution of Integral Proteins: Freeze-Fracture Analysis

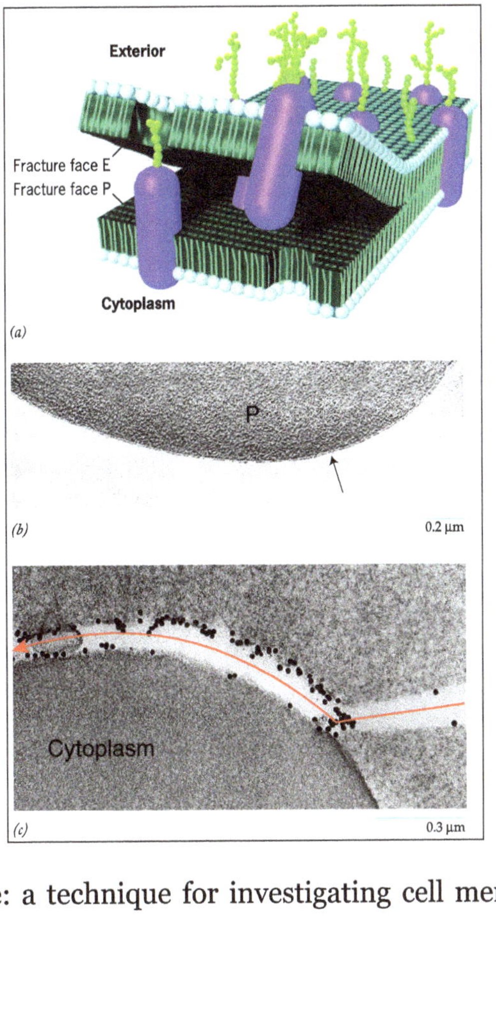

Figure shows freeze fracture: a technique for investigating cell membrane structure. (a) When a block of frozen tissue is struck by a knife blade, a fracture plane runs through the tissue, often following a path that leads it through the middle of the lipid bilayer. The fracture plane goes around the proteins rather than cracking them in half, and they segregate with one of the two halves of the bilayer. The exposed faces within the center of the bilayer can then be covered with a metal deposit to form a metallic replica. These exposed faces are referred to as the E, or ectoplasmic face, and the P, or protoplasmic face. (b) Replica of a freeze-fractured human erythrocyte. The P fracture face is seen to be covered with particles approximately 8 nm in diameter. A small ridge (arrow) marks the junction of the particulate face with the surrounding ice. (c) This micrograph shows the surface of an erythrocyte that was frozen and then fractured, but rather than preparing a replica, the cell was thawed, fixed, and labeled with a marker for the carbohydrate groups that project from the external surface of the integral protein glycophorin. Thin sections of the labeled, fractured cell reveal that glycophorin molecules (black particles) have preferentially segregated with the outer half of the membrane. The red line shows the path of the fracture plane.

The concept that proteins penetrate through membranes, rather than simply remaining external to the bilayer, was derived primarily from the results of a technique called freezefracture replication. In this procedure, tissue is frozen solid and then struck with a knife blade, which fractures the block into two pieces. As this occurs, the fracture plane often takes a path between the two leaflets of the lipid bilayer. Once the membranes are split in this manner, metals are deposited on their exposed surfaces to form a shadowed replica, which is viewed in the electron microscope. As shown in figure b, the replica resembles a road strewn with pebbles, which are called membrane-associated particles. Since the fracture plane passes through the center of the bilayer, most of these particles correspond to integral membrane proteins that extend at least halfway through the lipid core. When the fracture plane reaches a given particle, it goes around it rather than cracking it in half. Consequently, each protein (particle) separates with one half of the plasma membrane, leaving a corresponding pit in the other half. One of the great values of the freeze-fracturing technique is that it allows an investigation of the microheterogeneity of the membrane. Localized differences in parts of the membrane stand out in these replicas and can be identified (as illustrated by the replica of a gap junction shown in figure d). Biochemical analyses, in contrast, average out such differences.

Studying the Structure and Properties of Integral Membrane Proteins

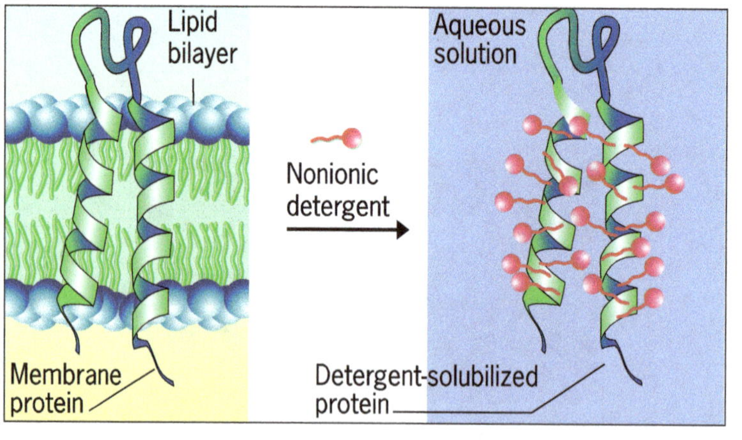

Figure shows solubilization of membrane proteins with detergents. The nonpolar ends of the detergent molecules associate with the nonpolar residues of the protein that had previously been in contact with the fatty acyl chains of the lipid bilayer. In contrast, the polar ends of the detergent molecules interact with the surrounding water molecules, keeping the protein in solution. Nonionic detergents, as shown here, solubilize membrane proteins without disrupting their structure.

Because of their hydrophobic transmembrane domains, integral membrane proteins are difficult to isolate in a soluble form. Removal of these proteins from the membrane normally requires the use of a detergent, such as the ionic (charged) detergent SDS (which denatures proteins) or the nonionic (uncharged) detergent Triton X-100 (which generally does not alter a protein's tertiary structure).

$$CH_3{-}(CH_2)_{11}{-}OSO_3^{-}Na^{+}$$

Sodium dodecyl sulfate (SDS)

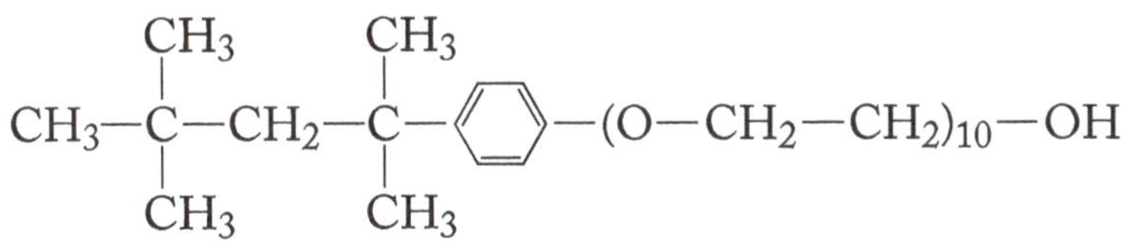

Triton X-100

Like membrane lipids, detergents are amphipathic, being composed of a polar end and a nonpolar hydrocarbon chain. As a consequence of their structure, detergents can substitute for phospholipids in stabilizing integral proteins while rendering them soluble in aqueous solution. Once the proteins have been solubilized by the detergent, various analyses can be carried out to determine the protein's amino acid composition, molecular mass, amino acid sequence, and so forth.

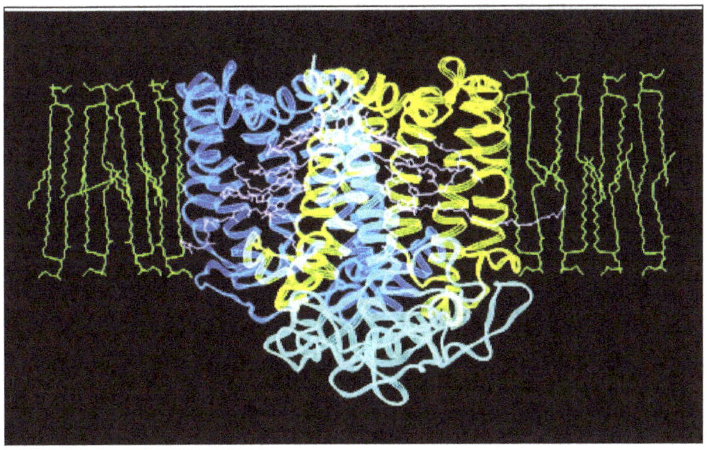

Figure shows an integral protein as it resides within the plasma membrane. Tertiary structure of the photosynthetic reaction center of a bacterium as determined by X-ray crystallography. The protein contains three different membrane-spanning polypeptides, shown in yellow, light blue, and dark blue. The helical nature of each of the transmembrane segments is evident.

Researchers have had great difficulty obtaining crystals of most integral membrane proteins for use in X-ray crystallography. In fact, fewer than 1 percent of the known high-resolution protein structures represent integral membrane proteins. Furthermore, most of these structures represent prokaryotic versions of a particular protein, which are often smaller than their eukaryotic homologues and easier to obtain in large quantity. Once the structure of one member of a membrane protein family is determined, researchers can usually apply a strategy called homology modeling to learn about the structure and activity of other members of the family. For example, solution of the structure of the bacterial K^+ channel KcsA provided a wealth of data that could be applied to the structure and mechanism of action of the much more complex eukaryotic K^+ channels.

One of the first membrane proteins whose entire three-dimensional structure was determined by X-ray crystallography is shown in figure. This protein—the bacterial photosynthetic reaction center—consists of three subunits containing 11 membrane-spanning α helices. Most membrane proteins have not been as amenable to crystallization as the photosynthetic reaction center. Among the obstacles, many membrane proteins are present at low numbers per cell, are unstable in the detergent-containing solutions required for their extraction, are prone to aggregation, and are heavily modified by glycosylation and cannot be expressed as recombinant proteins in other types of cells. Some of the technical difficulties in preparing membrane protein crystals have been overcome using new methodologies and laborious efforts. In one study, for example researchers were able to obtain high-quality crystals of a bacterial transporter after testing and refining more than 95,000 different conditions for crystallization. In some cases, mutant versions of a membrane protein are found to be better suited to forming the ordered arrays of molecules that make up a crystal lattice. In many other cases, crystallization has been achieved by covalently linking the membrane protein to other molecules, often small soluble proteins. These added elements stabilize the structure of the target protein and increase the number of sites on the surface of the protein that can participate in intermolecular contacts required for lattice formation. For example, the crystal structure of the β_2-adrenergic receptor was obtained by inserting a T4-lysozyme molecule (derived from a T4 bacteriophage) into one of the intracellular loops of the receptor. Crystallization of the β2-adrenergic receptor was also facilitated by the presence of a specific ligand bound to the receptor and the use of cholesterol in the lipid-phase medium in which the crystals were grown. Despite increasing success in protein crystallization, researchers still rely largely on indirect approaches for determining the three-dimensional organization of most membrane proteins. We will examine some of these approaches in the following paragraphs.

Identifying Transmembrane Domains

A great deal can be learned about the structure of a membrane protein and its orientation within the lipid bilayer from a computer-based (computational) analysis of its amino acid sequence, which is readily deduced from the nucleotide sequence of an isolated gene. The first question one might ask is: Which segments of the polypeptide chain are actually embedded in the lipid bi-layer? Those segments of a protein embedded within the membrane, which are described as the transmembrane domains, have a simple structure; they consist of a string of about 20 predom-inantly nonpolar amino acids that span the core of the lipid bilayer as an α helix. The chemical structure of a single transmembrane helix is shown in figure, which depicts the two-dimensional structure of glycophorin A, the major integral protein of the erythrocyte plasma membrane of the 20 amino acids that make up the lone a helix of a glycophorin monomer, all but three have hydrophobic side chains (or an H atom in the case of the glycine residues). The exceptions are serine and threonine, which are noncharged, polar residues. Figure a shows a portion of a transmembrane helix with a threonine residue, not unlike those of glycophorin A. The hydroxyl group of the residue's side chain can form a hydrogen bond with one of the oxygen atoms of the peptide backbone. Fully charged residues may also appear in transmembrane helices, but they tend to be near one of the ends of the helix and to be accommodated in ways that allow them to fit into their hydrophobic environment. This is illustrated in the model transmembrane helices shown in Figures Each of the helices in these figures contains a pair of charged residues whose side chains are able to reach out and interact with the innermost polar regions of the membrane, even if it requires distorting the helix to do so. Figure d shows two tyrosine residues with their hydrophobic aromatic side chains; each aromatic ring is oriented parallel with the hydrocarbon chains of the bilayer with which it has become integrated.

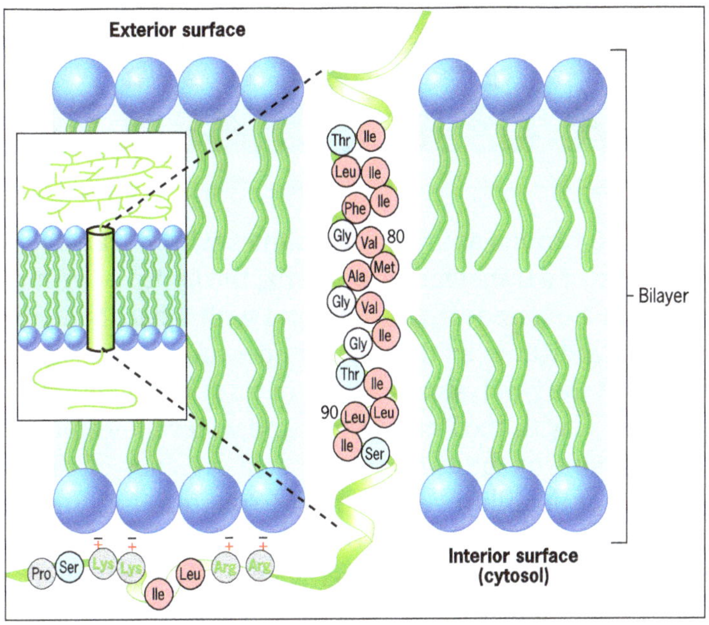

Figure shows Glycophorin A, an integral protein with a single transmembrane domain. The single α helix that passes through the membrane consists predominantly of hydrophobic residues (brown-colored circles). The four positively charged amino acid residues of the cytoplasmic domain of the membrane protein form ionic bonds with negatively charged lipid head groups. Carbohydrates are attached to a number of amino acid residues on the outer surface of the protein (shown in the inset). All but one of the 16 oligosaccharides are small O-linked chains (the exception is a larger oligosaccharide linked to the asparagine residue at position 26). Glycophorin molecules are present as homodimers within the erythrocyte membrane. The two helices of the dimer cross over one another in the region between residues 79 and 83. This Gly-X-X-X-Gly sequence is commonly found where transmembrane helices come into close proximity.

Knowing the amino acid sequence of an integral membrane protein, we can usually identify the transmembrane segments using a hydropathy plot, in which each site along a polypeptide is assigned a value that provides a measure of the hydrophobicity of the amino acid at that site as well as that of its neighbors. This approach provides a "running average" of the hydrophobicity of short sections of the polypeptide, and guarantees that one or a few polar amino acids in a sequence do not alter the profile of the entire stretch. Hydrophobicity of amino acids can be determined using various criteria, such as their lipid solubility or the energy that would be required to transfer them from a nonpolar medium into an aqueous medium. A hydropathy plot for glycophorin A is shown in figure. Transmembrane segments are usually identified as a jagged peak that extends well into the hydrophobic side of the spectrum. A reliable prediction concerning the orientation of the transmembrane segment within the bilayer can usually be made by examining the flanking amino acid residues. In most cases, as illustrated by glycophorin in figure, those parts of the polypeptide at the cytoplasmic flank of a transmembrane segment tend to be more positively charged than those at the extracellular flank.

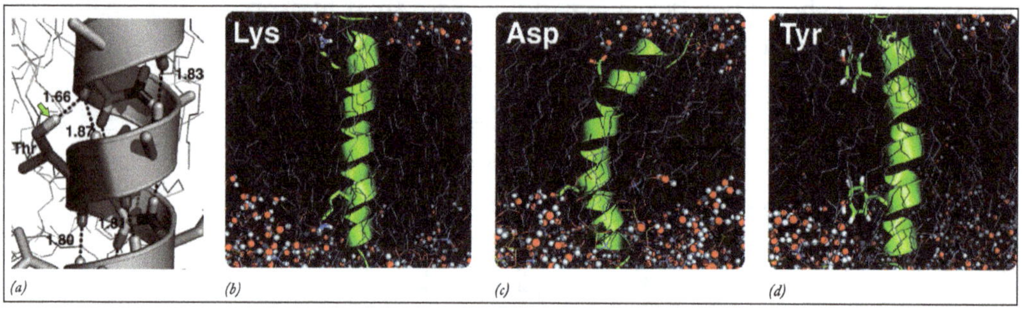

Figure shows accommodating various amino acid residues within transmembrane helices. (a) In this portrait of a small portion of a transmembrane helix, the hydroxyl group of the threonine side chain (arrow) is able to form a (shared) hydrogen bond with a backbone oxygen within the lipid bilayer. Hydrogen bonds are indicated by the dashed lines and their distances are shown in angstroms. (b) The side chains of the two lysine residues of this transmembrane helix are sufficiently long and flexible to form bonds with the head groups and water molecules of the polar face of the lipid bilayer. (c) The side chains of the two aspartic acid residues of this transmembrane helix can also reach the polar face of the bilayer but introduce distortion in the helix to do so. (d) The aromatic side chains of the two tyrosine residues of this transmembrane helix are oriented perpendicular to the axis of the membrane and parallel to the fatty acyl chains with which they interact.

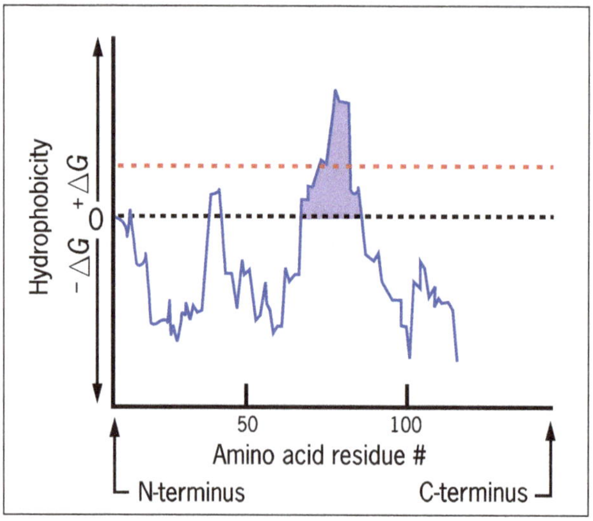

Figure shows hydropathy plot for glycophorin A, a single membrane-spanning protein.

Hydrophobicity is measured by the free energy required to transfer each segment of the polypeptide from a nonpolar solvent to an aqueous medium. Values above the o line are energy-requiring (+ ΔGs), indicating they consist of stretches of amino acids that have predominantly nonpolar side chains. Peaks that project above the red-colored line are interpreted as a transmembrane domain.

Not all integral membrane proteins contain transmembrane a helices. A number of membrane proteins contain a relatively large channel positioned within a circle of membranespanning b strands organized into a barrel as illustrated in figure. To date, aqueous channels constructed of b barrels have only been found in the outer membranes of bacteria, mitochondria, and chloroplasts.

Experimental Approaches to Identifying Conformational Changes within an Integral Membrane Protein Suppose you have isolated a gene for an integral membrane protein and, based on its nucleotide sequence, determined that it contains 12 apparent membrane-spanning α helices. You might want to know which sites within these transmembrane domains are accessible to the aqueous environment and how this changes as the protein carries out its function. This type of information is particularly important when working with a membrane channel or transporter that functions in the movement of hydrophilic substances across the membrane. Although these determinations are difficult to make without detailed structural models, considerable insight can be gained by sitedirected mutagenesis, that is, by introducing specific changes into the gene that encodes the protein. An experiment of this type, which was conducted on lactose permease, a sugar-transporting protein in bacterial cell membranes, is shown in figure. In this experiment, researchers prepared different versions of the permease in which nearly all of the individual amino acids of the protein were replaced, one at a time, by a cysteine residue. Each mutated membrane protein was then incubated with a water-soluble reagent (NEM) that is capable of adding an alkyl group to the cysteine side chain—as long as it has access to that side chain. The incubations with NEM were carried out 136 under two different conditions: either in the presence of the sugar to be transported or in its absence. The image in figure a shows those residues (depicted as red spheres) of the protein that become alkylated by NEM in the absence of the sugar. Many of these accessible residues are thought to line an inward-facing hydrophilic cavity in the permease that is open to the cytoplasm. Those cysteine residues that are not labeled, most notably on helices VI and XII, are presumed to either face the fatty acyl chains of the phospholipids or to be located in tightly packed regions of the molecule, thus making them inaccessible to NEM.

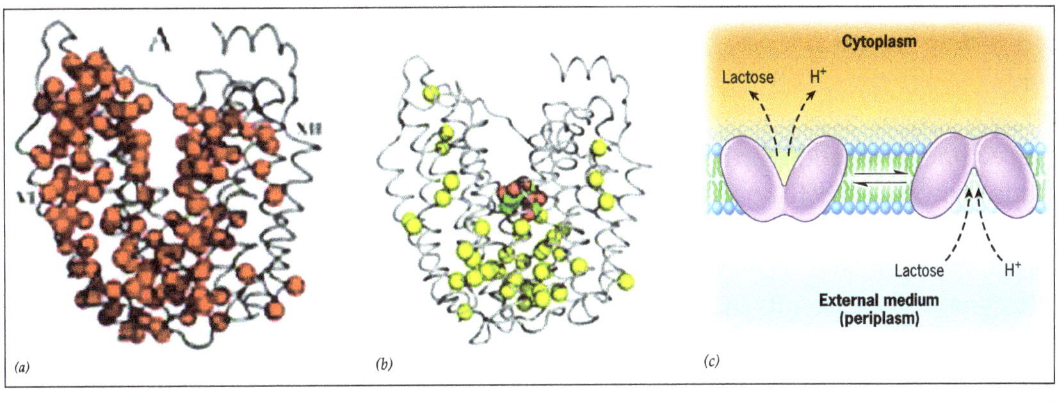

Figure shows an experiment employing site-directed mutagenesis to learn about dynamic changes in the conformation of a membrane protein as it carries out its activity. The cytoplasmic surface of the protein (lactose permease) is at the top. (a) The red spheres indicate the residues of the membrane protein that reacted with the alkylating agent NEM in the absence of a sugar to be transported. (b) The gold spheres denote the residues that are much more accessible to the alkylating agent when the protein is incubated with the sugar. The gold spheres in b are seen to cluster in a portion of the protein near the external medium (the periplasm in bacteria). The gold spheres correspond to residues that line an outward-facing cavity, i.e., one that is open to the external medium. The results support a model of alternating access to the medium as depicted in part c. (Note the structure that is depicted in parts a–b is that of the inward-facing conformation as determined by X-ray crystallography. The structure of the outward-facing conformation has not been directly determined, which is why these types of labeling experiments remain important).

The image in figure b shows those residues (depicted as gold spheres) that show a significant increase in reactivity to NEM when the permease is incubated with the sugar to be transported. The difference in NEM reactivity of the residues between the protein in the absence of the sugar (a) and the presence of the sugar (b) indicates that different parts of the protein are accessible to the aqueous medium under these two conditions. The results, in other words, suggest that addition of the sugar induces a conformational change in the permease. Careful analysis of the positions of the residues that become labeled in these two conformations supports the model shown in figure c. According to this model of alternating access, the sugar-binding site is open to the cytoplasm of the bacterial cell in one conformation and is open to the extracellular space in the other conformation. Alternation between the two conformations brings about transport of the sugar across the membrane.

Determining spatial relationships between specific amino acids in a membrane protein is another approach to learning about dynamic events that occur as a protein carries out its function. One way to learn about the distance between selected residues in a protein is to introduce chemical groups whose properties are sensitive to the distance that separates them. Nitroxides are chemical groups that contain an unpaired electron, which produces a characteristic spectrum when monitored by a technique called electron paramagnetic resonance (EPR) spectroscopy. A nitroxide group can be introduced at any site in a protein by first mutating that site to a cysteine and then attaching the nitroxide to the —SH group of the cysteine residue. Figure shows how this technique was used to discover the conformational changes that occur in a membrane protein as its channel is opened in response to changes in the pH of the medium. The protein in question, a bacterial K^+ channel, is a tetramer composed of four identical subunits. The cytoplasmic opening to the channel is bounded by four transmembrane helices, one from each subunit of the protein. Figure a shows the EPR spectra that were obtained when a nitroxide was introduced near the cytoplasmic end of each transmembrane helix. The red line shows the spectrum obtained at pH 6.5 when the channel is in the closed state, and the blue line shows the spectrum at pH 3.5 when the channel is open. The shape of each line depends on the proximity of the nitroxides to one another. The spectrum is broader at pH 6.5 because the nitroxide groups on the four subunits are closer together at this pH, which decreases the intensity of their EPR signals. These results indicate that the activation of the channel is accompanied by an increased separation between the labeled residues of the four subunits. An increase in diameter of the channel opening allows ions in the cytoplasm to reach the actual permeation pathway (shown in red) within the channel, which allows only the passage of K^+ ions. An alternate technique, called FRET, that can also be used to measure changes in the distance between labeled groups within a protein, is illustrated in figure.

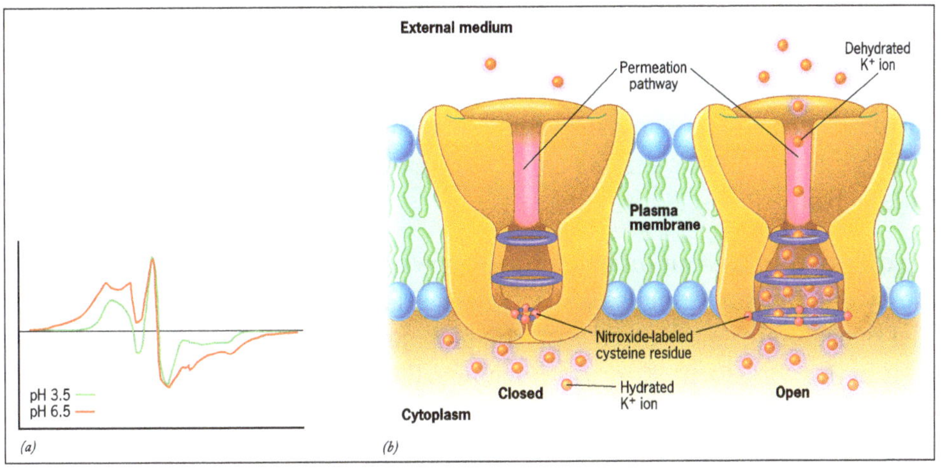

Figure shows use of EPR spectroscopy to monitor changes in conformation of a bacterial K^+ ion channel as it opens and closes. (a) EPR spectra from nitroxides that have been attached to cysteine residues near the cytoplasmic end of the four transmembrane helices that line the channel. The cysteine residue in each helix replaces a glycine residue that is normally at that position. The shapes of the spectra depend on the distances between unpaired electrons in the nitroxides on different subunits. (Nitroxides are described as "spin-labels," and this technique is known as site-directed spin labeling.) (b) A highly schematic model of the ion channel in the open and closed states based on the data from part a. Opening of the channel is accompanied by the movement of the four nitroxide groups apart from one another.

Peripheral Membrane Proteins

Peripheral proteins are associated with the membrane by weak electrostatic bonds. Peripheral proteins can usually be solubilized by extraction with high-concentration salt solutions that weaken the electrostatic bonds holding peripheral proteins to a membrane. In actual fact, the distinction between integral and peripheral proteins is blurred because many integral membrane proteins consist of several polypeptides, some that penetrate the lipid bilayer and others that remain on the periphery.

The best studied peripheral proteins are located on the internal (cytosolic) surface of the plasma membrane, where they form a fibrillar network that acts as a membrane "skeleton". These proteins provide mechanical support for the membrane and function as an anchor for integral membrane proteins. Other peripheral proteins on the internal plasma membrane surface function as enzymes, specialized coats, or factors that transmit transmembrane signals. Peripheral proteins typically have a dynamic relationship with the membrane, being recruited to the membrane or released from the membrane depending on prevailing conditions.

Lipid-Anchored Membrane Proteins

Several types of lipid-anchored membrane proteins can be distinguished. Numerous proteins present on the external face of the plasma membrane are bound to the membrane by a small, complex oligosaccharide linked to a molecule of phosphatidylinositol that is embedded in the outer leaflet of the lipid bilayer. Peripheral membrane proteins containing this type of glycosyl-phosphatidylinositol linkage are called GPI-anchored proteins. They were discovered when it was shown that certain membrane proteins could be released by a phospholipase that specifically recognized and cleaved inositol-containing phospholipids. The normal cellular scrapie protein PrP^C is a GPI-linked molecule, as are various receptors, enzymes, and cell-adhesion proteins. A rare type of anemia, paroxysmal nocturnal hemoglobinuria, results from a deficiency in GPI synthesis that makes red blood cells susceptible to lysis.

Another group of proteins present on the cytoplasmic side of the plasma membrane is anchored to the membrane by one or more long hydrocarbon chains embedded in the inner leaflet of the lipid bilayer. At least two proteins associated with the plasma membrane in this way (Src and Ras) have been implicated in the transformation of a normal cell to a malignant state.

Membrane Lipids and Membrane Fluidity

The physical state of the lipid of a membrane is described by its fluidity (or viscosity). Consider a simple artificial bilayer composed of phosphatidylcholine and phosphatidylethanolamine, whose fatty acids are largely unsaturated. If the temperature of the bilayer is kept relatively warm (e.g., 37 °C), the lipid exists in a relatively fluid state. At this temperature, the lipid bilayer is best described as a two-dimensional liquid crystal. As in a crystal, the molecules still retain a specified orientation; in this case, the long axes of the molecules tend toward a parallel arrangement, yet individual phospholipids can rotate around their axis or move laterally within the plane of the bilayer. If the temperature is slowly lowered, a point is reached where the bilayer distinctly changes. The lipid is converted from a liquid crystalline phase to a frozen crystalline gel in which the movement of the phospholipid fatty acid chains is greatly restricted. The temperature at which this change occurs is called the transition temperature.

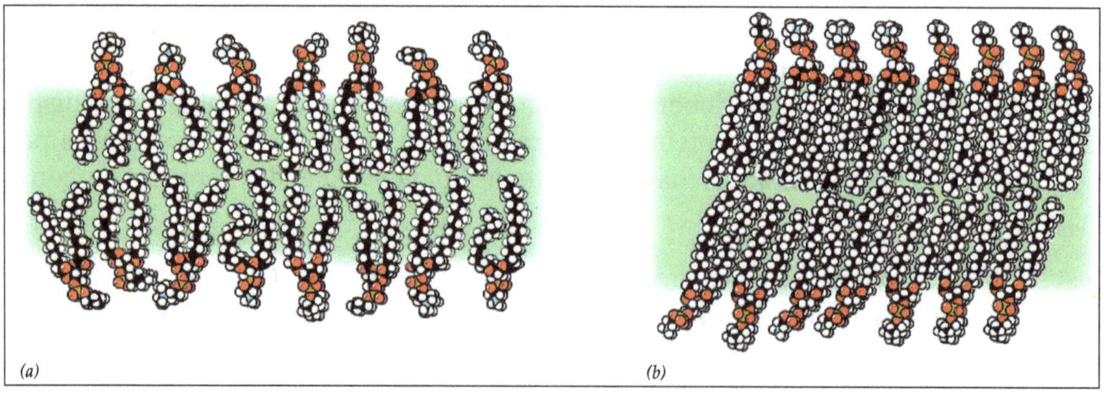

Figure shows the structure of the lipid bilayer depends on the temperature. The bilayer shown here is composed of two phospholipids: phosphatidylcholine and phosphatidylethanolamine. (a) Above the transition temperature, the lipid molecules and their hydrophobic tails are free to move in certain directions, even though they retain a considerable degree of order. (b) Below the transition temperature, the movement of the molecules is greatly restricted, and the entire bilayer can be described as a crystalline gel.

The transition temperature of a particular bilayer depends on the ability of the lipid molecules to be packed together, which depends in turn on the particular lipids of which it is constructed. Saturated fatty acids have the shape of a straight, flexible rod. Cis-unsaturated fatty acids, on the other hand, have crooks in the chain at the sites of a double bond. Consequently, phospholipids with saturated chains pack together more tightly than those containing unsaturated chains. The greater the degree of unsaturation of the fatty acids of the bilayer, the lower the temperature before the bilayer gels. The introduction of one double bond in a molecule of stearic acid lowers the melting temperature almost 60 °C. Another factor that influences bilayer fluidity is fatty acid chain length. The shorter the fatty acyl chains of a phospholipid, the lower its melting temperature. The physical state of the membrane is also affected by cholesterol. Because of their orientation within the bilayer, cholesterol molecules disrupt the close packing of fatty acyl chains and interfere with their mobility. The presence of cholesterol tends to abolish sharp transition temperatures and creates a condition of intermediate fluidity. In physiologic terms, cholesterol tends to increase the durability while decreasing the permeability of a membrane.

The Importance of Membrane Fluidity

What effect does the physical state of the lipid bilayer have on the biological properties of the membrane? Membrane fluidity provides a perfect compromise between a rigid, ordered structure in which mobility would be absent and a completely fluid, nonviscous liquid in which the components of the membrane could not be oriented and structural organization and mechanical support would be lacking. In addition, fluidity allows for interactions to take place within the membrane. For example, membrane fluidity makes it possible for clusters of membrane proteins to assemble at particular sites within the membrane and form specialized structures, such as intercellular junctions, light-capturing photosynthetic complexes, and synapses. Because of membrane fluidity, molecules that interact can come together, carry out the necessary reaction, and move apart.

Table: Melting Points of the Common 18-Carbon Fatty Acids.

Fatty acid	cis Double bonds	M.p.(°C)
Stearic acid	0	70
Oleic acid	1	19
Linoleic acid	2	-9
Linolenic acid	3	-17
Eicosapentanoic acid (EPA)	5	-54

Membranes arise only from preexisting membranes, and their growth is accomplished by the insertion of lipids and proteins into the fluid matrix of the membranous sheet. Many of the most basic cellular processes, including cell movement, cell growth, cell division, formation of intercellular junctions, secretion, and endocytosis, depend on the movement of membrane components and would probably not be possible if membranes were rigid, nonfluid structures.

Maintaining Membrane Fluidity

The internal temperature of most organisms (other than birds and mammals) fluctuates with the temperature of the external environment. Since it is essential for many activities that the membranes of a cell remain in a fluid state, cells respond to changing conditions by altering the types of phospholipids of which they are made. Maintenance of membrane fluidity is an example of homeostasis at the cellular level and can be demonstrated in various ways. For example, if the temperature of a culture of cells is lowered, the cells respond metabolically.

The initial "emergency" response is mediated by enzymes that remodel membranes, making the cell more cold resistant. Remodeling is accomplished by (1) desaturating single bonds in fatty acyl chains to form double bonds, and (2) reshuffling the chains between different phospholipid molecules to produce ones that contain two unsaturated fatty acids, which greatly lowers the melting temperature of the bilayer. Desaturation of single bonds to form double bonds is catalyzed by enzymes called desaturases. Reshuffling is accomplished by phospholipases, which split the fatty acid from the glycerol backbone, and acyltransferases, which transfer fatty acids between phospholipids. In addition, the cell changes the types of phospholipids being synthesized in favor of ones containing more unsaturated fatty acids. As a result of the activities of these various enzymes, the physical properties of a cell's membranes are matched to the prevailing environmental conditions. Maintenance of fluid membranes by adjustments in fatty acyl composition has been demonstrated in a variety of organisms, including hibernating mammals, pond-dwelling fish whose body temperature changes markedly from day to night, cold resistant plants, and bacteria living in hot springs. Prokaryotic cells that live at very high temperatures have plasma membranes that contain highly unusual lipids.

Lipid Rafts

Every so often an issue emerges that splits the community of cell biologists into believers and nonbelievers. The issue of lipid rafts falls into this category. When membrane lipids are extracted from cells and used to prepare artificial lipid bilayers, cholesterol and sphingolipids tend to self-assemble into microdomains that are more gelated and highly ordered than surrounding regions consisting primarily of phosphoglycerides. Because of their distinctive physical properties, such microdomains tend to float within the more fluid and disordered environment of the artificial bilayer. As a result, these patches of cholesterol and sphingolipid are referred to as lipid rafts. When added to these artificial bilayers, certain proteins tend to become concentrated in the lipid rafts, whereas others tend to remain outside their boundaries. GPI-anchored proteins show a particular fondness for the ordered regions of the bilayer.

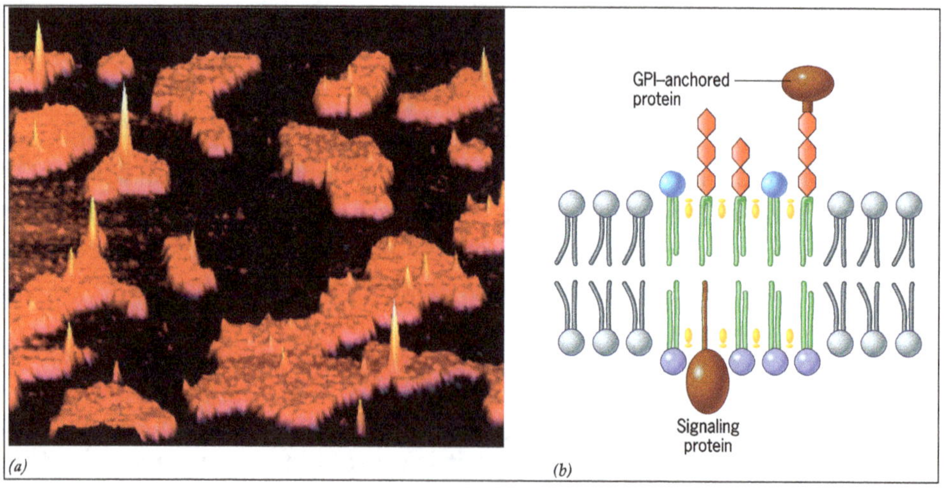

Figure shows Lipid rafts. (a) Image of the upper surface of an artificial lipid bilayer containing phosphatidylcholine, which appears as the black background, and sphingomyelin molecules, which organize themselves spontaneously into the orange-colored rafts. The yellow peaks show the positions of a GPI-anchored protein, which is almost exclusively raft-associated. This image is provided by an atomic force microscope, which measures the height of various parts of the specimen at the molecular level. (b) Schematic model of a lipid raft within a cell. The outer leaflet of the raft consists primarily of cholesterol (yellow) and sphingolipids (red head groups). Phosphatidylcholine molecules (blue head groups) with long saturated fatty acids also tend to concentrate in this region. GPI-anchored proteins are thought to become concentrated in lipid rafts. The lipids in the outer leaflet of the raft have an organizing effect on the lipids of the inner leaflet. As a result, the inner-leaflet raft lipids consist primarily of cholesterol and glycerophospholipids with long saturated fatty acyl tails. The inner leaflet tends to concentrate lipid-anchored proteins, such as Src kinase, that are involved in cell signaling.

The controversy arises over whether similar types of sphingolipid- and cholesterol-rich lipid rafts, such as that illustrated in figure b, exist within living cells. Most of the evidence in favor of lipid rafts is derived from studies that employ unnatural treatments, such as detergent extraction or cholesterol depletion, which makes the results difficult to interpret. Attempts to demonstrate the presence of lipid rafts in living cells have generally been unsuccessful, which can either mean that such rafts do not exist or they are so small (5 to 25 nm diameter) and short-lived as to be difficult to detect with current techniques. The concept of lipid rafts is very appealing because it provides a means to introduce order into a seemingly random sea of lipid molecules. Lipid rafts are postulated to serve as floating platforms that concentrate particular proteins, thereby organizing the membrane into functional compartments. For example, lipid rafts are thought to provide a favorable local environment for cell-surface receptors to interact with other membrane proteins that transmit signals from the extracellular space to the cell interior.

The Dynamic Nature of the Plasma Membrane

The lipid bilayer can exist in a relatively fluid state. As a result, a phospholipid can move laterally within the same leaflet with con siderable ease. The mobility of individual lipid molecules within the bilayer of the plasma membrane can be directly observed under the microscope by linking the polar heads of the lipids to gold particles or fluorescent compounds. It is estimated that a phospholipid can diffuse from one end of a bacterium to the other end in a second or two. In contrast, it takes a phospholipid molecule a matter of hours to days to move across to the other leaflet. Thus, of all the possible motions that a phospholipid can make, its flipflop to the other side of the membrane is the most restricted. This finding is not surprising. For flip-flop to occur, the hydrophilic head group of the lipid must pass through the internal hydrophobic sheet of the membrane, which is thermodynamically unfavorable. However, cells contain enzymes that actively move certain phospholipids from one leaflet to the other. These enzymes play a role in establishing lipid asymmetry and may also reverse the slow rate of passive transmembrane movement.

Because lipids provide the matrix in which integral proteins of a membrane are embedded, the physical state of the lipid is an important determinant of the mobility of integral proteins. The demonstration that integral proteins can move within the plane of the membrane was a cornerstone in the formulation of the fluid-mosaic model. The dynamic properties of membrane proteins have been revealed in several ways.

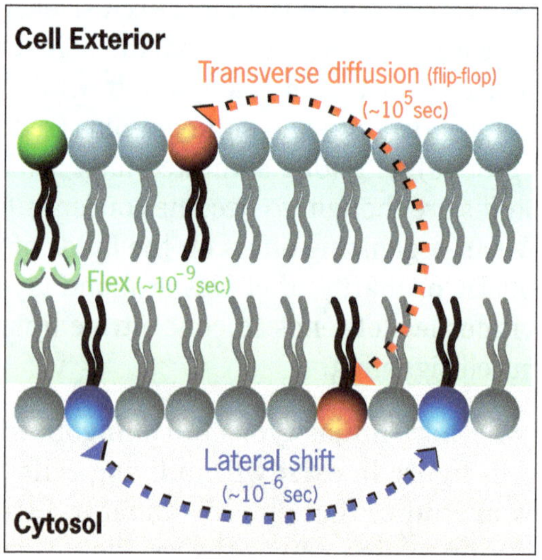

Figure shows the possible movements of phospholipids in a membrane. The types of movements in which membrane phospholipids can engage and the approximate time scales over which they occur. Whereas phospholipids move from one leaflet to another at a very slow rate, they diffuse laterally within a leaflet rapidly. Lipids lacking polar groups, such as cholesterol, can move across the bilayer quite rapidly.

The Diffusion of Membrane Proteins after Cell Fusion

Cell fusion is a technique whereby two different types of cells, or cells from two different species, can be fused to produce one cell with a common cytoplasm and a single, continuous plasma membrane. Cells are induced to fuse with one another by making the outer surface of the cells "sticky" so that their plasma membranes adhere to one another. Cells can be induced to fuse by addition of certain inactivated viruses that attach to the surface membrane, by adding the compound polyethylene glycol, or by a mild electric shock. Cell fusion has played an important role in cell biology and is currently used in an invaluable technique to prepare specific antibodies.

The first experiments to demonstrate that membrane proteins could move within the plane of the membrane utilized cell fusion. In their experiments, mouse and human cells were fused, and the locations of specific proteins of the plasma membrane were followed once the two membranes had become continuous. To follow the distribution of either the mouse membrane proteins or the human membrane proteins at various times after fusion, antibodies against one or the other type of protein were prepared and covalently linked to fluorescent dyes. The antibodies against the mouse proteins were complexed with a dye that fluoresces green and the antibodies against human proteins with one that fluoresces red. When the antibodies were added to fused cells, they bound to the human or mouse proteins and could be located under a fluorescence light microscope. At the time of fusion, the plasma membrane appeared half human and half mouse; that is, the two protein types remained segregated in their own hemisphere. As the time after fusion increased, the membrane proteins were seen to move laterally within the membrane into the opposite hemisphere. By about 40 minutes, each species' proteins were uniformly distributed around the entire hybrid cell membrane. If the same experiment was performed at lower temperature, the viscosity of the lipid bilayer increased, and the mobility of the membrane proteins decreased. These early cell fusion experiments gave the impression that integral membrane proteins were capable of virtually unrestricted movements. As we will see shortly, subsequent studies made it apparent that membrane dynamics was a much more complex subject than first envisioned.

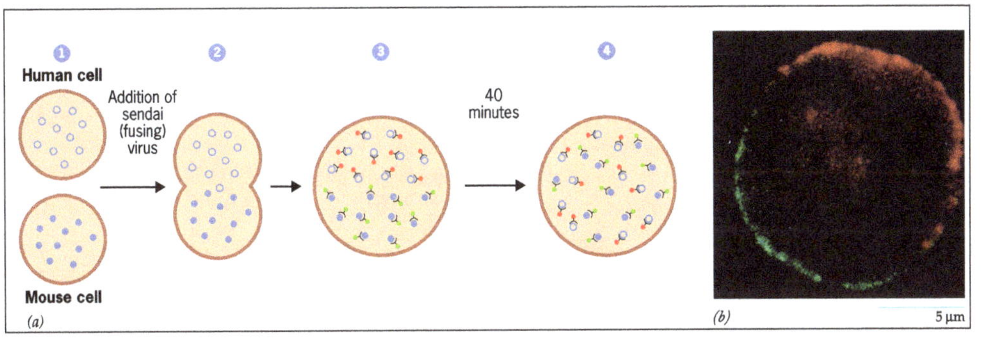

Figure shows the use of cell fusion to reveal mobility of membrane proteins. (a) Outline of the experiment in which human and mouse cells were fused (steps 1–2) and the distribution of the proteins on the surface of each cell were followed in the hybrids over time (steps 3–4). Mouse membrane proteins are indicated by solid circles, human membrane proteins by open circles. Locations of human and mouse proteins in the plasma membranes of the hybrid cells were monitored by interaction with fluorescent red and fluorescent green antibodies, respectively. (b) Micrograph showing a fused cell in which mouse and human proteins are still in their respective hemispheres (equivalent to the hybrid in step 3 of part a).

Restrictions on Protein and Lipid Mobility

Several techniques allow researchers to follow the movements of molecules in the membranes of living cells using the light microscope. In a technique called fluorescence recovery after photo-bleaching (FRAP), which is illustrated in figure a, integral membrane components in cultured cells are first labeled by linkage to a fluorescent dye. A particular membrane protein can be labeled using a specific probe, such as a fluorescent antibody. Once labeled, cells are placed under the microscope and irradiated by a sharply focused laser beam that bleaches the fluorescent molecules in its path, leaving a circular spot (typically about 1 mm diameter) on the surface of the cell that is largely devoid of fluorescence. If the labeled proteins in the membrane are mobile, then the random movements of these molecules should produce a gradual reappearance of fluorescence in the irradiated circle. The rate of fluorescence recovery provides a direct measure of the rate of diffusion (expressed as a diffusion coefficient, D) of the mobile molecules. The extent of fluorescence recovery (expressed as a percentage of the original intensity) provides a measure of the percentage of the labeled molecules that are free to diffuse.

Early studies utilizing FRAP suggested that (1) membrane proteins moved much more slowly in a plasma membrane than they would in a pure lipid bilayer and (2) a significant fraction of membrane proteins (30 to 70 percent) were not free to diffuse back into the irradiated circle. But the FRAP technique has its drawbacks. FRAP can only follow the average movement of a relatively large number of labeled molecules (hundreds to thousands) as they diffuse over a relatively large distance (e.g., 1 mm). As a result, researchers using FRAP cannot distinguish between proteins that are truly immobile and ones that can only diffuse over a limited distance in the time allowed. To get around these limitations, alternate techniques have been developed that allow researchers to observe the movements of individual protein molecules over very short distances and to determine how they might be restrained.

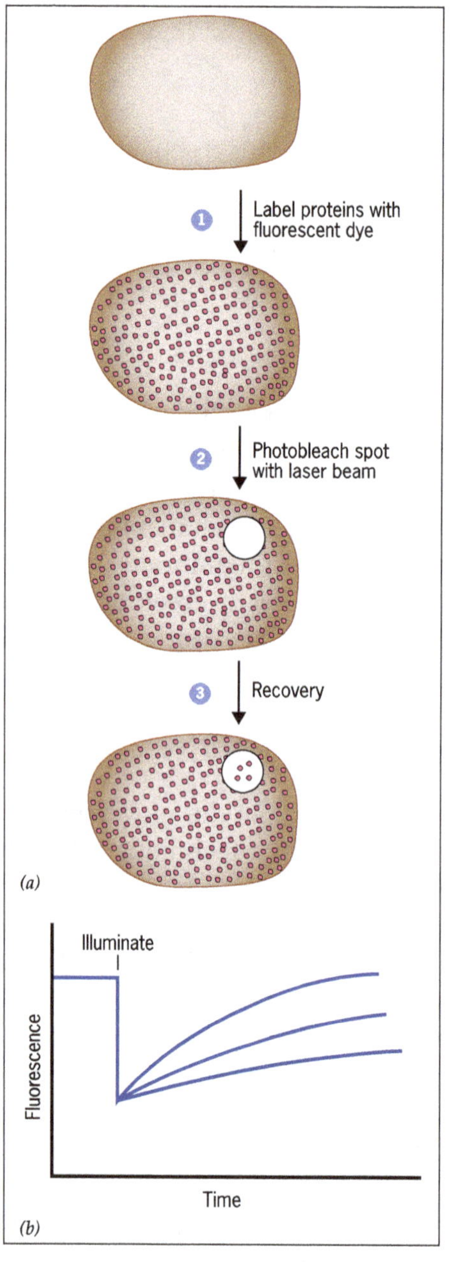

Figure shows measuring the diffusion rates of membrane proteins by fluorescence recovery after photobleaching (FRAP). (a) In this technique, a particular component of the membrane is first labeled with a fluorescent dye (step 1). A small region of the surface is then irradiated to bleach the dye molecules (step 2), and the recovery of fluorescence in the bleached region is followed over time (step 3). (b) The rate of fluorescence recovery within the illuminated spot can vary depending on the protein(s) being followed. The rate of recovery is related to the diffusion coefficient of the fluorescently labeled protein.

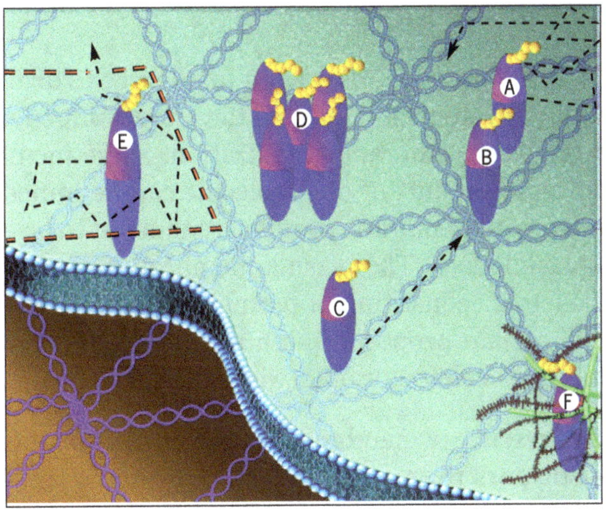

Figure shows patterns of movement of integral membrane proteins. Depending on the cell type and the conditions, integral membrane proteins can exhibit several different types of mobility. Protein A is capable of diffusing randomly throughout the membrane, though its rate of movement may be limited; protein B is immobilized as the result of its interaction with the underlying membrane skeleton; protein C is being moved in a particular direction as the result of its interaction with a motor protein at the cytoplasmic surface of the membrane; movement of protein D is restricted by other integral proteins of the membrane; movement of protein E is restricted by fences formed by proteins of the membrane skeleton, but it can hop into adjacent compartments through transient openings in a fence; movement of protein F is restrained by extracellular materials.

In single-particle tracking (SPT), individual membrane protein molecules are labeled, usually with antibody-coated gold particles (approximately 40 nm in diameter), and the movements of the labeled molecules are followed by computer-enhanced video microscopy. The results of these studies often depend on the particular protein being investigated. For example:

- Some membrane proteins move randomly throughout the membrane, though generally at rates considerably less than would be measured in an artificial lipid bilayer. (If protein mobility were based strictly on physical parameters such as lipid viscosity and protein size, one would expect proteins to migrate with diffusion coefficients of approximately 10^{-8} to 10^{-9} cm^2/sec rather than 10^{-10} to 10^{-12} cm^2/sec, as is observed for molecules of this group.) The reasons for the reduced diffusion coefficient have been debated.
- Some membrane proteins fail to move and are considered to be immobilized.
- In some cases, a protein is found to move in a highly directed (i.e., nonrandom) manner toward one part of the cell or another. For example, a particular membrane protein may tend to move toward the leading or the trailing edge of a moving cell.
- In most studies, the largest fraction of protein species exhibit random (Brownian) movement within the membrane at rates consistent with free diffusion (diffusion coefficients about 5×10^{-9} cm^2/sec), but the molecules are unable to migrate freely more than a few tenths of a micrometer. Long-range diffusion occurs but at slower rates, apparently because of the presence of a system of barriers.

Control of Membrane Protein Mobility

It is apparent that plasma membrane proteins are not totally free to drift around randomly on the lipid "sea." Instead, this class of proteins is subjected to various influences that affect their mobility and, in doing so, promote intramembrane organization. Some membranes are crowded with proteins, so that the random movements of one molecule can be impeded by its neighbors. The strongest influences on an integral membrane protein are thought to be exerted from just beneath the membrane on its cytoplasmic face. The plasma membranes of many cells possess a fibrillar network, or "membrane skeleton," consisting of peripheral proteins situated on the cytoplasmic surface of the membrane. A certain proportion of a membrane's integral protein molecules are either tethered to the membrane skeleton or otherwise restricted by it.

Information concerning the presence of membrane barriers has been obtained using an innovative technique that allows investigators to trap integral proteins and drag them through the plasma membrane with a known force. This technique, which uses an apparatus referred to as optical tweezers, takes advantage of the tiny optical forces that are generated by a focused laser beam. The integral proteins to be studied are tagged with antibody-coated beads, which serve as handles that can be gripped by the laser beam. It is generally found that optical tweezers can drag an integral protein for a limited distance before the protein encounters a barrier that causes it to be released from the laser's grip. As it is released, the protein typically springs backward, suggesting that the barriers are elastic structures.

One approach to studying factors that affect membrane protein mobility is to genetically modify cells so that they produce altered membrane proteins. Integral proteins whose cytoplasmic portions have been genetically deleted often move much greater distances than their intact counterparts, indicating that barriers reside on the cytoplasmic side of the membrane. These findings suggest that the membrane's underlying skeleton forms a network of "fences" around portions of the membrane, creating compartments that restrict the distance an integral protein can travel. Proteins move across the boundaries from one compartment to another through breaks in the fences. Such openings are thought to appear and disappear along with the dynamic disassembly and reassembly of parts of the meshwork. Membrane compartments may keep specific combinations of proteins in close enough proximity to facilitate their interaction.

Integral proteins lacking that portion that would normally project into the extracellular space typically move at a much faster rate than the wild-type version of the protein. This finding suggests that the movement of a transmembrane protein through the bilayer is slowed by extracellular materials that can entangle the external portion of the protein molecule.

Membrane Lipid Mobility

Proteins are huge molecules, so it isn't surprising that their movement within the lipid bilayer might be restricted. Phospholipids, by comparison, are small molecules that make up the very fabric of the lipid bilayer. One might expect their movements to be completely unfettered, yet a number of studies have suggested that phospholipid diffusion is also restricted. When individual phospholipid molecules of a plasma membrane are tagged and followed under the microscope using ultra high-speed cameras, they are seen to be confined for very brief periods and then hop from one confined area to another. Figure a shows the path taken by an individual phospholipid within the plasma membrane over a period of 56 milliseconds. Computer analysis indicates that the phospholipid diffuses freely within one compartment (shaded in purple) before it jumps the "fence" into a neighboring compartment (shaded in blue) and then over another fence into an adjacent compartment (shaded in green), and so forth. Treatment of the membrane with agents that disrupt the underlying membrane skeleton removes some of the fences that restrict phospholipid diffusion. But if the membrane skeleton lies beneath the lipid bilayer, how can it interfere with phospholipid movement? The fences are constructed of rows of integral membrane proteins whose cytoplasmic domains are attached to the membrane skeleton. This is not unlike the confinement of horses or cows by a picket fence whose posts are embedded in the underlying soil.

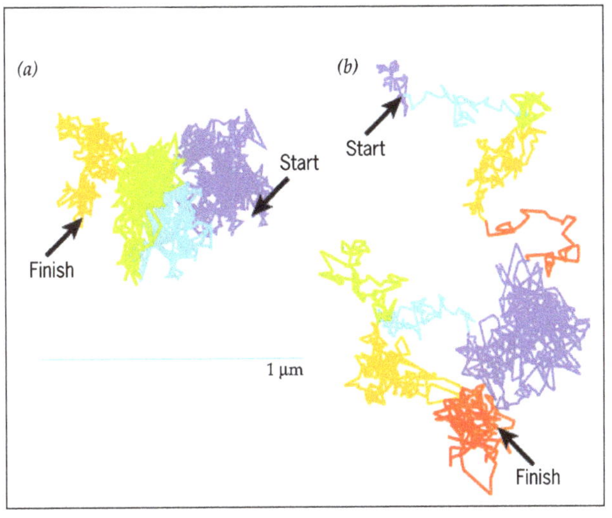

Figure shows experimental demonstration that diffusion of phospholipids within the plasma membrane is confined.(a) The track of a single labeled unsaturated phospholipid is followed for 56 ms as it diffuses within the plasma membrane of a rat fibroblast. Phospholipids diffuse freely within a confined compartment before hopping into a neighboring compartment. The rate of diffusion within a compartment is as rapid as that expected by unhindered Brownian movement. However, the overall rate of diffusion of the phospholipid appears slowed because the molecule must hop a barrier to continue its movement. The movement of the phospholipid within each compartment is represented by a single color. (b) The same experiment shown in a is carried out for 33 ms in an artificial bilayer, which lacks the "picket fences" present in a cellular membrane. The much more open, extended trajectory of the phospholipid can now be explained by simple, unconfined Brownian movement. For the sake of comparison, fake compartments were assigned in b and indicated by different colors.

Membrane Domains and Cell Polarity

For the most part, studies of membrane dynamics, such as those discussed above, are carried out on the relatively homogeneous plasma membrane situated at the upper or lower surface of a cell residing on a culture dish. Most membranes, however, exhibit distinct variations in protein composition and mobility, especially in cells whose various surfaces display distinct functions.

For example, the epithelial cells that line the intestinal wall or make up the microscopic tubules of the kidney are highly polarized cells whose different surfaces carry out different functions. Studies indicate that the apical plasma membrane, which selectively absorbs substances from the lumen, possesses different enzymes than the lateral plasma membrane, which interacts with neighboring epithelial cells, or the basal membrane, which adheres to an underlying extracellular substrate (a basement membrane). In other examples, the receptors for neurotransmitter substances are concentrated into regions of the plasma membrane located within synapses, and receptors for low-density lipoproteins are concentrated into patches of the plasma membrane specialized to facilitate their internalization.

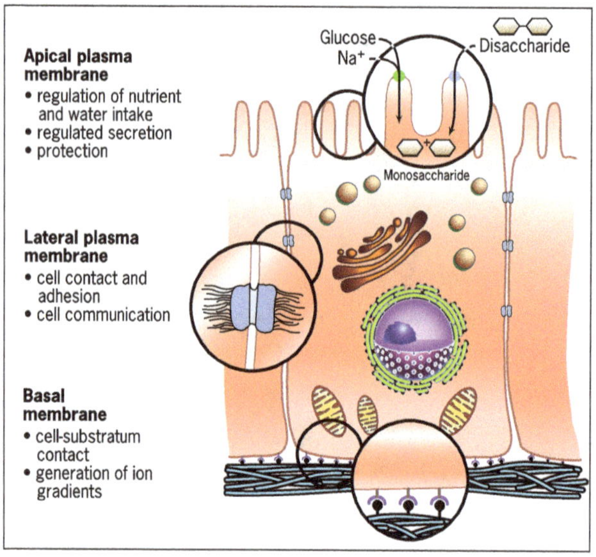

Figure shows differentiated functions of the plasma membrane of an epithelial cell. The apical surface of this intestinal epithelial cell contains integral proteins that function in ion transport and hydrolysis of disaccharides, such as sucrose and lactose; the lateral surface contains integral proteins that function in intercellular interaction; and the basal surface contains integral proteins that function in the association of the cell with the underlying basement membrane.

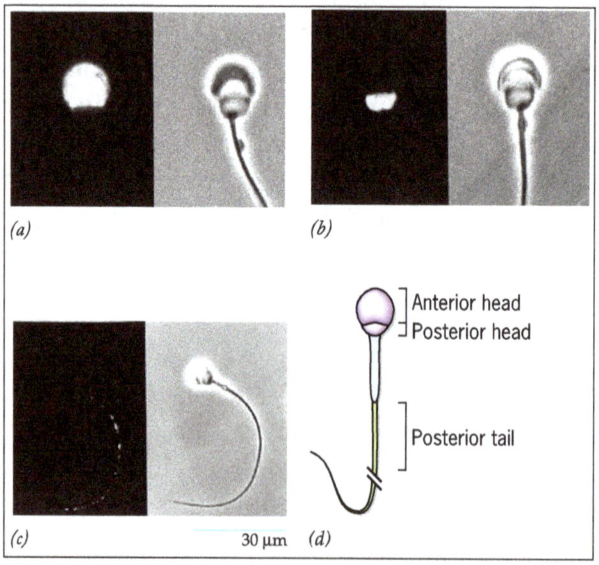

Figure shows differentiation of the mammalian sperm plasma membrane as revealed by fluorescent antibodies. (a–c) Three pairs of micrographs, each showing the distribution of a particular protein at the cell surface as revealed by a bound fluorescent antibody. The three proteins are localized in different parts of the continuous sperm membrane. Each pair of photographs shows the fluoresence pattern of the bound antibody and a phase contrast micrograph of the same cell. (d) Diagram summarizing the distribution of the proteins.

Of all the various types of mammalian cells, sperm may have the most highly differentiated structure. A mature sperm can be divided into head, midpiece, and tail, each having its own specialized functions. Although divided into a number of distinct parts, a sperm is covered by a continuous plasma membrane which, as revealed by numerous techniques, consists of a mosaic of different types of localized domains. For example, when sperm are treated with a variety of antibodies, each antibody combines with the surface of the cell in a unique topographic pattern that reflects the distribution within the plasma membrane of the particular protein antigen recognized by that antibody.

The Red Blood Cell: An Example of Plasma Membrane Structure

Of all the diverse types of membranes, the plasma membrane of the human erythrocyte (red blood cell) is the most studied and best understood. There are several reasons for the popularity of this membrane. The cells are inexpensive to obtain and readily available in huge numbers from whole blood. They are already present as single cells and need not be dissociated from a complex tissue. The cells are simple by comparison with other cell types, lacking nuclear and cytoplasmic membranes that inevitably contaminate plasma membrane preparations from other cells. In addition, purified, intact erythrocyte plasma membranes can be obtained simply by placing the cells in a dilute (hypotonic) salt solution. The cells respond to this osmotic shock by taking up water and swelling, a phenomenon termed hemolysis. As the surface area of each cell increases, the cell becomes leaky, and the contents, composed almost totally of dissolved hemoglobin, flow out of the cell leaving behind a plasma membrane "ghost".

Once erythrocyte plasma membranes are isolated, the proteins can be solubilized and separated from one another (fractionated), providing a better idea of the diversity of proteins within the membrane. Fractionation of membrane proteins can be accomplished using polyacrylamide gel electrophoresis (PAGE) in the presence of the ionic detergent sodium dodecyl sulfate (SDS). The SDS keeps the integral proteins soluble and, in addition, adds a large number of negative charges to the proteins with which it associates. Because the number of charged SDS molecules per unit weight of protein tends to be relatively constant, the molecules separate from one another according to their molecular weight. The largest proteins move most slowly through the molecular sieve of the gel. The major proteins of the erythrocyte membrane are separated into about a dozen conspicuous bands by SDS–PAGE. Among the proteins are a variety of enzymes (including glyceraldehyde 3-phosphate dehydrogenase, one of the enzymes of glycolysis), transport proteins (for ions and sugars), and skeletal proteins (e.g., spectrin).

Integral Proteins of the Erythrocyte Membrane

A model of the erythrocyte plasma membrane showing its major proteins is seen in figure. The most abundant integral proteins of this membrane are a pair of carbohydrate-containing, membrane-spanning proteins, called band 3 and glycophorin A. The high density of these proteins within the membrane is evident in the freeze-fracture micrographs of figure. Band 3, which gets its name from its position in an electrophoretic gel, is present as a dimer composed of two identical subunits (a homodimer). Each subunit spans the membrane at least a dozen times and contains a relatively small amount of carbohydrate (6–8 percent of the molecule's weight). Band 3 protein serves as a channel for the passive exchange of anions across the membrane. As blood circulates through the tissues, carbon dioxide becomes dissolved in the fluid of the bloodstream (the plasma) and undergoes the following reaction:

$$H_2O + CO_2 \rightarrow H_2CO_3 \rightarrow HCO^-_3 + H^+$$

The bicarbonate ions (HCO^-_3) enter the erythrocyte in exchange for chloride ions, which leave the cell. In the lungs, where carbon dioxide is released, the reaction is reversed and bicarbonate ions leave the erythrocyte in exchange for chloride ions. The reciprocal movement of HCO^-_2 and Cl^- occurs through a channel in the center of each band 3 dimer.

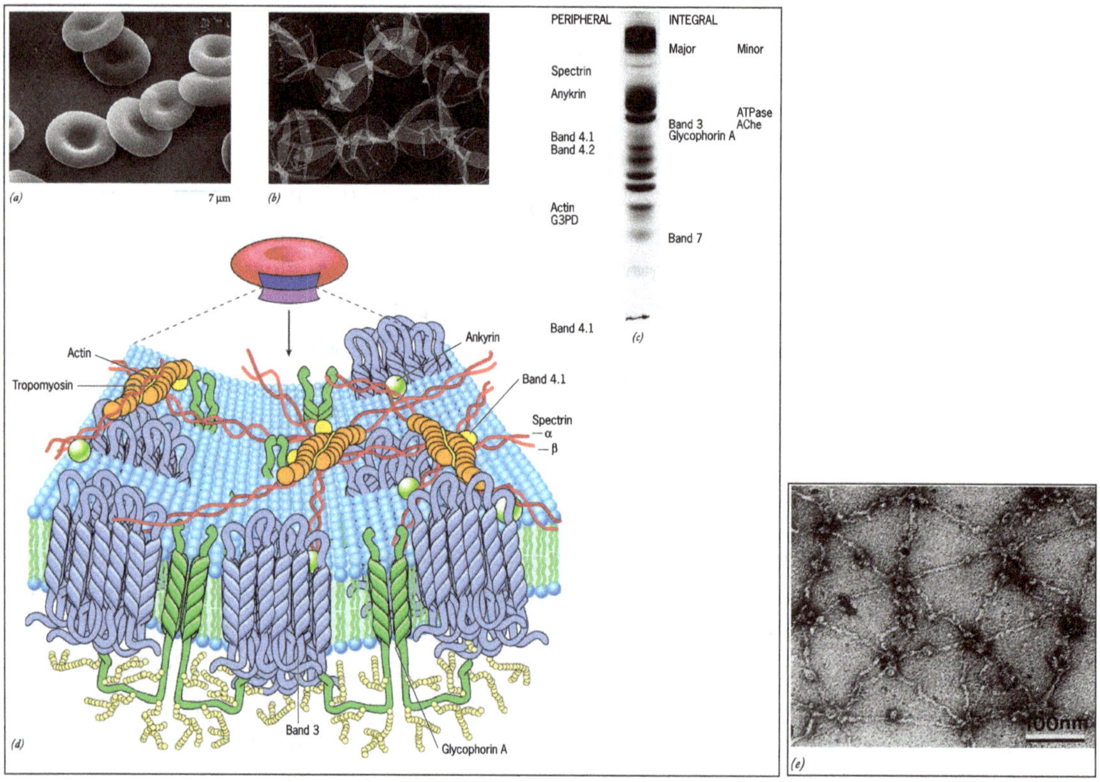

Figure shows the plasma membrane of the human erythrocyte. (a) Scanning electron micrograph of human erythrocytes. (b) Micrograph showing plasma membrane ghosts, which were isolated by allowing erythrocytes to swell and hemolyze as described in the text. (c) The results of SDS–polyacrylamide gel electrophoresis (SDS–PAGE) used to fractionate the proteins of the erythrocyte membrane, which are identified at the sides of the gel. (d) A model of the erythrocyte plasma membrane as viewed from the internal surface, showing the integral proteins embedded in the lipid bilayer and the arrangement of peripheral proteins that make up the membrane's internal skeleton. The band 3 dimer shown here is simplified. The band 4.1 protein stabilizes actin–spectrin complexes. (e) Electron micrograph showing the arrangement of the proteins of the inner membrane skeleton.

Glycophorin A was the first membrane protein to have its amino acid sequence determined. The arrangement of the polypeptide chain of glycophorin A within the plasma membrane is shown in Figure. (Other related glycophorins, B, C, D, and E, are also present in the membrane at much lower concentrations.) Like band 3, glycophorin A is also present in the membrane as a dimer. Unlike band 3, each glycophorin A subunit spans the membrane only once, and it contains a bushy carbohydrate cover consisting of 16 oligosaccharide chains that together make up about 60 percent of the molecule's weight. It is thought that the primary function of glycophorin derives from the large number of negative charges borne on sialic acid, the sugar residue at the end of each carbohydrate chain. Because of these charges, red blood cells repel each other, which prevents the cells from clumping as they circulate through the body's tiny vessels. It is noteworthy that persons who lack both glycophorin A and B in their red blood cells show no ill-effects from their absence. At the same time, the band 3 proteins in these individuals are more heavily glycosylated, which apparently compensates for the otherwise missing negative charges required to prevent cell–cell interaction. Glycophorin also happens to be the receptor utilized by the protozoan that causes malaria, providing a path for entry into the blood cell. Consequently, individuals whose erythrocytes lack glycophorin A and B are thought to be protected from acquiring malaria. Differences in glycophorin amino acid sequence determine whether a person has an MM, MN, or NN blood type.

The Erythrocyte Membrane Skeleton

The peripheral proteins of the erythrocyte plasma membrane are located on its internal surface and constitute a fibrillar membrane skeleton that plays a major role in determining the biconcave shape of the erythrocyte. The membrane skeleton can establish domains within the membrane that enclose particular groups of membrane proteins and may greatly restrict the movement of these proteins. The major component of the skeleton is an elongated fibrous protein, called spectrin. Spectrin is a heterodimer approximately 100 nm long, consisting of an α and β subunit that curl around one another. Two such dimeric molecules are linked at their head ends to form a 200-nm-long filament that is both flexible and elastic. Spectrin is attached to the internal surface of the membrane by means of noncovalent bonds to another peripheral protein, ankyrin, which in turn is linked noncovalently to the cytoplasmic domain of a band 3 molecule. As is evident in figures d and e, spectrin filaments are organized into hexagonal or pentagonal arrays. This two-dimensional network is constructed by linking both ends of each spectrin filament to a cluster of proteins that include a short filament of actin and tropomyosin, proteins typically involved in contractile activities. A number of genetic diseases (hemolytic anemias) characterized by fragile, abnormally shaped erythrocytes have been traced to mutations in ankyrin or spectrin.

If the peripheral proteins are removed from erythrocyte ghosts, the membrane becomes fragmented into small vesicles, indicating that the inner protein network is required to maintain the integrity of the membrane. Erythrocytes are circulating cells that are squeezed under pressure through microscopic capillaries whose diameter is considerably less than that of the erythrocytes themselves. To traverse these narrow passageways, and to do so day after day, the red blood cell must be highly deformable, durable, and capable of withstanding shearing forces that tend to pull it apart. The spectrin–actin network gives the cell the strength, elasticity, and pliability necessary to carry out its demanding function.

When first discovered, the membrane skeleton of the erythrocyte was thought to be a unique structure suited to the unique shape and mechanical needs of this cell type. However, as other cells were examined, similar types of membrane skeletons containing members of the spectrin and ankyrin families have been revealed, indicating that inner membrane skeletons are widespread. Dystrophin, for example, is a member of the spectrin family that is found in the membrane skeleton of muscle cells. Mutations in dystrophin are responsible for causing muscular dystrophy, a devastating disease that cripples and kills children. As in the case of cystic fibrosis, the most debilitating mutations are ones that lead to a complete absence of the protein in the cell. The plasma membranes of muscle cells lacking dystrophin are apparently destroyed as a consequence of the mechanical stress exerted on them as the muscle contracts. As a result, the muscle cells die and eventually are no longer replaced.

The Movement of Substances Across Cell Membranes

Because the contents of a cell are completely surrounded by its plasma membrane, all communication between the cell and the extracellular medium must be mediated by this structure. In a sense, the plasma membrane has a dual function. On one hand, it must retain the dissolved materials of the cell so that they do not simply leak out into the environment, while on the other hand, it must allow the necessary exchange of materials into and out of the cell. The lipid bilayer of the membrane is ideally suited to prevent the loss of charged and polar solutes from a cell. Consequently, some special provision must be made to allow the movement of nutrients, ions, waste products, and other compounds, in and out of the cell. There are basically two means for the movement of substances through a membrane: passively by diffusion or actively by an energy-coupled transport process. Both types of movements lead to the net flux of a particular ion or compound. The term net flux indicates that the movement of the substance into the cell (influx) and out of the cell (efflux) is not balanced, but that one exceeds the other.

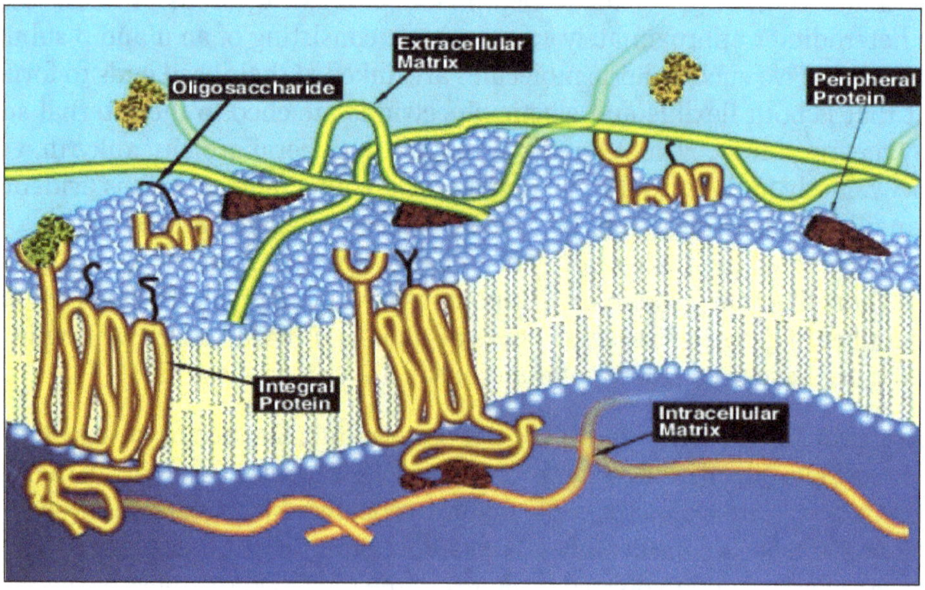

Several different processes are known by which substances move across membranes: simple diffusion through the lipid bilayer; simple diffusion through an aqueous, proteinlined channel; diffusion that is facilitated by a protein transporter; and active transport, which requires an energy-driven protein "pump" capable of moving substances against a concentration gradient. We will consider each in turn, but first we will discuss the energetics of solute movement.

The Energetics of Solute Movement

Diffusion is a spontaneous process in which a substance moves from a region of high concentration to a region of low concentration, eventually eliminating the concentration difference between the two regions. Diffusion depends on the random thermal motion of solutes and is an exergonic process driven by an increase in entropy. We will restrict the following discussion to diffusion of substances across membranes.

The free-energy change when an uncharged solute (a nonelectrolyte) diffuses across a membrane depends on the magnitude of the concentration gradient, that is, the difference in concentration on each side of the membrane. The following relationship describes the movement of a nonelectrolyte into the cell:

$$\Delta G = RT \ln \frac{[C_i]}{[C_o]}$$

$$\Delta G = 2.303 RT \log_{10} \frac{[C_i]}{[C_o]}$$

where DG is the free-energy change, R is the gas constant, T is the absolute temperature, and $[C_i]/[C_o]$ is the ratio of the concentration of the solute on the inside (i) and outside (o) surfaces of the membrane. At 25 °C,

$$\Delta G = 1.4 \text{ kcal/mol} \cdot \log_{10} \frac{[C_i]}{[C_o]}$$

If the ratio of $[C_i]/[C_o]$ is less than 1.0, then the log of the ratio is negative, ΔG is negative, and the net influx of solute is thermodynamically favored (exergonic). If, for example, the external concentration of solute is 10 times the internal concentration, ΔG = -1.4 kcal/mole. Thus, the maintenance of a tenfold concentration gradient represents a storage of 1.4 kcal/mol. As solute moves into the cell, the concentration gradient decreases, the stored energy is dissipated, and ΔG decreases until, at equilibrium, ΔG is zero. (To calculate ΔG for movement of a solute out of the cell, the term for concentration ratios becomes $[C_o]/[C_i]$).

If the solute is an electrolyte (a charged species), the overall charge difference between the two compartments must also be considered. As a result of the mutual repulsion of ions of like charges, it is thermodynamically unfavorable for an electrolyte to move across a membrane from one compartment into another compartment having a net charge of the same sign. Conversely, if the charge of the electrolyte is opposite in sign to the compartment into which it is moving, the process is thermodynamically favored. The greater the difference in charge (the potential difference or voltage) between the two compartments, the greater the difference in free energy. Thus, the tendency of an electrolyte to diffuse between two compartments depends on two gradients: a chemical gradient, determined by the concentration difference of the substance between the two compartments, and the electric potential gradient, determined by the difference in charge. Together these differences are combined to form an electrochemical gradient. The free-energy change for the diffusion of an electrolyte into the cell is,

$$\Delta G = RT \ln \frac{[C_i]}{[C_o]} + zF\Delta E_m$$

where z is the charge of the solute, F is the Faraday constant (23.06 kcal/V ? equivalent, where an equivalent is the amount of the electrolyte having one mole of charge), and ΔE_m is the potential difference (in volts) between the two compartments. A tenfold difference in concentration of a nonelectrolyte across a membrane at 25 °C generates a ΔG of -1.4 kcal/mol. Suppose the concentration gradient consisted of Na^+ ions, which were present at tenfold higher concentration outside the cell than in the cytoplasm. Because the voltage across the membrane of a cell is typically about -70 mV, the free-energy change for the movement of a mole of Na1 ions into the cell under these conditions would be,

$$\Delta G = -1.4\,\text{kcal}/\text{mol} + zF\Delta E_m$$
$$\Delta G = -1.4\,\text{kcal}/\text{mol} + (1)(23.06\,\text{kcal}/\text{V}\cdot\text{mol})(-0.07\,\text{V})$$
$$= -3.1\,\text{kcal}/\text{mol}$$

Thus under these conditions, the concentration difference and the electric potential make similar contributions to the storage of free energy across the membrane.

The interplay between concentration and potential differences is seen in the diffusion of potassium ions (K^+) out of a cell. The efflux of the ion is favored by the K^+ concentration gradient, which has a higher K^+ concentration inside the cell, but hindered by the electrical gradient that its diffusion creates, which leaves a higher negative charge inside the cell.

Diffusion of Substances through Membranes

Two qualifications must be met before a nonelectrolyte can diffuse passively across a plasma membrane. The substance must be present at higher concentration on one side of the membrane than the other, and the membrane must be permeable to the substance. A membrane may be permeable to a given solute either (1) because that solute can pass directly through the lipid bilayer, or (2) because that solute can traverse an aqueous pore that spans the membrane. Let us begin by considering the former route in which a substance must dissolve in the lipid bilayer on its way through the membrane.

Discussion of simple diffusion leads us to consider the polarity of a solute. One simple measure of the polarity (or nonpolarity) of a substance is its partition coefficient, which is the ratio of its solubility in a nonpolar solvent, such as octanol or a vegetable oil, to that in water under conditions where the nonpolar solvent and water are mixed together. Figure shows the relationship between partition coefficient and membrane permeability of a variety of chemicals and drugs. It is evident that the greater the lipid solubility, the faster the penetration.

Another factor determining the rate of penetration of a compound through a membrane is its size. If two molecules have approximately equivalent partition coefficients, the smaller molecule tends to penetrate the lipid bilayer of a membrane more rapidly than the larger one. Very small, uncharged molecules penetrate very rapidly through cellular membranes. Consequently, membranes are highly permeable to small inorganic molecules, such as O_2, CO_2, NO, and H_2O, which are thought to slip between adjacent phospholipids. In contrast, larger polar molecules, such as sugars, amino acids, and phosphorylated intermediates, exhibit poor membrane penetrability. As a result, the lipid bilayer of the plasma membrane provides an effective barrier that keeps these essential metabolites from diffusing out of the cell. Some of these molecules (e.g., sugars and amino acids) must enter cells from the bloodstream, but they cannot do so by simple diffusion. Instead, special mechanisms must be available to mediate their penetration through the plasma membrane. The use of such mechanisms allows a cell to regulate the movement of substances across its surface barrier.

The Diffusion of Water through Membranes

Water molecules move much more rapidly through a cell membrane than do dissolved ions or small polar organic solutes, which are essentially non-penetrating. Because of this difference in the penetrability of water versus solutes, membranes are said to be semipermeable. Water moves readily through a semipermeable membrane from a region of lower solute concentration to a region of higher solute concentration. This process is called osmosis, and it is readily demonstrated by placing a cell into a solution containing a nonpenetrating solute at a concentration different than that present within the cell itself.

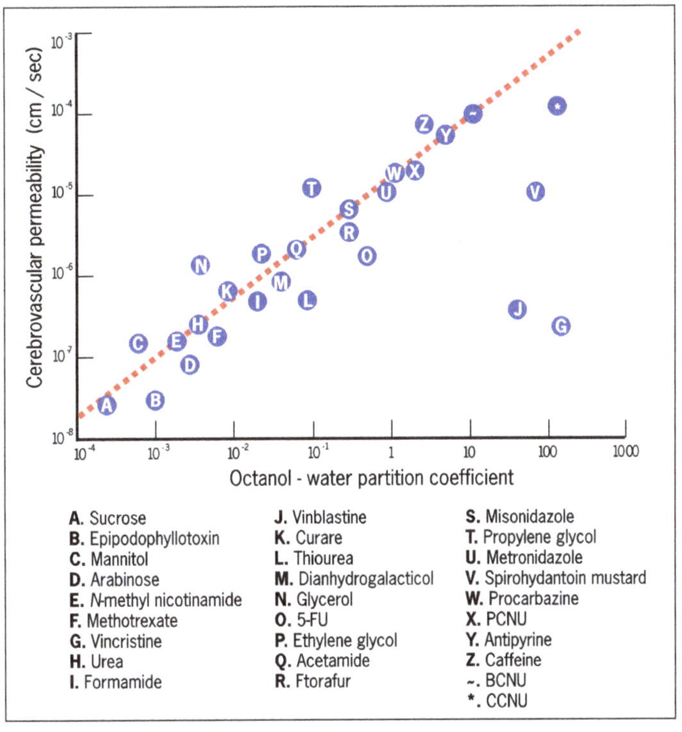

Figure shows the relationship between partition coefficient and membrane permeability.In this case, measurements were made of the penetration of a variety of chemicals and drugs across the plasma membranes of the cells that line the capillaries of the brain. Substances penetrate by passage through the lipid bilayer of these cells. The partition coefficient is expressed as the ratio of solubility of a solute in octanol to its solubility in water. Permeability is expressed as penetrance (P) in cm/sec. For all but a few compounds, such as vinblastine and vincristine, penetrance is directly proportional to lipid solubility.

When two compartments of different solute concentration are separated by a semipermeable membrane, the compartment of higher solute concentration is said to be hypertonic (or hyperosmotic) relative to the compartment of lower solute concentration, which is described as being hypotonic (or hypoosmotic). When a cell is placed into a hypotonic solution, the cell rapidly gains water by osmosis and swells. Conversely, a cell placed into a hypertonic solution rapidly loses water by osmosis and shrinks. These simple observations show that a cell's volume is controlled by the difference between the solute concentration inside the cell and that in the extracellular medium. The swelling and shrinking of cells in slightly hypotonic and hypertonic media are usually only temporary events. Within a few minutes, the cells recover and return to their original volume. In a hypotonic medium, recovery occurs as the cells lose ions, thereby reducing their internal osmotic pressure. In a hypertonic medium, recovery occurs as the cells gain ions from the medium. Once the internal solute concentration (which includes a high concentration of dissolved proteins) equals the external solute concentration, the internal and external fluids are isotonic (or isosmotic), and no net movement of water into or out of the cells occurs.

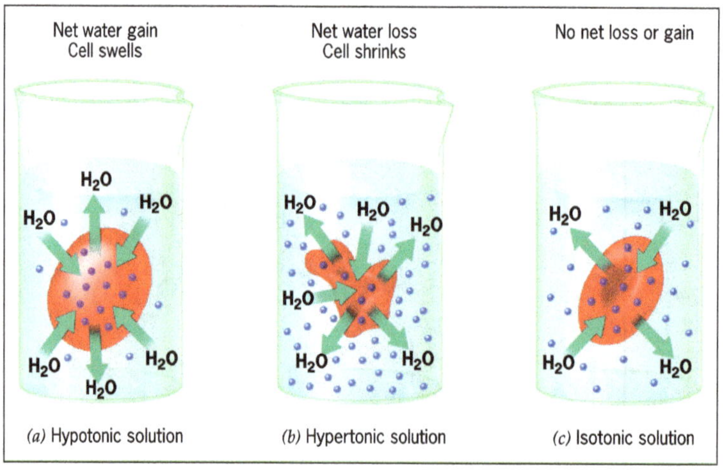

Figure shows the effects of differences in the concentration of solutes on opposite sides of the plasma membrane. (a) A cell placed in a hypotonic solution (one having a lower solute concentration than the cell) swells because of a net gain of water by osmosis. (b) A cell in a hypertonic solution shrinks because of a net loss of water by osmosis. (c) A cell placed in an isotonic solution maintains a constant volume because the inward flux of water is equal to the outward flux.

Osmosis is an important factor in a multitude of bodily functions. Your digestive tract, for example, secretes several liters of fluid daily, which is reabsorbed osmotically by the cells that line your intestine. If this fluid weren't reabsorbed, as happens in cases of extreme diarrhea, you would face the prospect of rapid dehydration. Plants utilize osmosis in different ways. Unlike animal cells, which are generally isotonic with the medium in which they are bathed, plant cells are generally hypertonic compared to their fluid environment. As a result, there is a tendency for water to enter the cell, causing it to develop an internal (turgor) pressure that pushes against its surrounding wall. Turgor pressure provides support for nonwoody plants and for the nonwoody parts of trees, such as the leaves. If a plant cell is placed into a hypertonic medium, its volume shrinks as the plasma membrane pulls away from the surrounding cell wall, a process called plasmolysis. The loss of water due to plasmolysis causes plants to lose their support and wilt.

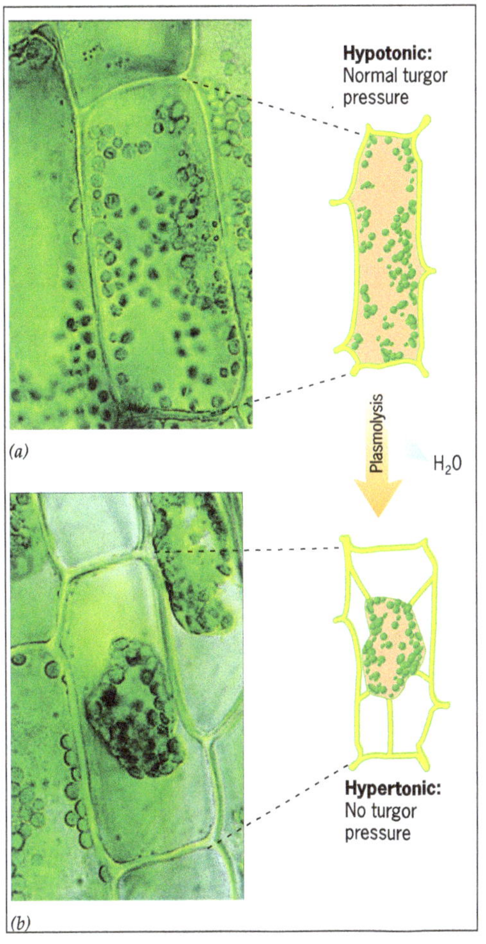

Figure shows the effects of osmosis on a plant cell. (a) Aquatic plants living in freshwater are surrounded by a hypotonic environment. Water therefore tends to flow into the cells, creating turgor pressure. (b) If the plant is placed in a hypertonic solution, such as seawater, the cell loses water, and the plasma membrane pulls away from the cell wall.

Not all cells are equally permeable to water. In fact, many cells are much more permeable to water than can be explained by simple diffusion through the lipid bilayer. In the early 1990s, Peter Agre and colleagues at Johns Hopkins University were attempting to isolate and purify the membrane proteins responsible for the Rh antigen on the surface of red blood cells. During this pursuit, they identified a protein they thought might be the long-sought water channel of the erythrocyte membrane. To test their hypothesis, they engineered frog oocytes to incorporate the newly discovered protein into their plasma membranes and then placed the oocytes in a hypotonic medium. Just as predicted, the oocytes swelled due to the influx of water and eventually burst. The team had discovered a family of small integral proteins, called aquaporins, that allow the passive movement of water from one side of the plasma membrane to the other. Each aquaporin subunit (in the four-subunit protein) contains a central channel that is lined primarily by hydrophobic amino acid residues and is highly specific for water molecules. A billion or so water molecules can pass—in single file—through each channel every second. At the same time, H^+ ions, which normally hop along a chain of water molecules, are not able to penetrate these open pores. The apparent mechanism by which these channels are able to exclude protons has been suggested by a combination of X-ray crystallographic studies, which has revealed the structure of the protein, and molecular dynamics simulations, which has put this protein structure into operation. A model based on such simulations is shown in figure a. Very near its narrowest point, the wall of an aquaporin channel contains a pair of precisely positioned positive charges (residues N203 and N68 in figure b) that attract the oxygen atom of each water molecule as it speeds through the constriction in the protein. This interaction reorients the central water molecule in a position that prevents it from maintaining the hydrogen bonds that normally link it to its neighboring water molecules. This removes the bridge that would normally allow protons to move from one water molecule to the next.

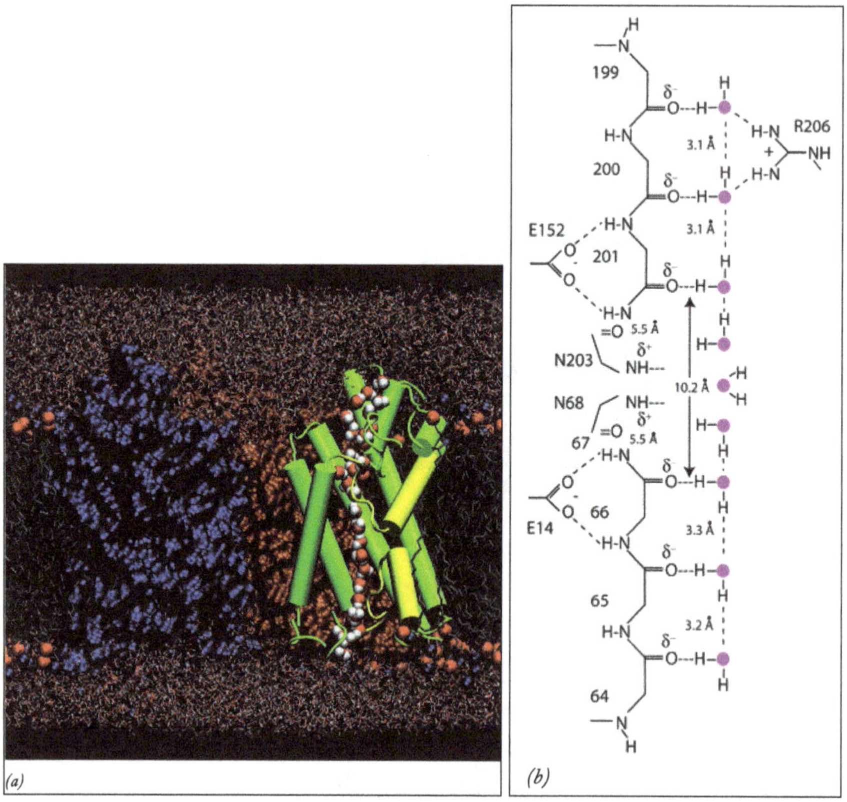

Figure shows passage of water molecules through an aquaporin channel. (a) Snapshot from a molecular dynamics simulation of a stream of water molecules (red and white spheres) passing in single file through the channel in one of the subunits of an aquaporin molecule residing within a membrane. (b) A model describing the mechanism by which water molecules pass through an aquaporin channel with the simultaneous exclusion of protons. Nine water molecules are shown to be lined up in single file along the wall of the channel. Each water molecule is depicted as a red circular O atom with two associated Hs. In this model, the four water molecules at the top and bottom of the channel are oriented, as the result of their interaction with the carbonyl (CPO) groups of the protein backbone, with their H atoms pointed away from the center of the channel. These water molecules are able to form hydrogen bonds (dashed lines) with their neighbors. In contrast, the single water molecule in the center of the channel is oriented in a position that prevents it from forming hydrogen bonds with other water molecules, which has the effect of interrupting the flow of protons through the channel.

Aquaporins are particularly prominent in cells, such as those of a kidney tubule or plant root, where the passage of water plays a crucial role in the tissue's physiologic activities. The hormone vasopressin, which stimulates water retention by the collecting ducts of the kidney, acts by way of one of these proteins (AQP2). Some cases of the inherited disorder congenital nephrogenic diabetes insipidus arise from mutations in this aquaporin channel. Persons suffering from this disease excrete large quantities of urine because their kidneys do not respond to vasopressin.

The Diffusion of Ions through Membranes

The lipid bilayer that constitutes the core of biological membranes is highly impermeable to charged substances, including small ions such as Na^+, K^+, Ca^{2+} and Cl^-. Yet the rapid movement (conductance) of these ions across membranes plays a critical role in a multitude of cellular activities, including formation and propagation of a nerve impulse, secretion of substances into the extracellular space, muscle contraction, regulation of cell volume, and the opening of stomatal pores on plant leaves.

In 1955, Alan Hodgkin and Richard Keynes of Cambridge University first proposed that cell membranes contain ion channels, that is, openings in the membrane that are permeable to specific ions. During the late 1960s and 1970s, Bertil Hille of the University of Washington and Clay Armstrong of the University of Pennsylvania began to obtain evidence for the existence of such channels. The final "proof " emerged through the work of Bert Sakmann and Erwin Neher at the Max-Planck Institute in Germany in the late 1970s and early 1980s who developed techniques to monitor the ionic current passing through a single ion channel. This is accomplished using very fine micropipette-electrodes made of polished glass that are placed on the outer cell surface and sealed to the membrane by suction. The voltage across the membrane can be maintained (clamped) at any particular value, and the current originating in the small patch of membrane surrounded by the pipette can be measured. These landmark studies marked the first successful investigations into the activities of individual protein molecules. Today, biologists have identified a bewildering variety of ion channels, each formed by integral membrane proteins that enclose a central aqueous pore. As might be predicted, mutations in the genes encoding ion channels can lead to many serious diseases.

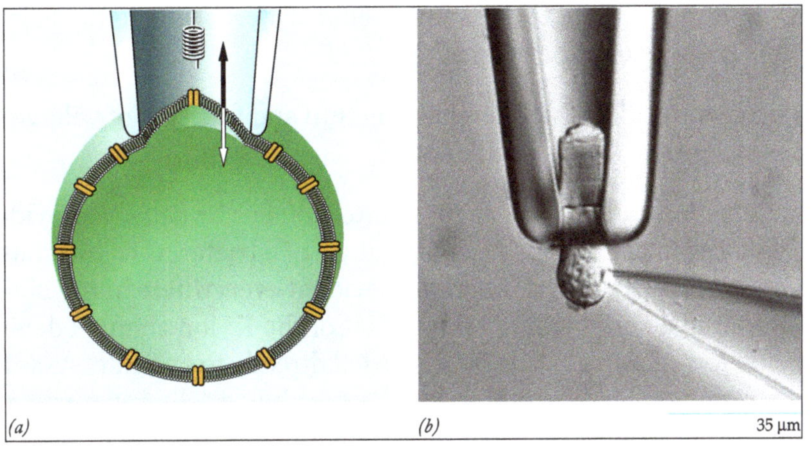

Figure shows measuring ion channel conductance by patch-clamp recording. (a) In this technique, a highly polished glass micropipette is placed against a portion of the outer surface of a cell, and suction is applied to seal the rim of the pipette against the plasma membrane. Because the pipette is wired as an electrode (a microelectrode), a voltage can be applied across the patch of membrane enclosed by the pipette, and the responding flow of ions through the membrane channels can be measured. As indicated in the figure, the micropipette can enclose a patch of membrane containing a single ion channel, which allows investigators to monitor the opening and closing of a single gated channel, as well as its conductance at different applied voltages. (b) The micrograph shows patch-clamp recordings being made from a single photoreceptor cell of the retina of a salamander. One portion of the cell is drawn into a glass micropipette by suction, while a second micropipette-electrode (lower right) is sealed against a small patch of the plasma membrane on another portion of the cell.

Most ion channels are highly selective in allowing only one particular type of ion to pass through the pore. As with the passive diffusion of other types of solutes across membranes, the diffusion of ions through a channel is always downhill, that is, from a state of higher energy to a state of lower energy. Most of the ion channels that have been identified can exist in either an open or a closed conformation; such channels are said to be gated. The opening and closing of the gates are subject to complex physiologic regulation and can be induced by a variety of factors depending on the particular channel. Three major categories of gated channels are distinguished:

- Voltage-gated channels whose conformational state depends on the difference in ionic charge on the two sides of the membrane.
- Ligand-gated channels whose conformational state depends on the binding of a specific molecule (the ligand), which is usually not the solute that passes through the channel. Some ligand-gated channels are opened (or closed) following the binding of a molecule to the outer surface of the channel; others are opened (or closed) following the binding of a ligand to the inner surface of the channel. For example, neurotransmitters, such as acetylcholine, act on the outer surface of certain cation channels, while cyclic nucleotides, such as cAMP, act on the inner surface of certain calcium ion channels.
- Mechano-gated channels whose conformational state depends on mechanical forces (e.g., stretch tension) that are applied to the membrane. Members of one family of cation channels, for example, are opened by the movements of stereocilia on the hair cells of the inner ear in response to sound or motions of the head.

We will focus in the following discussion on the structure and function of voltage-gated potassium ion channels because these are the best understood.

In 1998, Roderick MacKinnon and his colleagues at Rockefeller University provided the first atomic-resolution image of an ion channel protein, in this case, a bacterial K^+ ion channel called KcsA. The relationship between structure and function is evident everywhere in the biological world, but it would be difficult to find a better example than that of the K^+ ion channel depicted in figure. As we will see shortly, the formulation of this structure led directly to an understanding of the mechanism by which these remarkable molecular machines are able to select overwhelmingly for K^+ ions over Na^+ ions, yet at the same time allow an incredibly rapid conductance of K^+ ions through the membrane. We will also see that the mechanisms of ion selectivity and conductance in this bacterial channel are virtually identical to those operating in the much larger mammalian channels. Evidently, the basic challenges in operating an ion channel were solved relatively early in evolution, although many refinements appeared over the following one or two billion years.

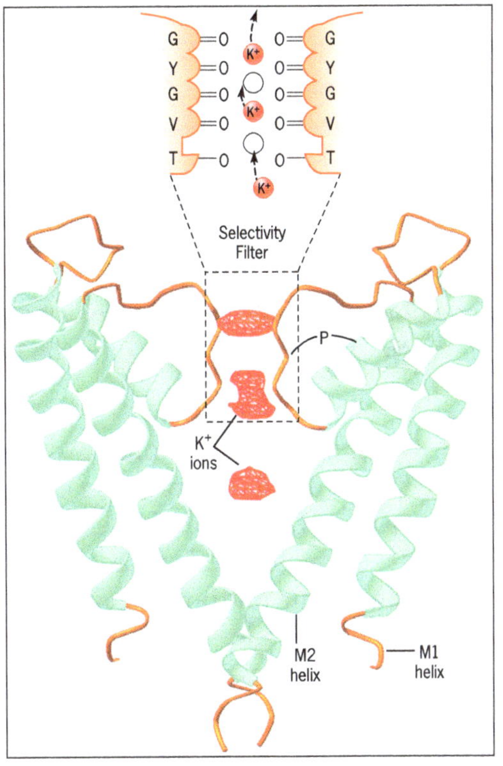

Figure shows three-dimensional structure of the bacterial KcsA channel and the selection of K^+ ions. This K^+ ion channel consists of four subunits, two of which are shown here. Each subunit is comprised of M1 and M2 helices joined by a P (pore) segment consisting of a short helix and a nonhelical portion that lines the channel through which the ions pass. A portion of each P segment contains a conserved pentapeptide (GYGVT) whose residues line the selectivity filter that screens for K^+ ions. The oxygen atoms of the carbonyl groups of these residues project into the channel where they can interact selectively with K^+ ions (indicated by the red mesh objects) within the filter. As indicated in the top inset, the selectivity filter contains four rings of carbonyl O atoms and one ring of threonyl O atoms; each of these five rings contains four O atoms, one donated by each subunit. The diameter of the rings is just large enough so that eight O atoms can coordinate a single K^+ ion, replacing its normal water of hydration. Although four K^+ binding sites are shown, only two are occupied at one time.

The KcsA channel consists of four subunits, two of which are shown in figure. Each subunit of Figure is seen to contain two membrane-spanning helices (M1 and M2) and a pore region (P) at the extracellular end of the channel. P consists of a short pore helix that extends approximately onethird the width of the channel and a nonhelical loop that forms the lining of a narrow selectivity filter, so named because of its role in allowing only the passage of K^+ ions.

The lining of the selectivity filter contains a highly conserved pentapeptide—Gly-Tyr-Gly-Val-Thr (or GYGVT in single-letter nomenclature). The X-ray crystal structure of the KcsA channel shows that the backbone carbonyl (CPO) groups from the conserved pentapeptide create five successive rings of oxygen atoms (four rings are made up of carbonyl oxygens from the polypeptide backbone, and one ring consists of oxygen atoms from the threonine side chain). Each ring contains four oxygen atoms (one from each subunit) and has a diameter of approximately 3 Å, which is slightly larger than the 2.7 Å diameter of a K^+ ion that has lost its normal shell of hydration. Consequently, the electronegative O atoms that line the selectivity filter can substitute for the shell of water molecules that are displaced as each K^+ ion enters the pore. In this model, the selectivity filter contains four potential K^+ ion binding sites. As indicated in the top inset of figure, a K^+ ion bound at any of these four sites would occupy the center of a "box" having four O atoms in a plane above the ion and four O atoms in a plane below the atom. As a result, each K^+ ion in one of these sites could coordinate with eight O atoms of the selectivity filter. Whereas the selectivity filter is a precise fit for a dehydrated K^+ ion, it is much larger than the diameter of a dehydrated Na^+ ion (1.9 Å). Consequently, a Na^+ ion cannot interact optimally with the eight oxygen atoms necessary to stabilize it in the pore. As a result, the smaller Na^+ ions cannot overcome the higher energy barrier required to penetrate the pore.

Although there are four potential K^+ ion binding sites, only two are occupied at any given time. Potassium ions are thought to move, two at a time—from sites 1 and 3 to sites 2 and 4—as indicated in the top inset of figure. The entry of a third K^+ ion into the selectivity filter creates an electrostatic repulsion that ejects the ion bound at the opposite end of the line. Studies indicate that there is virtually no energy barrier for an ion to move from one binding site to the next, which accounts for the extremely rapid flow of ions across the membrane. Taken together, these conclusions concerning K^+ ion selectivity and conductance provide a superb example of how much can be learned about biological function through an understanding of molecular structure.

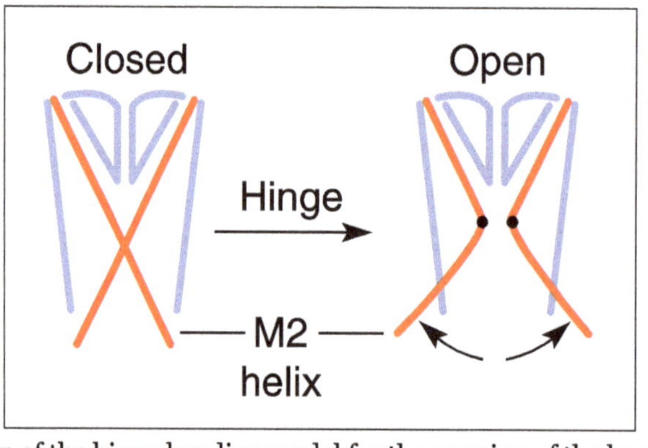

Schematic illustration of the hinge-bending model for the opening of the bacterial KcsA channel. The M2 helices from each subunit bend outward at a specific glycine residue, which opens the gate at the intracellular end of the channel to K^+ ions.

The KcsA channel depicted in figure has a gate, just like eukaryotic channels. The opening of the gate of the KcsA channel in response to very low pH was illustrated in figure. The structure of KcsA shown in figure is actually the closed conformation of the protein (despite the fact that it contains ions in its channel). It has not been possible to crystallize the KcsA channel in its open conformation, but the structure of a homologous prokaryotic K^+ channel (called MthK) in the open conformation has been crystallized and its structure determined. Comparison of the open structure of MthK and the closed structure of the homologous protein KcsA strongly suggested that gating of these molecules is accomplished by conformational changes of the cytoplasmic ends of the inner (M2) helices. In the closed conformation, as seen in figure and figure, left drawing, the M2 helices are straight and cross over one another to form a "helix bundle" that seals the cytoplasmic face of the pore. In the model shown in figure, the channel opens when the M2 helices bend at a specific hinge point where a glycine residue is located.

Now that we have seen how these prokaryotic K^+ channels operate, we are in a better position to understand the structure and function of the more complex eukaryotic versions, which are thought to perform in a similar manner. Genes that encode a variety of distinct voltage-gated K^+ (or Kv) channels have been isolated and the molecular anatomy of their proteins scrutinized. The Kv channels of plants play an important role in salt and water balance and in regulation of cell volume. The Kv channels of animals are best known for their role in muscle and nerve function. Eukaryotic Kv channel subunits contain six membrane-associated helices, named S1–S6, which are shown two-dimensionally in figure. These six helices can be grouped into two functionally distinct domains:

- A pore domain, which has the same basic architecture as that of the entire bacterial channel illustrated in figure and contains the selectivity filter that permits the selective passage of K^+ ions. Helices M1 and M2 and the P segment of the KcsA channel of figure are homologous to helices S5 and S6 and the P segment of the voltage-gated eukaryotic channel illustrated in figure. Like the four M2 helices of KcsA, the four S6 helices line much of the pore, and their configuration determines whether the gate to the channel is open or closed.
- A voltage-sensing domain consisting of helices S1–S4 that senses the voltage across the plasma membrane.

The three-dimensional crystal structure of a complete eukaryotic Kv channel purified from rat brain is shown in figure. Determination of this structure was made possible by use of a mixture of detergent and lipid throughout the purification and crystallization process. The presence of negatively charged phospholipids is thought to be important in maintaining the native structure of the membrane protein and promoting its function as a voltage-gated channel. Like the KcsA channel, a single eukaryotic Kv channel consists of four homologous subunits arranged symmetrically around the central ion-conducting pore. The selectivity filter, and thus the presumed mechanism of K^+ ion selection, is virtually identical in the prokaryotic KcsA and eukaryotic Kv proteins. The gate leading into a Kv channel is formed by the inner ends of the S6 helices and is thought to open and close in a manner roughly similar to that of the M2 helices of the bacterial channel. The protein depicted in figure represents the open state of the channel.

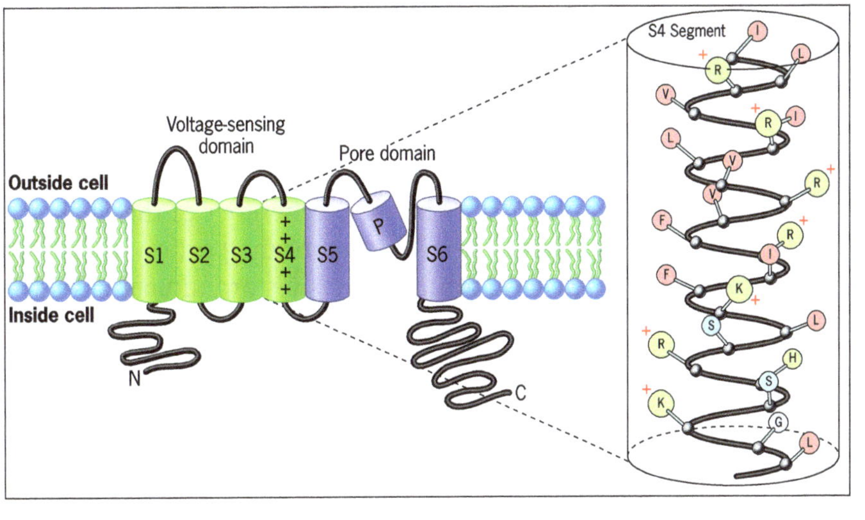

Figure shows the structure of one subunit of a eukaryotic, voltage-gated K^+ channel. A two-dimensional portrait of a K^+ channel subunit showing its six transmembrane helices and a portion of the polypeptide (called the pore helix or P) that dips into the protein to form part of the channel's wall. The inset shows the sequence of amino acids of the positively charged S4 helix of the Drosophila K^+ Shaker ion channel, which serves as a voltage sensor. The positively charged side chains are situated at every third residue along the otherwise hydrophobic helix. This member of the Kv family is called a Shaker channel because flies with certain mutations in the protein shake vigorously when anesthesized with ether. The Shaker channel was the first K^+ channel to be identified and cloned in 1987.

The S4 helix, which contains several positively charged amino acid residues spaced along the polypeptide chain, acts as the key element of the voltage sensor. The voltage-sensing domain is seen to be connected to the pore domain by a short linker helix denoted as S4–S5 in the model in figure. Under resting conditions, the negative potential across the membrane keeps the gate closed. A change in the potential to a more positive value exerts an electric force on the S4 helix. This force is thought to cause the transmembrane S4 helix to move in such a way that its positively charged residues shift from a position where they were exposed to the cytoplasm to a new position where they are exposed to the outside of the cell. Voltage sensing is a dynamic process whose mechanism cannot be resolved by a single static view of the protein such as that shown in figure. In fact, several competing models describing the mechanism of action of the voltage sensor are currently debated. However it occurs, the movement of the S4 helix in response to membrane depolarization initiates a series of conformational changes within the protein that opens the gate at the cytoplasmic end of the channel.

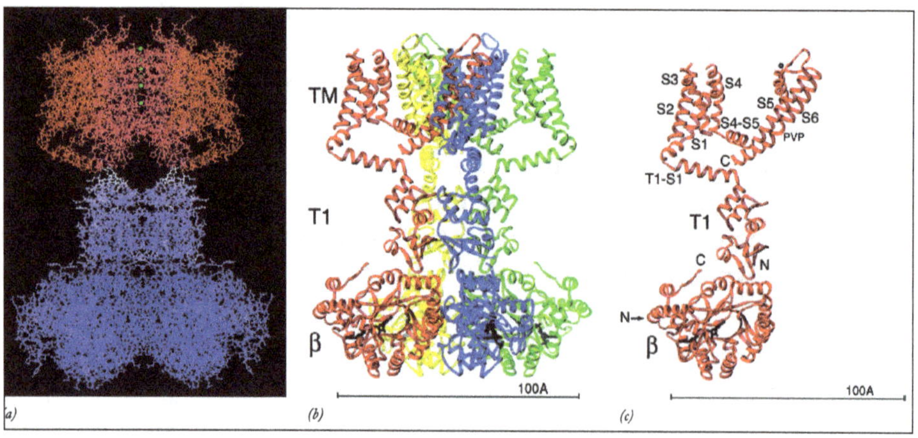

Figure shows three-dimensional structure of a voltage-gated mammalian K^+ channel. (a) The crystal structure of the entire tetrameric Kv1.2 channel, a member of the Shaker family of K^+ ion channels found in nerve cells of the brain. The transmembrane portion is shown in red, and the cytoplasmic portion in blue. The potassium ion binding sites are indicated in green. (b) Ribbon drawing of the same channel shown in a, with the four subunits that make up the channel shown in different colors. If you focus on the red subunit, you can see (1) the spatial separation between the voltage-sensing and pore domains of the subunit and (2) the manner in which the voltage-sensing domain from each subunit is present on the outer edge of the pore domain of a neighboring subunit. The cytoplasmic portion of this particular channel consists of a T1 domain, which is part of the channel polypeptide itself, and a separate b polypeptide. (c) Ribbon drawing of a single subunit showing the spatial orientation of the six membrane-spanning helices (S1–S6) and also the presence of the S4–S5 linker helix, which connects the voltage-sensing and pore domains. This linker transmits the signal from the S4 voltage sensor that opens the channel. The inner surface of the channel below the pore domain is lined by the S6 helix (roughly similar to the M2 helix of the bacterial channel shown in figure). The channel shown here is present in the open configuration with the S6 helices curved outward at the site marked PVP (standing for Pro-Val-Pro, which is likely the amino acid sequence of the "hinge").

Once opened, more than ten million potassium ions can pass through the channel per second, which is nearly the rate that would occur by free diffusion in solution. Because of the large ion flux, the opening of a relatively small number of K^+ channels has significant impact on the electrical properties of the membrane. After the channel is open for a few milliseconds, the movement of K^+ ions is "automatically" stopped by a process known as inactivation. To understand channel inactivation, we have to consider an additional portion of a Kv channel besides the two transmembrane domains.

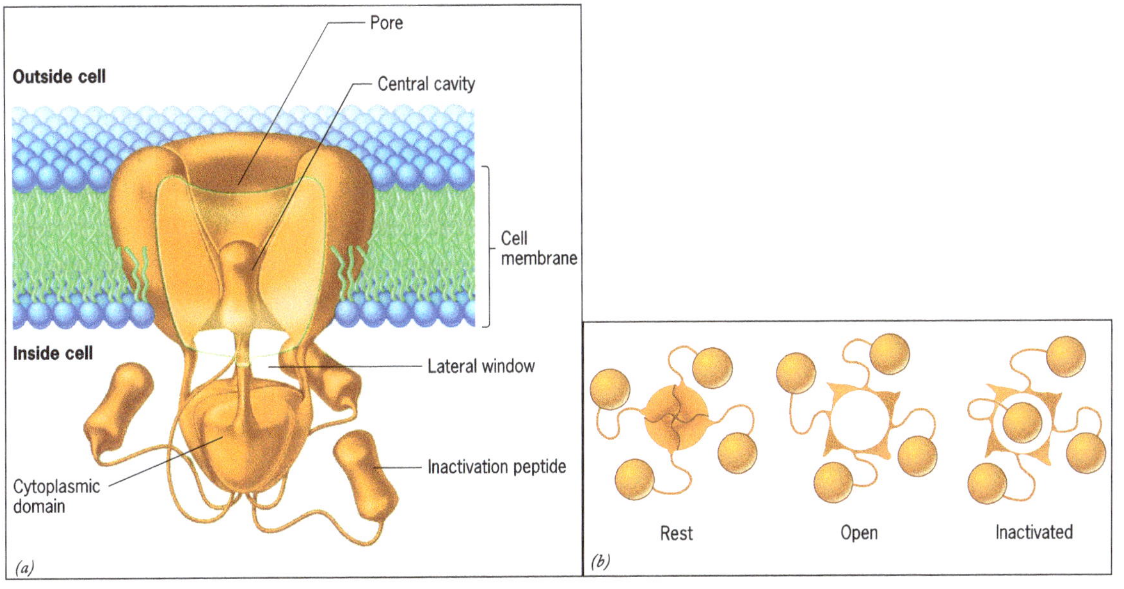

Figure shows conformational states of a voltage-gated K^+ ion channel. (a) Three-dimensional model of a eukaryotic K^+ ion channel. Inactivation of channel activity occurs as one of the inactivation peptides, which dangle from the cytoplasmic portion of the complex, fits into the cytoplasmic opening of the channel. (b) Schematic representation of a view into a K^+ ion channel, perpendicular to the membrane from the cytoplasmic side, showing the channel in the closed (resting), open, and inactivated state.

Eukaryotic Kv channels typically contain a large cytoplasmic structure whose composition varies among different channels. As indicated in figure a inactivation of the channel is accomplished by movement of a small inactivation peptide that dangles from the cytoplasmic portion of the protein. The inactivation peptide is thought to gain access to the cytoplasmic mouth of the pore by snaking its way through one of four "side windows" indicated in the figure. When one of these dangling peptides moves up into the mouth of the pore, the passage of ions is blocked, and the channel is inactivated. At a subsequent stage of the cycle, the inactivation peptide is released, and the gate to the channel is closed. It follows from this discussion that the potassium channel can exist in three different states—open, inactivated, and closed— which are illustrated schematically in figure b.

Potassium channels come in many different varieties. It is remarkable that C. elegans, a nematode worm whose body consists of only about 1000 cells, contains approximately 80 different genes that encode K^+ channels. It is evident that a single cell—whether in a nematode, human, or plant—is likely to possess a variety of different K^+ channels that open and close in response to different voltages. In addition, the voltage required to open or close a particular K^+ channel can vary depending on whether or not the channel protein is phosphorylated, which in turn is regulated by hormones and other factors. It is apparent that ion channel function is under the control of a diverse and complex set of regulatory agents. The structure and function of a very different type of ion channel, the ligand-gated nicotinic acetylcholine receptor.

Facilitated Diffusion

Substances always diffuse across a membrane from a region of higher concentration on one side to a region of lower concentration on the other side, but they do not always diffuse through the lipid bilayer or through a channel. In many cases, the diffusing substance first binds selectively to a membrane spanning protein, called a facilitative transporter that facilitates the diffusion process. The binding of the solute to the facilitative transporter on one side of the membrane is thought to trigger a conformational change in the protein, exposing the solute to the other surface of the membrane, from where it can diffuse down its concentration gradient. This mechanism is illustrated in figure. Because they operate passively, that is, without being coupled to an energy-releasing system, facilitated transporters can mediate the movement of solutes equally well in both directions. The direction of net flux depends on the relative concentration of the substance on the two sides of the membrane.

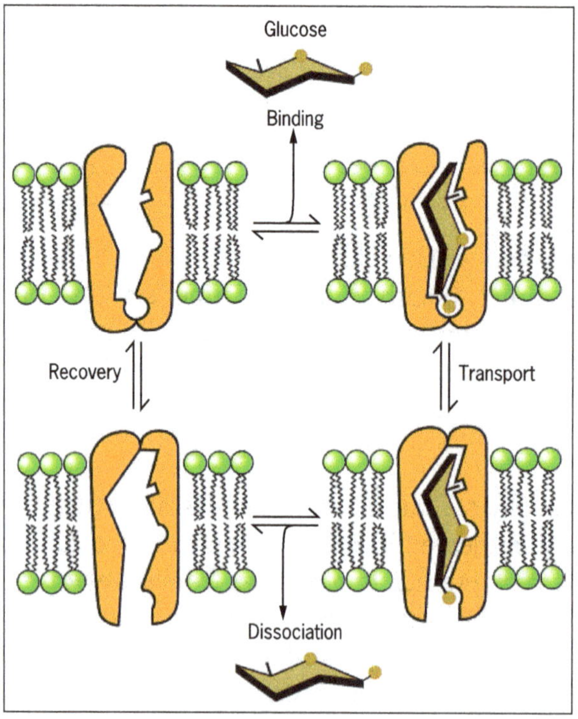

Facilitated diffusion. A schematic model for the facilitated diffusion of glucose depicts the alternating conformation of a carrier that exposes the glucose binding site to either the inside or outside of the membrane.

Facilitated diffusion, as this process is called, is similar in many ways to an enzyme-catalyzed reaction. Like enzymes, facilitative transporters are specific for the molecules they transport, discriminating, for example, between D and L stereoisomers. In addition, both enzymes and transporters exhibit saturation-type kinetics. Unlike ion channels, which can conduct millions of ions per second, most facilitative transporters can move only hundreds to thousands of solute molecules per second across the membrane. Another important feature of facilitative transporters is that, like enzymes and ion channels, their activity can be regulated. Facilitated diffusion is particularly important in mediating the entry and exit of polar solutes, such as sugars and amino acids that do not penetrate the lipid bilayer.

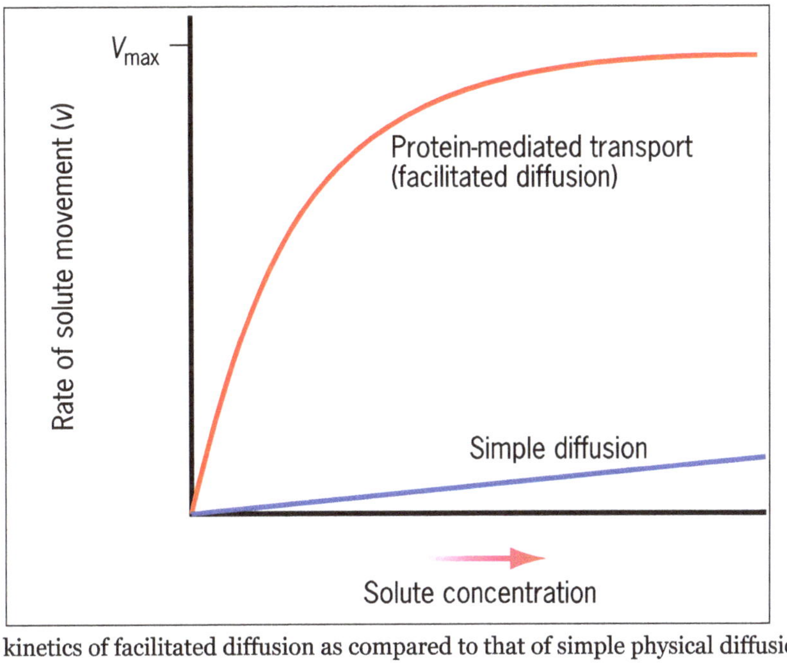

The kinetics of facilitated diffusion as compared to that of simple physical diffusion.

The Glucose Transporter: An Example of Facilitated Diffusion

Glucose is the body's primary source of direct energy, and most mammalian cells contain a membrane protein that facilitates the diffusion of glucose from the bloodstream into the cell. A gradient favoring the continued diffusion of glucose into the cell is maintained by phosphorylating the sugar after it enters the cytoplasm, thus lowering the intracellular glucose concentration. Humans have at least five related proteins (isoforms) that act as facilitative glucose transporters. These isoforms, termed GLUT1 to GLUT5, are distinguished by the tissues in which they are located, as well as their kinetic and regulatory characteristics.

Insulin is a hormone produced by endocrine cells of the pancreas and plays a key role in maintaining proper blood sugar levels. An increase in blood glucose levels triggers the secretion of insulin, which stimulates the uptake of glucose into various target cells, most notably skeletal muscle and fat cells (adipocytes). Insulin-responsive cells share a common isoform of the facilitative glucose transporter, specifically GLUT4. When insulin levels are low, these cells contain relatively few glucose transporters on their plasma membrane. Instead, the transporters are present within the membranes of cytoplasmic vesicles. Rising insulin levels act on target cells to stimulate the fusion of the cytoplasmic vesicles to the plasma membrane, which moves transporters to the cell surface where they can bring glucose into the cell.

Active Transport

Life cannot exist under equilibrium conditions. Nowhere is this more apparent than in the imbalance of ions across the plasma membrane. The differences in concentration of the major ions between the outside and inside of a typical mammalian cell are shown in table. The ability of a cell to generate such steep concentration gradients across its plasma membrane cannot occur by either simple or facilitated diffusion. Rather, these gradients must be generated by active transport.

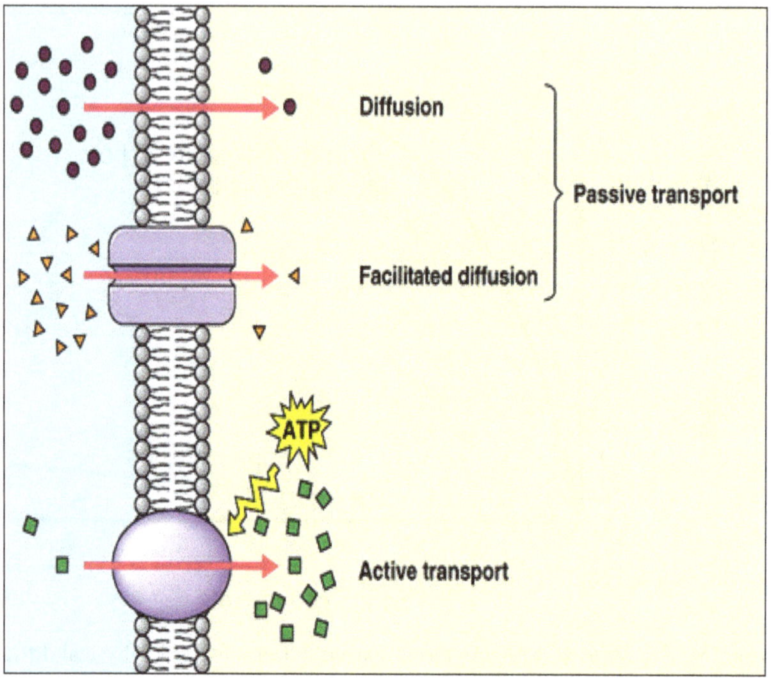

Like facilitated diffusion, active transport depends on integral membrane proteins that selectively bind a particular solute and move it across the membrane in a process driven by changes in the protein's conformation. Unlike facilitated diffusion, however, movement of a solute against a gradient requires the coupled input of energy. Consequently, the endergonic movement of ions or other solutes across the membrane against a concentration gradient is coupled to an exergonic process, such as the hydrolysis of ATP, the absorbance of light, the transport of electrons, or the flow of other substances down their gradients. Proteins that carry out active transport are often referred to as "pumps."

Primary Active Transport: Coupling Transport to ATP Hydrolysis

In 1957, Jens Skou, a Danish physiologist, discovered an ATP-hydrolyzing enzyme in the nerve cells of a crab that was only active in the presence of both Na^+ and K^+ ions. Skou proposed, and correctly so, that this enzyme, which was responsible for ATP hydrolysis, was the same protein that was active in transporting the two ions; the enzyme was called the Na^+/K^+-ATPase, or the sodium–potassium pump.

Unlike the protein-mediated movement of a facilitated diffusion system, which will carry the substance equally well in either direction, active transport drives the movement of ions in only one direction. It is the Na^+/K^+-ATPase that is responsible for the large excess of Na^+ ions outside of the cell and the large excess of K^+ ions inside the cell. The positive charges carried by these two cations are balanced by negative charges carried by various anions so that the extracellular and intracellular compartments are, for the most part, electrically neutral. Cl^- ions are present at greater concentration outside of cells, where they balance the extracellular Na^+ ions. The abundance of intracellular K^+ ions is balanced primarily by excess negative charges carried by proteins and nucleic acids.

Table: Ion Concentrations Inside and Outside of a Typical Mammalian Cell.

	Extracellular concentration	Intracellular concentration	Ionic gradient
Na^+	150 mM	10 mM	15x
K^+	5 mM	140 mM	28x
Cl^-	120 mM	10 mM	12x
Ca^{2+}	10^{-3} M	10^{-7} M	10,000x
H^+	$10^{-7.4}$ M (pH of 7.4)	$10^{-7.2}$ M (pH of 7.2)	Nearly 2x

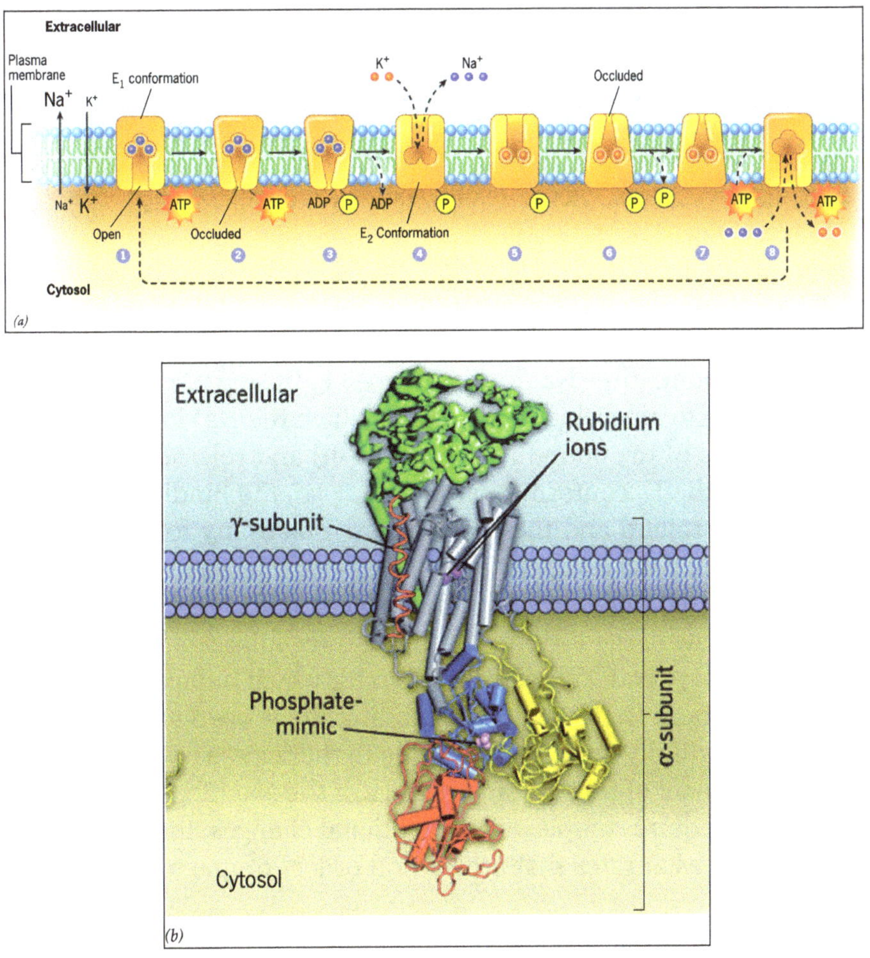

Figure shows the Na^+/K^+-ATPase. (a) Simplified schematic model of the transport cycle as described in the text. Note that the actual Na^+/K^+-ATPase is composed of at least two different membrane-spanning subunits: a larger α subunit, which carries out the transport activity, and a smaller β subunit, which functions primarily in the maturation and assembly of the pump within the membrane. A third (γ) subunit may also be present. (b) A model of the E_2 conformation of the protein based on an X-ray crystallographic study. The cation binding sites are located deep within the transmembrane domain, which consists of 10 membrane spanning helices. The two rubidium ions are located where the potassium ions would normally be bound.

The ratio of Na^+:K^+ pumped by the Na^+/K^+-ATPase is not 1:1, but 3:2. In other words, for each ATP hydrolyzed, three sodium ions are pumped out as two potassium ions are pumped in. Because of this pumping ratio, the Na^+/K^+-ATPase is electrogenic, which means that it contributes directly to the separation of charge across the membrane. The Na^+/K^+-ATPase is an example of a P-type ion pump. The "P" stands for phosphorylation, indicating that, during the pumping cycle, the hydrolysis of ATP leads to the transfer of the released phosphate group to an aspartic acid residue of the transport protein. As denoted by the name of the protein, this reaction is catalyzed by the pump itself.

Consider the activity of the protein. It must pick up sodium or potassium ions from a region of low concentration, which means that the protein must have a relatively high affinity for the ions. Then the protein must release the ions on the other side of the membrane into a much greater concentration of each ion. To do this, the affinity of the protein for that ion must decrease. Thus, the affinity for each ion on the two sides of the membrane must be different. This is achieved by ATP hydrolysis and the subsequent release of ADP, which induces a large conformational change within the protein molecule. The change in protein structure—between the E_1 and E_2 conformations—also serves to alternately expose the ion binding sites to the opposite side of the membrane.

A proposed scheme for the pumping cycle of the Na^+/K^+-ATPase is shown in figure a, which is based on recent X-ray crystallographic structures of P-type ion pumps. In step 1, Figure a, the pump is in the E_1 conformation, and the ion binding sites are accessible to the inside of the cell. In this step, the protein has bound three Na^+ ions and an ATP. In step 2, a gate within the protein has closed, shifting the pump into an occluded E_1 state in which the Na^+ ions can no longer flow back into the cytosol. Hydrolysis of the bound ATP (step 2→3) and release of ADP (step 3→4) cause the change from the E_1 to the E_2 conformation. In doing so, the binding site becomes accessible to the extracellular compartment, and the protein loses its affinity for Na1 ions, which are then released outside the cell. Once the three Na^+ ions have been released, the protein picks up two K^+ ions (step 5).The closing of another gate within the protein shifts the pump into an occluded state (step 6), preventing back flow of the K^+ ions into the extracellular space. Occlusion is followed by dephosphorylation (step 6→7) and ATP binding (step 8), which induces the return of the protein to the original E_1 conformation. In this state, the binding site is open to the internal surface of the membrane and has lost its affinity for K^+ ions, leading to the release of these ions into the cell. The cycle is then repeated. A model of the crystal structure of the Na^+/K^+ - ATPase structure is shown in Figure b. Because they require complex conformational changes, these active transport pumps move ions across membranes at rates that are several orders of magnitude lower than their flow through ion channels.

The sodium–potassium pump is found only in animal cells. This protein is thought to have evolved in primitive animals as the primary means to maintain cell volume and as the mechanism to generate the steep Na^+ and K^+ gradients that play such a key role in the formation of impulses in nerve and muscle cells. These same ionic gradients are used in nonexcitable cells to power the movement of other solutes. The importance of the sodium–potassium pump becomes evident when one considers that it consumes approximately one-third of the energy produced by most animal cells and two-thirds of the energy produced by nerve cells. Digitalis, a steroid obtained from the foxglove plant that has been used for 200 years as a treatment for congestive heart disease, binds to the Na^+ /K^+ - ATPase. Digitalis strengthens the heart's contraction by inhibiting the Na^+ /K^+ pump, which leads to a chain of events that increases Ca^{2+} availability inside the muscle cells of the heart.

Other Primary Ion Transport Systems

The best studied P-type pump is the Ca^{2+} -ATPase whose three-dimensional structure has been determined at several stages of the pumping cycle. The calcium pump is present in the membranes of the endoplasmic reticulum, where it actively transports calcium ions out of the cytosol into the lumen of this organelle. Plant cells have a H^+ -transporting, P-type, plasma membrane pump. In plants, this proton pump plays a key role in the secondary transport of solutes, in the control of cytosolic pH, and possibly in control of cell growth by means of acidification of the plant cell wall.

The epithelial lining of the stomach also contains a P-type pump, the H^+/K^+ -ATPase, which secretes a solution of concentrated acid (up to 0.16 N HCl) into the stomach chamber. In the resting state, these pump molecules are situated in cytoplasmic membranes of the parietal cells of the stomach lining and are nonfunctional. When food enters the stomach, a hormonal message is transmitted to the parietal cells that causes the pump-containing membranes to move to the apical cell surface, where they fuse with the plasma membrane and begin secreting acid. In addition to functioning in digestion, stomach acid can also lead to heartburn. Prilosec and related drugs are widely used to prevent heartburn by inhibiting the stomach's H^+/K^+ -ATPase. Other acid-blocking heartburn medications (e.g., Zantac, Pepcid, and Tagamet) do not inhibit the H^+/K^+ -ATPase directly, but block a receptor on the surface of the parietal cells, thereby stopping the cells from becoming activated by the hormone.

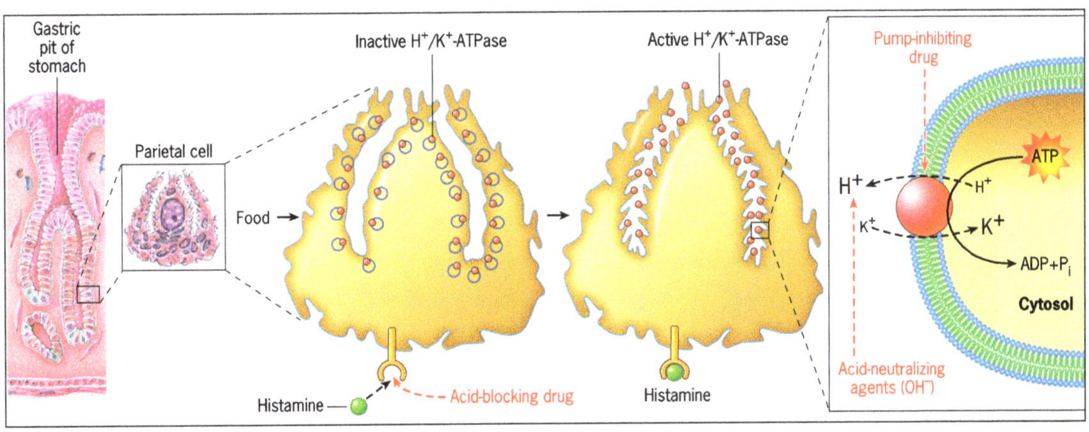

Figure shows control of acid secretion in the stomach. In the resting state, the H^+/K^+ -ATPase molecules are present in the walls of cytoplasmic vesicles. Food entering the stomach triggers a cascade of hormone-stimulated reactions in the stomach wall leading to the release of histamine, which binds to a receptor on the surface of the acid-secreting parietal cells. Binding of histamine to its receptor stimulates a response that causes the H^+/K^+ -ATPase-containing vesicles to fuse to the plasma membrane, forming deep folds, or canaliculi. Once at the surface, the transport protein is activated and pumps protons into the stomach cavity against a concentration gradient (indicated by the size of the letters). The heartburn drugs Prilosec, Nexium, and Prevacid block acid secretion by directly inhibiting the H^+/K^+ - ATPase, whereas several other acid-blocking medications interfere with activation of the parietal cells. Acid-neutralizing medications provide basic anions that combine with the secreted protons.

Unlike P-type pumps, V-type pumps utilize the energy of ATP without forming a phosphorylated protein intermediate. V-type pumps actively transport hydrogen ions across the walls of cytoplasmic organelles and vacuoles (hence the designation V-type). They occur in the membranes that line lysosomes, secretory granules, and plant cell vacuoles where they maintain the low pH of the contents. V-type pumps have also been found in the plasma membranes of a variety of cells. For example, a V-type pump in the plasma membranes of kidney tubules helps maintain the body's acid–base balance by secreting protons into the forming urine. V-type pumps are large multi-subunit complexes and similar in structure to that of the ATP synthase shown in figure.

Another diverse group of proteins that actively transport ions is the ATP-binding cassette (ABC) transporters, so called because all of the members of this superfamily share a homologous ATP-binding domain. The best studied ABC transporter is described in the accompanying Human Perspective.

Using Light Energy to Actively Transport Ions

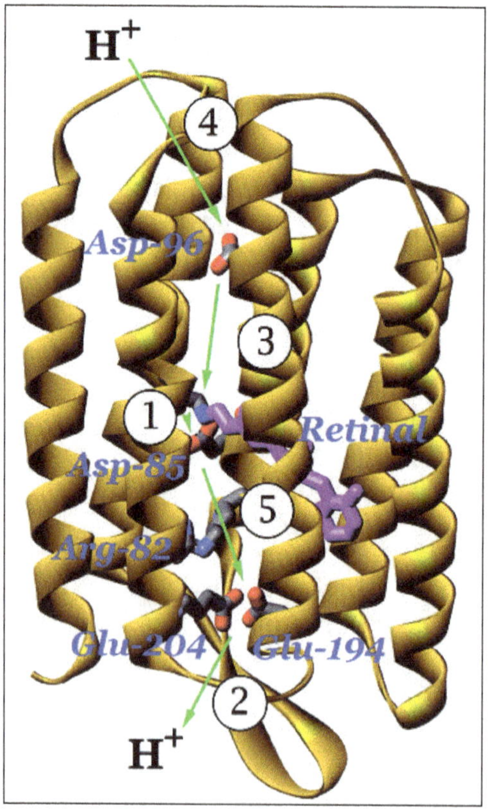

Figure shows bacteriorhodopsin: a light-driven proton pump. The protein contains seven membrane-spanning helices and a centrally located retinal group (shown in purple), which serves as the lightabsorbing element (chromophore). Absorption of a photon of light causes a change in the electronic structure of retinal, leading to the transfer of a proton from the $-NH^+$ group to a closely associated, negatively charged aspartic acid residue (#85) (step 1). The proton is then released to the extracellular side of the membrane (step 2) by a relay system consisting of several amino acid residues (Asp82, Glu204, and Glu194). The spaces between these residues are filled with hydrogen-bonded water molecules that help shuttle protons along the pathway. The deprotonated retinal is returned to its original state (step 3) when it accepts a proton from an undissociated aspartic acid residue (Asp96) located near the cytoplasmic side of the membrane. Asp96 is then reprotonated by a H^+ from the cytoplasm (step 4). Asp85 is deprotonated (step 5) prior to receiving a proton from retinal in the next pumping cycle. As a result of these events, protons move from the cytoplasm to the cell exterior through a central channel in the protein.

Halobacterium salinarium (or H. halobium) is an archaebacterium that lives in extremely salty environments, such as that found in the Great Salt Lake. When grown under anaerobic conditions, the plasma membranes of these prokaryotes take on a purple color due to the presence of one particular protein, bacteriorhodopsin, which acts a light-driven proton pump. As shown in figure, bacteriorhodopsin contains retinal, the same prosthetic group present in rhodopsin, the light-absorbing protein of the rods of the vertebrate retina. The absorption of light energy by the retinal group induces a series of conformational changes in the protein that cause a proton to move from the retinal group, through a channel in the protein, to the cell exterior. The proton donated by the photoexcited retinal is replaced by another proton transferred to the protein from the cytoplasm. In effect, this process results in the translocation of protons from the cytoplasm to the external environment, thereby generating a steep H1 gradient across the plasma membrane. This gradient is subsequently used by an ATP-synthesizing enzyme to phosphorylate ADP.

Secondary Active Transport (or Cotransport): Coupling Transport to Existing Ion Gradients

The establishment of concentration gradients, such as those of Na^+, K^+, and H^+, provides a means by which free energy can be stored in a cell. The potential energy stored in ionic gradients is utilized by a cell in various ways to perform work, including the transport of other solutes. Consider the physiologic activity of the intestine. Within its lumen, enzymes hydrolyze high-molecularweight polysaccharides into simple sugars, which are absorbed by the epithelial cells that line the intestine. The movement of glucose across the apical plasma membrane of the epithelial cells, against a concentration gradient, occurs by cotransport with sodium ions, as illustrated in figure. The Na^+ concentration is kept very low within the cells by the action of a primary active transport system (the Na^+/K^+ -ATPase), located in the basal and lateral plasma membrane, which pumps sodium ions out of the cell against a concentration gradient. The tendency for sodium ions to diffuse back across the apical plasma membrane down their concentration gradient is "tapped" by the epithelial cells to drive the cotransport of glucose molecules into the cell against a concentration gradient. The glucose molecules are said to be driven by secondary active transport. In this case, the transport protein, called a Na^+/glucose cotransporter, moves two sodium ions and one glucose molecule with each cycle. Once inside, the glucose molecules diffuse through the cell and are moved across the basal membrane by facilitated diffusion.

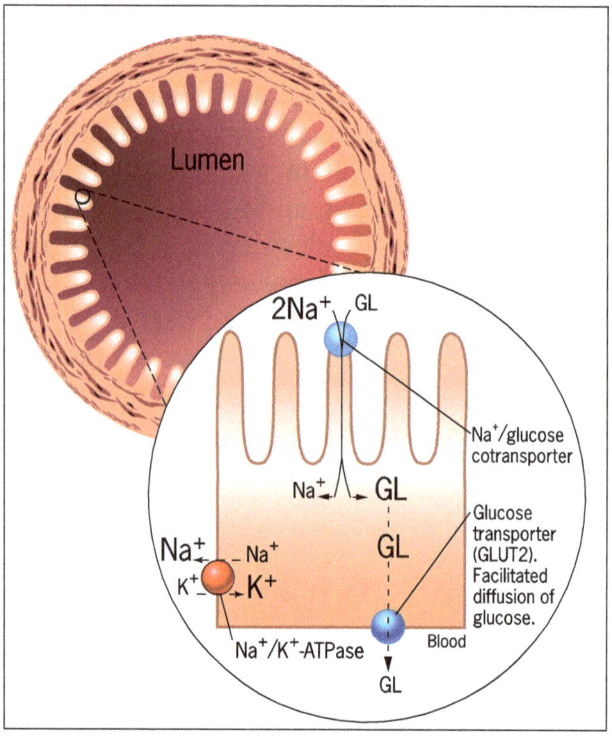

Figure shows secondary transport: the use of energy stored in an ionic gradient. The Na^+/K^+-ATPase residing in the plasma membrane of the lateral surface maintains a very low cytosolic concentration of Na^+. The Na^+ gradient across the plasma membrane represents a storage of energy that can be tapped to accomplish work, such as the transport of glucose by a Na^+/glucose cotransporter located in the apical plasma membrane. Once transported across the apical surface into the cell, the glucose molecules diffuse to the basal surface where they are carried by a glucose facilitative transporter out of the cell and into the bloodstream. The relative size of the letters indicates the directions of the respective concentration gradients. Two Na^+ ions are transported for each glucose molecule; the 2;1 Na^+/glucose provides a much greater driving force for moving glucose into the cell than a 1;1 ratio.

To appreciate the power of an ion gradient in accumulating other types of solutes in cells, we can briefly consider the energetics of the Na^+/glucose cotransporter. The free-energy change for the movement of a mole of Na^+ ions into the cell is equal to -3.1 kcal/mol, and thus 6.2 kcal for two moles of Na^+ ions, which would be available to transport one mole of glucose uphill into the cell. The equation for the movement of a nonelectrolyte, such as glucose, across the membrane is,

$$\Delta G = RT \ln \frac{[C_i]}{[C_o]}$$

$$\Delta G = 2.303 RT \log_{10} \frac{[C_i]}{[C_o]}$$

Using this equation, we can calculate how steep a concentration gradient of glucose (X) that this cotransporter can generate. At 25 °C,

$$-6.2\ \text{kcal/mol} = 1.4\ \text{kcal/mol} \cdot \log_{10} X$$
$$\log_{10} X = -4.43$$
$$X = \frac{1}{23{,}000}$$

This calculation indicates that the Na^+/glucose cotransporter is capable of transporting glucose into a cell against a concentration gradient greater than 20,000-fold.

Plant cells rely on secondary active transport systems to take up a variety of nutrients, including sucrose, amino acids, and nitrate. In plants, uptake of these compounds is coupled to the downhill, inward movement of H^+ ions rather than Na^+ ions. The secondary active transport of glucose into the epithelial cells of the intestine and the transport of sucrose into a plant cell are examples of symport, in which the two transported species (Na^+ and glucose or H^+ and sucrose) move in the same direction. Numerous cotransporters have been isolated that engage in antiport, in which the two transported species move in opposite directions. For example, cells often maintain a proper cytoplasmic pH by coupling the inward, downhill movement of Na^+ with the outward movement of H^+. Cotransporters that mediate antiport are usually called exchangers. The three-dimensional structures of a number of secondary transporters have been solved in recent years, and, like the Na^+/K^+-ATPase, they exhibit a transport cycle in which the protein's binding sites gain alternating access to the cytoplasm and the extracellular space. A proposed model of the transport cycle of one of the major families of secondary transporters is shown in figure.

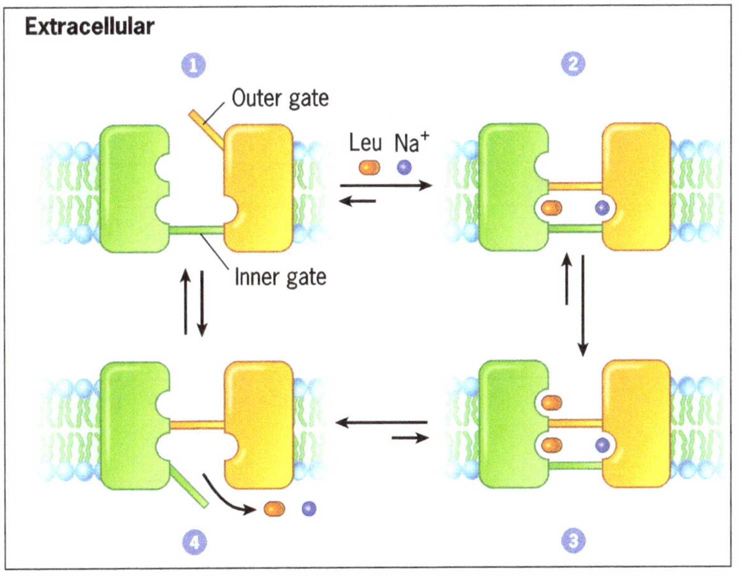

Figure shows a schematic model of the transport cycle of a secondary transporter. Four different conformational states during the transport cycle of a bacterial symporter of the LeuT family are shown. The protein actively transports the amino acid leucine into the cell using the established Na^+ ion gradient as its source of energy. In step 1, the outer gate in the protein is open, whch allows both Na^+ and leucine to reach their binding sites from the extracellular space. In step, 2, the outer gate closes, occluding the substrates within the protein. In step 3, a second leucine molecule binds to another site just outside the outer gate. In step 4, the inner gate opens and the substrates are released into the cytoplasm. The protein returns to its original state when the inner gate is closed and the outer gate is opened.

The Human Perspective

Defects in Ion Channels and Transporters as a Cause of Inherited Disease

Several severe, inherited disorders have been traced to mutations in genes that encode ion channel proteins. Most of the disorders listed in table affect the movement of ions across the plasma membranes of excitable cells (i.e., muscle, nerve, and sensory cells), reducing the ability of these cells to develop or transmit impulses. In contrast, cystic fibrosis, the best studied and most common inherited ion channel disorder, results from a defect in the ion channels of epithelial cells.

Inherited disorder	Type of channel	Gene	Clinical consequences
Familial hemiplegic migraine (FHM)	Ca^{2+}	CACNL1A4	Migraine headaches
Episodic ataxia type-2 (EA-2)	Ca^{2+}	CACNL1A4	Ataxia (lack of balance and coordination)
Hypokalemic periodic paralysis	Ca^{2+}	CACNL1A3	Periodic myotonia (muscle stiffness) and paralysis
Episodic ataxia type-1	K^+	KCNA1	Ataxia
Benign familial neonatal convulsions	K^+	KCNQ2	Epileptic convulsions
Nonsyndromic dominant deafness	K^+	KCNQ4	Deafness
Long QT syndrome	K^+ Na^+	HERG KCNQ1, or SCN5A	Dizziness, sudden death from ventricular fibrillation
Hyperkalemic periodic paralysis	Na^+	SCN4A	Periodic myotonia and paralysis
Liddle Syndrome	Na^+	B-ENaC	Hypertension (high blood pressure)
Myasthenia gravis	Na^+	nAChR	Muscle weakness
Dent's disease	Cl^-	CLCN5	Kidney stones
Myotonia congenital	Cl^-	CLC-1	Periodic myotonia
Bartter's syndrome type IV	Cl^-	CLC-Kb	Kidney dysfunction, deafness
Cystic fibrosis	Cl^-	CFTR	Lung congestion and infections
Cardiac arrhythmias	Na^+ K^+ Ca^{2+}	Many different genes	Irregular or rapid heartbeat

On average, 1 out of every 25 persons of Northern European descent carries one copy of the mutant gene that can cause cystic fibrosis. Because they show no symptoms of the mutant gene, most heterozygotes are unaware that they are carriers. Consequently, approximately 1 out of every 2500 infants in this Caucasian population (1/25 x 1/25 x 1/4) is homozygous recessive at this locus and born with cystic fibrosis (CF). Although cystic fibrosis affects various organs, including the intestine, pancreas, sweat glands, and reproductive tract, the respiratory tract usually exhibits the most severe effects. Victims of CF produce a thickened, sticky mucus that is very hard to propel out of the airways. Afflicted individuals typically suffer from chronic lung infections and inflammation, which progressively destroy pulmonary function.

The gene responsible for cystic fibrosis was isolated in 1989. Once the sequence of the CF gene was determined and the amino acid sequence of the corresponding polypeptide was deduced, it was apparent that the polypeptide was a member of the ABC transporter superfamily. The protein was named cystic fibrosis transmembrane conductance regulator (CFTR), an ambiguous term that reflected the fact that researchers weren't sure of its precise function. The question was thought to be answered after the protein was purified, incorporated into artificial lipid bilayers, and shown to act as a cyclic AMP-regulated chloride channel, not a transporter. But subsequent studies have added numerous complications to the story as it has been shown that, in addition to functioning as a chloride channel, CFTR also (1) conducts bicarbonate (HCO_3^-) ions, (2) suppresses the activity of an epithelial Na^+ ion channel (ENaC), and (3) stimulates the activity of a family of epithelial chloride/bicarbonate exchangers. As the role of CFTR has become more complex, it has become difficult to establish precisely how a defect in this protein leads to the development of chronic lung infections.

Because the movement of water out of epithelial cells by osmosis follows the movement of salts, abnormalities in the flux of Cl^-, HCO_3^-, and/or Na^+ caused by CFTR deficiency leads to a decrease in the fluid that bathes the epithelial cells of the airways. A reduction in volume of the surface liquid, and a resulting increase in viscosity of the secreted mucus, impair the function of the cilia that push mucus and bacteria out of the respiratory tract. Many CF patients are helped by inhaling a mist of hypertonic saline, which helps to draw more water into the airways and reduce the viscosity of the mucus. Clinical trials are also being carried out on compounds that have the potential to increase the volume of surface fluid by altering the movements of ions in and out of the epithelial cells. These compounds include Na^+-channel inhibitors (e.g., GS-9411) with the potential to reduce Na^+ ion absorption from the airway fluid into the epithelium and activators of Cl^- channels (other than CFTR) (e.g., Moli1901) with the potential to increase Cl^- ion conductance from the epithelium into the airway fluid.

In the past decade, researchers have identified more than 1000 different mutations that give rise to cystic fibrosis. However, approximately 70 percent of the alleles responsible for cystic fibrosis in the United States contain the same genetic alteration (designated ΔF508)—they are all missing three base pairs of DNA that encode a phenylalanine at position 508, within one of the cytoplasmic domains of the CFTR polypeptide. Subsequent research has revealed that CFTR polypeptides lacking this particular amino acid fail to be processed normally within the membranes of the endoplasmic reticulum and, in fact, never reach the surface of epithelial cells. As a result, CF patients who are homozygous for the DF508 allele completely lack the CFTR channel in their plasma membranes and have a severe form of the disease. When cells from these patients are grown in culture at lower temperature, the mutant protein is transported to the plasma membrane where it functions quite well. This finding has prompted a number of drug companies to screen for small molecules that can bind to these mutant CFTR molecules, preventing their destruction in the cytoplasm and allowing them to reach the cell surface. One of these candidate drugs, VX-809, is in phase II clinical trials. VX-809 is given in combination with another drug, Kalydeco, which can keep the ion channel open if it arrives at the cell surface. Kalydeco has recently been approved as a solo treatment for the approximately 4 percent of CF patients whose CFTR protein carries a single amino acid substitution, G551D. While this mutation does not interfere with the protein reaching the cell surface, it does prevent the ion channel from opening properly.

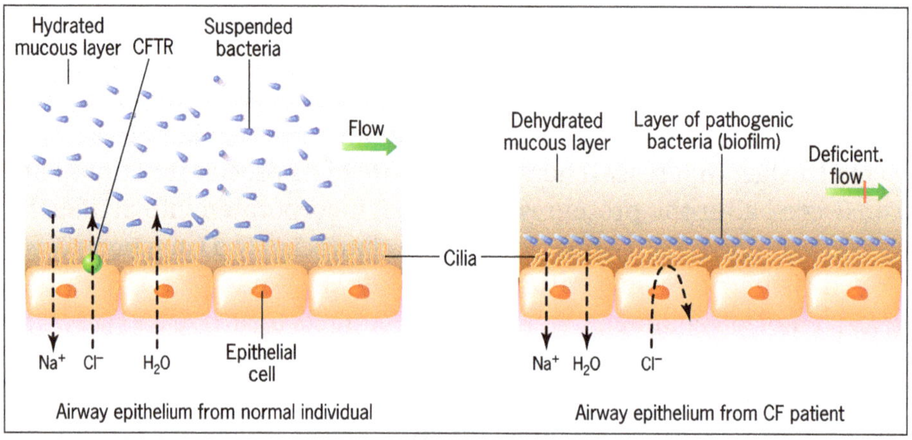

Figure shows an explanation for the debilitating effects on lung function from the absence of the CFTR protein. In the airway epithelium of a normal individual, water flows out of the epithelial cells in response to the outward movement of ions, thus hydrating the surface mucous layer. The hydrated mucous layer, with its trapped bacteria, is readily moved out of the airways. In the airway epithelium of a person with cystic fibrosis, the abnormal movement of ions causes water to flow in the opposite direction, thus dehydrating the mucous layer. As a result, trapped bacteria cannot be moved out of the airways, which allows them to proliferate as a biofilm and cause chronic infections.

According to one estimate, the DF508 mutation had to have originated more than 50,000 years ago to have reached such a high frequency in the population. The fact that the CF gene has reached this frequency suggests that heterozygotes may receive some selective advantage over those lacking a copy of the defective gene. It has been proposed that CF heterozygotes may be protected from the effects of cholera, a disease that is characterized by excessive fluid secretion by the wall of the intestine. One difficulty with this proposal is that there is no record of cholera epidemics in Europe until the 1820s. An alternate proposal suggests that heterozygotes are protected from typhoid fever because the bacterium responsible for this disease adheres poorly to the wall of an intestine having a reduced number of CFTR molecules.

Ever since the isolation of the gene responsible for CF, the development of a cure by gene therapy—that is, by replacement of the defective gene with a normal version—has been a major goal of CF researchers. Cystic fibrosis is a good candidate for gene therapy because the worst symptoms of the disease result from the defective activities of epithelial cells that line the airways and, therefore, are accessible to agents that can be delivered by inhalation of an aerosol. Clinical trials have been conducted using several different types of delivery systems. In one group of trials, the normal CFTR gene was incorporated into the DNA of a defective adenovirus, a type of virus that normally causes upper respiratory tract infections. The recombinant virus particles were then allowed to infect the cells of the airway, delivering the normal gene to the genetically deficient cells. The primary disadvantage in using adenovirus is that the viral DNA (along with the normal CFTR gene) does not become integrated into the chromosomes of the infected host cell so that the virus must be readministered frequently. As a result, the procedure often induces an immune response within the patient that eliminates the virus and leads to lung inflammation. Researchers are hesitant to employ viruses that integrate their genomes for fear of initiating the formation of cancers. In other trials, the DNA encoding the normal CFTR gene has been linked to positively charged liposomes that can fuse with the plasma membranes of the airway cells, delivering their DNA contents into the cytoplasm. Lipid-based delivery has an advantage over viruses in being less likely to stimulate a destructive immune response following repeated treatments, but has the disadvantage of being less effective in achieving genetic modification of target cells. To date, none of the clinical trials of gene therapy has resulted in significant improvement of either physiologic processes or disease symptoms. The development of more effective DNA delivery systems, which are capable of genetically altering a greater percentage of airway cells, will be required if a treatment for CF based on gene therapy is to be achieved.

Membrane Potentials and Nerve Impulses

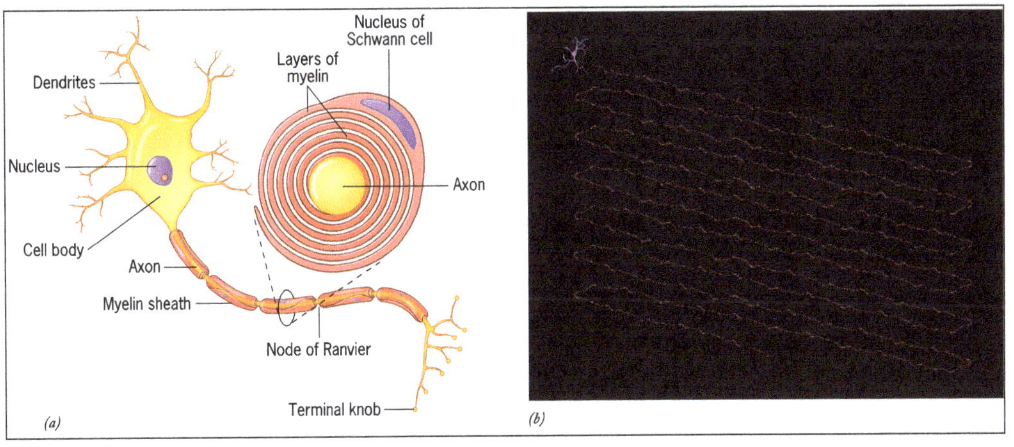

Figure shows the structure of a nerve cell. (a) Schematic drawing of a simple neuron with a myelinated axon. As the inset shows, the myelin sheath comprises individual Schwann cells that have wrapped themselves around the axon. The sites where the axon lacks myelin wrapping are called nodes of Ranvier. (Myelin-forming cells within the central nervous system are called oligodendrocytes rather than Schwann cells.) (b) A composite micrograph of a single rat hippocampal neuron with cell body and dendrites (purple) and an axon 1 cm in length (red).

All organisms respond to external stimulation, a property referred to as irritability. Even a single-celled ameba, if poked with a fine glass needle, responds by withdrawing its pseudopodia, rounding up, and moving off in another direction. Irritability in an ameba depends on the same basic properties of membranes that lead to the formation and propagation of nerve impulses.

Nerve cells (or neurons) are specialized for the collection, conduction, and transmission of information, which is coded in the form of fast-moving electrical impulses. The basic parts of a typical neuron are illustrated in figure. The nucleus of the neuron is located within an expanded region called the cell body, which is the metabolic center of the cell and the site where most of its material contents are manufactured. Extending from the cell bodies of most neurons are a number of fine extensions, called dendrites, which receive incoming information from external sources, typically other neurons. Also emerging from the cell body is a single, more prominent extension, the axon, which conducts outgoing impulses away from the cell body and toward the target cell(s). Although some axons may be only a few micrometers in length, others extend for many meters in the body of a large vertebrate, such as a giraffe or whale. Most axons split near their ends into smaller processes, each ending in a terminal knob—a specialized site where impulses are transmitted from neuron to target cell. Many neurons in the brain end in thousands of terminal knobs, allowing these brain cells to communicate with thousands of potential targets.

The Resting Potential

A voltage (or electric potential difference) between two points, such as the inside and outside of the plasma membrane, results when there is an excess of positive ions at one point and an excess of negative ions at the other point. Voltages across plasma membranes can be measured by inserting one fine glass electrode (or microelectrode) into the cytoplasm of a cell, placing another electrode in the extracellular fluid outside the cell, and connecting the electrodes to a voltmeter, an instrument that measures a difference in charge between two points. When this experiment was first carried out on a giant axon of the squid, a potential difference of approximately 70 millivolts (mV) was recorded, the inside being negative with respect to the outside (indicated with a minus sign, 270 mV). The presence of a membrane potential is not unique to nerve cells; such potentials are present in all types of cells, the magnitude varying between about 215 and 2100 mV. When a nerve or muscle cell is in an unexcited state, the membrane potential is referred to as the resting potential because it is subject to dramatic change, as discussed in the following section. The magnitude and direction of the voltage across the plasma membrane are determined by the differences in concentrations of ions on either side of the membrane and their relative permeabilities. The Na^+/K^+-ATPase pumps Na^+ out of the cell and K^+ into the cell, thereby establishing steep gradients of these two ions across the plasma membrane. Because of these gradients, you might expect that potassium ions would leak out of the cell and sodium ions would leak inward through their respective ion channels. However, the vast majority of the ion channels in the plasma membrane of a resting nerve cell are closed.

Those that are open are selective for K^+; they are often referred to as K^+ leak channels. K^+ leak channels are members of the K2P family of K^+ channels, which lack the S4 voltage sensor and fail to respond to changes in voltage.

Because K^+ ions are the only charged species with significant permeability in a resting nerve cell, their outflow through the membrane leaves an excess of negative charges on the cytoplasmic side of the membrane. Although the concentration gradient across the membrane favors continued efflux of K^+, the electrical gradient resulting from the excess negative charge on the inside of the membrane favors the retention of K^+ ions inside the cell. When these two opposing forces are balanced, the system is at equilibrium, and there is no further net movement of K^+ ions across the membrane. Using the following equation, which is called the Nernst equation, one can calculate the membrane potential (V_m) that would be measured at equilibrium if the plasma membrane of a nerve cell were permeable only to K^+ ions. In this case, V_m would be equal to the potassium equilibrium potential (E_K),

$$E_K = 2.303\frac{RT}{zF} \bullet \log_{10}\frac{[K_o^+]}{[K_i^+]}$$

For a squid giant axon, the internal $[K_i^+]$ is approximately 350 mM, while the external $[K_o^+]$ is approximately 10 mM; thus at 25 °C (298 K) and z = +1 (for the univalent K^+ ion),

$$E_K = 59\log_{10} 0.028 = -91\,\text{mV}$$

A similar calculation of the Na^+ equilibrium potential (E_{Na}) would produce a value of approximately 155 mV. Because measurements of the voltage across the resting nerve membrane are similar in sign and magnitude (270 mV) to the potassium equilibrium potential just calculated, the movement of potassium ions across the membrane is considered the most important factor in determining the resting potential. The difference between the calculated K^+ equilibrium potential (-91 mV) and the measured resting potential (-70 mV) is due to a slight permeability of the membrane to Na^+ through a recently described Na^+ leak channel.

Cell Wall

A cell wall is an outer layer surrounding certain cells that is outside of the cell membrane. All cells have cell membranes, but generally only plants, fungi, algae, most bacteria, and archaea have cells with cell walls. The cell wall provides strength and structural support to the cell, and can control to some extent what types and concentrations of molecules enter and leave the cell. The materials that make up the cell wall differ depending on the type of organism. The cell wall has evolved many different times among different groups of organisms.

Cell Wall Functions

The cell wall has a few different functions. It is flexible, but provides strength to the cell, which helps protect the cell against physical damage. It also gives the cell its shape and allows the organism to maintain a certain shape overall. The cell wall can also provide protection from pathogens such as bacteria that are trying to invade the cell. The structure of the cell wall allows many small molecules to pass through it, but not larger molecules that could harm the cell.

Cell Wall Structure

Plant Cell Walls

The main component of the plant cell wall is cellulose, a carbohydrate that forms long fibers and gives the cell wall its rigidity. Cellulose fibers group together to form bundles called microfibrils. Other important carbohydrates include hemicellulose, pectin, and liginin. These carbohydrates form a network along with structural proteins to form the cell wall. Plant cells that are in the process of growing have primary cell walls, which are thin. Once the cells are fully grown, they develop secondary cell walls. The secondary cell wall is a thick layer that is formed on the inside of the primary cell wall. This layer is what is usually meant when referring to a plant's cell wall. There is also another layer in between plant cells called the middle lamella; it is pectin-rich and helps plant cells stick together.

The cell walls of plant cells help them maintain turgor pressure, which is the pressure of the cell membrane pressing against the cell wall. Ideally, plants cells should have lots of water within them, leading to high turgidity. Whereas a cell without a cell wall, such as an animal cell, can swell and burst of too much water diffuses into it, plants need to be in hypotonic solutions (more water inside than outside, leading to lots of water entering the cell) to maintain turgor pressure and their structural shape. The cell wall efficiently holds water in so that the cell does not burst. When turgor pressure is lost, a plant will begin to wilt. Turgor pressure is what gives plant cells their characteristic square shape; the cells are full of water, so they fill up the space available and press against each other.

This diagram of a plant cell depicts the cell wall in green, surrounding the contents of the cell.

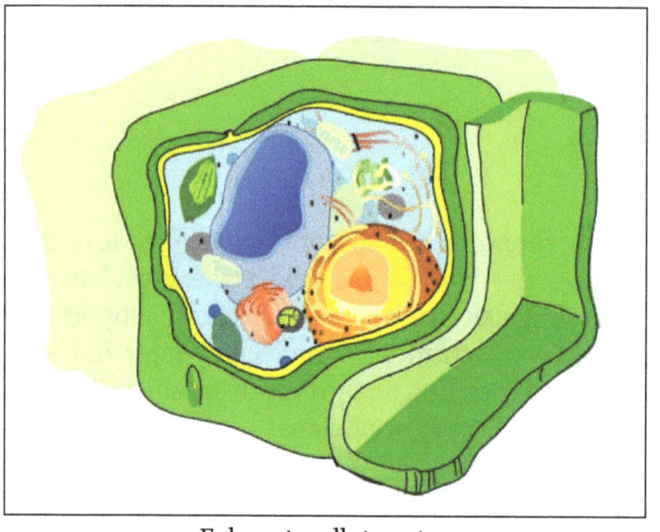

Eukaryota cell strucutre.

Algae Cell Walls

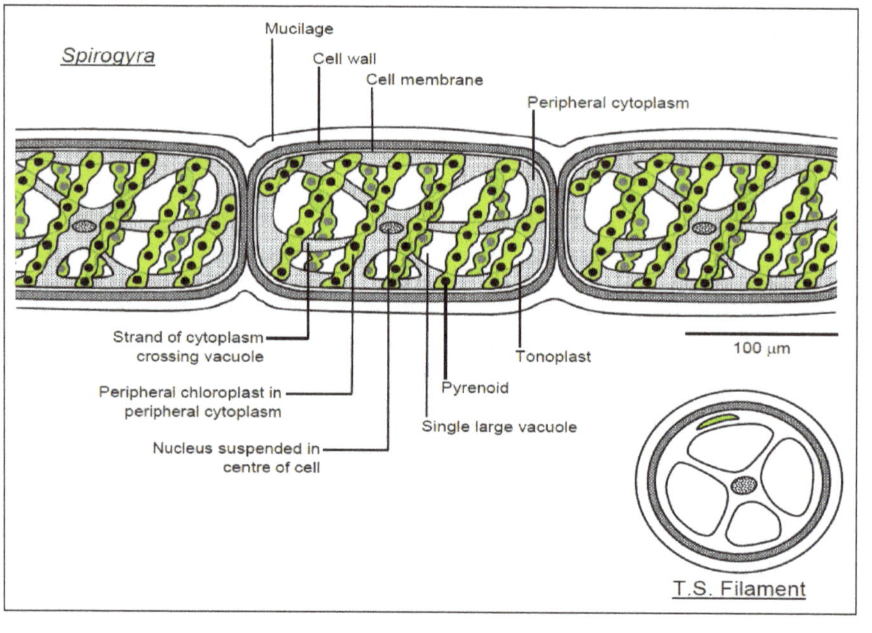

Algae are a diverse group, and the diversity in their cell walls reflects this. Some algae, such as green algae, have cell walls that are similar in structure to those of plants. Other algae, such as brown algae and red algae, have cellulose along with other polysaccharides or fibrils. Diatoms have cell walls that are made from silicic acid. Other important molecules in algal cell walls include mannans, xylans, and alginic acid.

Fungi Cell Walls

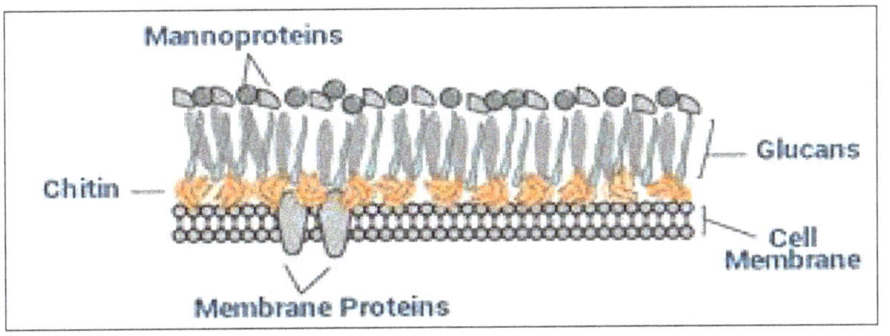

The cell walls of fungi contain chitin, which is a glucose derivative that is similar in structure to cellulose. Layers of chitin are very tough; chitin is the same molecule found in the rigid exoskeletons of animals such as insects and crustaceans. Glucans, which are other glucose polymers, are also found in the fungal cell wall along with lipids and proteins. Fungi have proteins called hydrophobins in their cell walls. Found only in fungi, hydrophobins give the cells strength, help them adhere to surfaces, and help control the movement of water into the cells. In fungi, the cell wall is the most external layer, and surrounds the cell membrane.

References

- What-is-a-cell-14023083: nature.com, Retrieved 11, July 2020
- Prokaryotic-cells: microbenotes.com, Retrieved 23, February 2020
- Characteristics-of-eukaryotic-cells: microbenotes.com, Retrieved 15, June 2020
- Building-the-cell-design-principles-of-cellular-13997438: nature.com, Retrieved 07, March 2020
- What-are-cell-prganelles, biology: byjus.com, Retrieved 30, January 2020

Chapter 3

Primary Components of a Cell

A cell is mainly composed of a membrane, cytoplasm, nucleus and the mitochondrion suspended in the cytoplasm. The cytoplasm contains all the organelles of the cell and forms the non-nuclear content. Contained within the cytoplasm is the nucleus, separated from the cytoplasm by an enclosing membrane. This chapter covers these topics in a clear and concise manner.

Cytoplasm

In cell biology, the cytoplasm is the material within a living cell, excluding the cell nucleus. It comprises cytosol (the gel-like substance enclosed within the cell membrane) and the organelles – the cell's internal sub-structures. All of the contents of the cells of prokaryote organisms (such as bacteria, which lack a cell nucleus) are contained within the cytoplasm. Within the cells of eukaryote organisms the contents of the cell nucleus are separated from the cytoplasm, and are then called the nucleoplasm. The cytoplasm is about 80% water and usually colorless.

The submicroscopic ground cell substance or cytoplasmatic matrix which remains after exclusion the cell organelles and particles is groundplasm. It is the hyaloplasm of light microscopy, and high complex, polyphasic system in which all of resolvable cytoplasmic elements of are suspended, including the larger organelles such the ribosomes, mitochondria, the plant plastids, lipid droplets, and vacuoles.

It is within the cytoplasm that most cellular activities occur, such as many metabolic pathways including glycolysis, and processes such as cell division. The concentrated inner area is called the endoplasm and the outer layer is called the cell cortex or the ectoplasm.

Movement of calcium ions in and out of the cytoplasm is a signaling activity for metabolic processes.

In plants, movement of the cytoplasm around vacuoles is known as cytoplasmic streaming.

History

The term was introduced by Rudolf von Kölliker in 1863, originally as a synonym for protoplasm, but later it has come to mean the cell substance and organelles outside the nucleus.

There has been certain disagreement on the definition of cytoplasm, as some authors prefer to exclude from it some organelles, especially the vacuoles and sometimes the plastids.

Physical Nature

The physical properties of the cytoplasm have been contested in recent years. It remains uncertain how the varied components of the cytoplasm interact to allow movement of particles and organelles while maintaining the cell's structure. The flow of cytoplasmic components plays an important role in many cellular functions which are dependent on the permeability of the cytoplasm. An example of such function is cell signalling, a process which is dependent on the manner in which signaling molecules are allowed to diffuse across the cell. While small signaling molecules like calcium ions are able to diffuse with ease, larger molecules and subcellular structures often require aid in moving through the cytoplasm. The irregular dynamics of such particles have given rise to various theories on the nature of the cytoplasm.

As a Sol-gel

There has long been evidence that the cytoplasm behaves like a sol-gel. It is thought that the component molecules and structures of the cytoplasm behave at times like a disordered colloidal solution (sol) and at other times like an integrated network, forming a solid mass (gel). This theory thus proposes that the cytoplasm exists in distinct fluid and solid phases depending on the level of interaction between cytoplasmic components, which may explain the differential dynamics of different particles observed moving through the cytoplasm.

As a Glass

Recently it has been proposed that the cytoplasm behaves like a glass-forming liquid approaching the glass transition. In this theory, the greater the concentration of cytoplasmic components, the less the cytoplasm behaves like a liquid and the more it behaves as a solid glass, freezing larger cytoplasmic components in place (it is thought that the cell's metabolic activity is able to fluidize the cytoplasm to allow the movement of such larger cytoplasmic components). A cell's ability to vitrify in the absence of metabolic activity, as in dormant periods, may be beneficial as a defence strategy. A solid glass cytoplasm would freeze subcellular structures in place, preventing damage, while allowing the transmission of very small proteins and metabolites, helping to kickstart growth upon the cell's revival from dormancy.

Other Perspectives

There has been research examining the motion of cytoplasmic particles independent of the nature of the cytoplasm. In such an alternative approach, the aggregate random forces within the cell caused by motor proteins explain the non-Brownian motion of cytoplasmic constituents.

Constituents

The three major elements of the cytoplasm are the cytosol, organelles and inclusions.

Cytosol

The cytosol is the portion of the cytoplasm not contained within membrane-bound organelles. Cytosol makes up about 70% of the cell volume and is a complex mixture of cytoskeleton filaments, dissolved molecules, and water. The cytosol's filaments include the protein filaments such as actin filaments and microtubules that make up the cytoskeleton, as well as soluble proteins and small structures such as ribosomes, proteasomes, and the mysterious vault complexes. The inner, granular and more fluid portion of the cytoplasm is referred to as endoplasm.

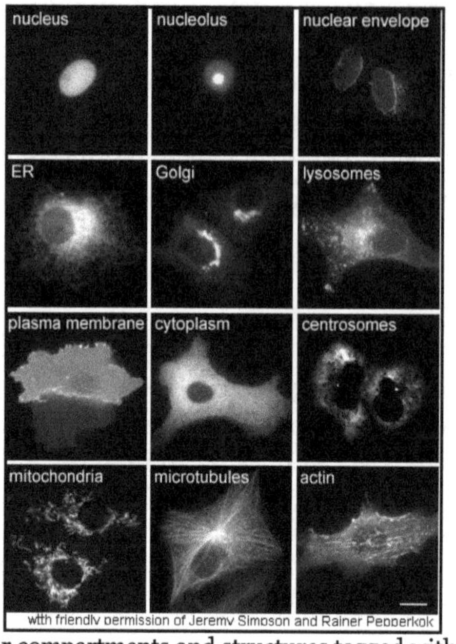

Proteins in different cellular compartments and structures tagged with green fluorescent protein.

Due to this network of fibres and high concentrations of dissolved macromolecules, such as proteins, an effect called macromolecular crowding occurs and the cytosol does not act as an ideal solution. This crowding effect alters how the components of the cytosol interact with each other.

Organelles

Organelles (literally "little organs"), are usually membrane-bound structures inside the cell that have specific functions. Some major organelles that are suspended in the cytosol are the mitochondria, the endoplasmic reticulum, the Golgi apparatus, vacuoles, lysosomes, and in plant cells, chloroplasts.

Cytoplasmic Inclusions

The inclusions are small particles of insoluble substances suspended in the cytosol. A huge range of inclusions exist in different cell types, and range from crystals of calcium oxalate or silicon dioxide in plants, to granules of energy-storage materials such as starch, glycogen, or polyhydroxybutyrate. A particularly widespread example are lipid droplets, which are spherical droplets composed of lipids and proteins that are used in both prokaryotes and eukaryotes as a way of storing lipids such as fatty acids and sterols. Lipid droplets make up much of the volume of adipocytes, which are specialized lipid-storage cells, but they are also found in a range of other cell types.

Controversy and Research

The cytoplasm, mitochondria and most organelles are contributions to the cell from the maternal gamete. Contrary to the older information that disregards any notion of the cytoplasm being active, new research has shown it to be in control of movement and flow of nutrients in and out of the cell by viscoplastic behavior and a measure of the reciprocal rate of bond breakage within the cytoplasmic network.

The material properties of the cytoplasm remain an ongoing investigation. Recent measurements using force spectrum microscopy reveal that the cytoplasm can be likened to an elastic solid, rather than a viscoelastic fluid.

Cytosol

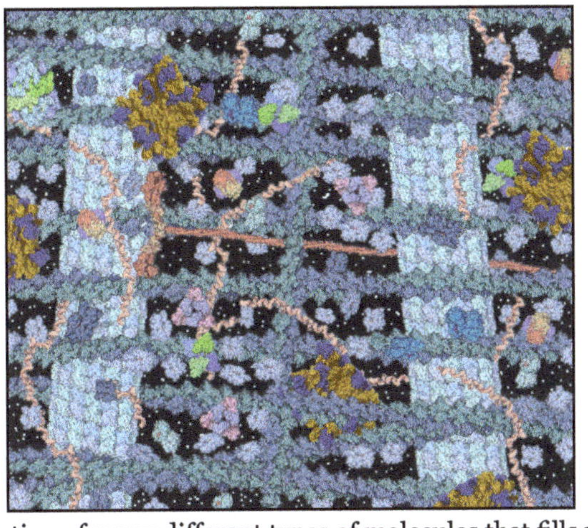

The cytosol is a crowded solution of many different types of molecules that fills much of the volume of cells.

The cytosol or cytoplasmic matrix is the liquid found inside cells. It constitutes most of the intracellular fluid (ICF). It is separated into compartments by membranes. For example, the mitochondrial matrix separates the mitochondrion into many compartments.

In the eukaryotic cell, the cytosol is within the cell membrane and is part of the cytoplasm, which also comprises the mitochondria, plastids, and other organelles (but not their internal fluids and structures); the cell nucleus is separate. The cytosol is thus a liquid matrix around the organelles. In prokaryotes, most of the chemical reactions of metabolism take place in the cytosol, while a few take place in membranes or in the periplasmic space. In eukaryotes, while many metabolic pathways still occur in the cytosol, others are contained within organelles.

The cytosol is a complex mixture of substances dissolved in water. Although water forms the large majority of the cytosol, its structure and properties within cells is not well understood. The concentrations of ions such as sodium and potassium are different in the cytosol than in the extracellular fluid; these differences in ion levels are important in processes such as osmoregulation, cell signaling, and the generation of action potentials in excitable cells such as endocrine, nerve and muscle cells. The cytosol also contains large amounts of macromolecules, which can alter how molecules behave, through macromolecular crowding.

Although it was once thought to be a simple solution of molecules, the cytosol has multiple levels of organization. These include concentration gradients of small molecules such as calcium, large complexes of enzymes that act together to carry out metabolic pathways, and protein complexes such as proteasomes and carboxysomes that enclose and separate parts of the cytosol.

Definition

The term "cytosol" was first introduced in 1965 by H. A. Lardy, and initially referred to the liquid that was produced by breaking cells apart and pelleting all the insoluble components by ultracentrifugation. Such a soluble cell extract is not identical to the soluble part of the cell cytoplasm and is usually called a cytoplasmic fraction.

The term *cytosol* is now used to refer to the liquid phase of the cytoplasm in an intact cell. This excludes any part of the cytoplasm that is contained within organelles. Due to the possibility of confusion between the use of the word "cytosol" to refer to both extracts of cells and the soluble part of the cytoplasm in intact cells, the phrase "aqueous cytoplasm" has been used to describe the liquid contents of the cytoplasm of living cells.

Prior to this, other terms were used for the cell fluid, not always synonymously, as its nature was not very clear.

Properties and Composition

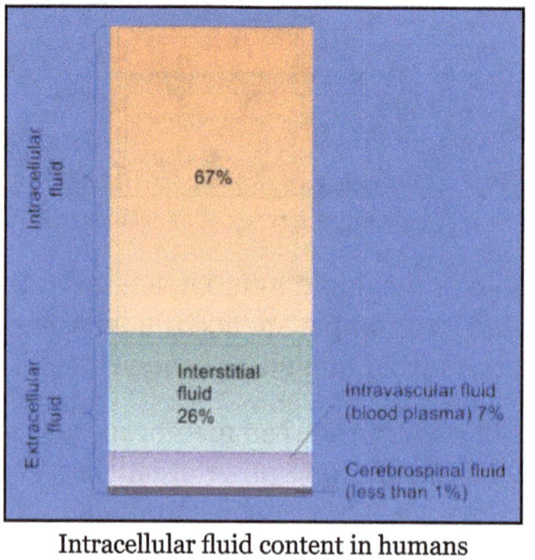

Intracellular fluid content in humans

The proportion of cell volume that is cytosol varies: for example while this compartment forms the bulk of cell structure in bacteria, in plant cells the main compartment is the large central vacuole. The cytosol consists mostly of water, dissolved ions, small molecules, and large water-soluble molecules (such as proteins). The majority of these non-protein molecules have a molecular mass of less than 300 Da. This mixture of small molecules is extraordinarily complex, as the variety of molecules that are involved in metabolism (the metabolites) is immense. For example, up to 200,000 different small molecules might be made in plants, although not all these will be present in the same species, or in a single cell. Estimates of the number of metabolites in single cells such as *E. coli* and baker's yeast predict that under 1,000 are made.

Water

Most of the cytosol is water, which makes up about 70% of the total volume of a typical cell. The pH of the intracellular fluid is 7.4. while human cytosolic pH ranges between 7.0 - 7.4, and is usually higher if a cell is growing. The viscosity of cytoplasm is roughly the same as pure water, although diffusion of small molecules through this liquid is about fourfold slower than in pure water, due mostly to collisions with the large numbers of macromolecules in the cytosol. Studies in the brine shrimp have examined how water affects cell functions; these saw that a 20% reduction in the amount of water in a cell inhibits metabolism, with metabolism decreasing progressively as the cell dries out and all metabolic activity halting when the water level reaches 70% below normal.

Although water is vital for life, the structure of this water in the cytosol is not well understood, mostly because methods such as nuclear magnetic resonance spectroscopy only give information on the average structure of water, and cannot measure local variations at the microscopic scale. Even the structure of pure water is poorly understood, due to the ability of water to form structures such as water clusters through hydrogen bonds.

The classic view of water in cells is that about 5% of this water is strongly bound in by solutes or macromolecules as water of solvation, while the majority has the same structure as pure water. This water of solvation is not active in osmosis and may have different solvent properties, so that some dissolved molecules are excluded, while others become concentrated. However, others argue that the effects of the high concentrations of macromolecules in cells extend throughout the cytosol and that water in cells behaves very differently from the water in dilute solutions. These ideas include the proposal that cells contain zones of low and high-density water, which could have widespread effects on the structures and functions of the other parts of the cell. However, the use of advanced nuclear magnetic resonance methods to directly measure the mobility of water in living cells contradicts this idea, as it suggests that 85% of cell water acts like that pure water, while the remainder is less mobile and probably bound to macromolecules.

Ions

The concentrations of the other ions in cytosol are quite different from those in extracellular fluid and the cytosol also contains much higher amounts of charged macromolecules such as proteins and nucleic acids than the outside of the cell structure.

Typical ion concentrations in mammalian cytosol and blood.		
Ion	Concentration in cytosol **(millimolar)**	Concentration in blood **(millimolar)**
Potassium	139	4
Sodium	12	145
Chloride	4	116
Bicarbonate	12	29
Amino acids in proteins	138	9
Magnesium	0.8	1.5
Calcium	<0.0002	1.8

In contrast to extracellular fluid, cytosol has a high concentration of potassium ions and a low concentration of sodium ions. This difference in ion concentrations is critical for osmoregulation, since if the ion levels were the same inside a cell as outside, water would enter constantly by osmosis - since the levels of macromolecules inside cells are higher than their levels outside. Instead, sodium ions are expelled and potassium ions taken up by the Na^+/K^+-ATPase, potassium ions then flow down their concentration gradient through potassium-selection ion channels, this loss of positive charge creates a negative membrane potential. To balance this potential difference, negative chloride ions also exit the cell, through selective chloride channels. The loss of sodium and chloride ions compensates for the osmotic effect of the higher concentration of organic molecules inside the cell.

Cells can deal with even larger osmotic changes by accumulating osmoprotectants such as betaines or trehalose in their cytosol. Some of these molecules can allow cells to survive being completely dried out and allow an organism to enter a state of suspended animation called cryptobiosis. In this state the cytosol and osmoprotectants become a glass-like solid that helps stabilize proteins and cell membranes from the damaging effects of desiccation.

The low concentration of calcium in the cytosol allows calcium ions to function as a second messenger in calcium signaling. Here, a signal such as a hormone or an action potential opens calcium channels so that calcium floods into the cytosol. This sudden increase in cytosolic calcium activates other signalling molecules, such as calmodulin and protein kinase C. Other ions such as chloride and potassium may also have signaling functions in the cytosol, but these are not well understood.

Macromolecules

Protein molecules that do not bind to cell membranes or the cytoskeleton are dissolved in the cytosol. The amount of protein in cells is extremely high, and approaches 200 mg/ml, occupying about 20-30% of the volume of the cytosol. However, measuring precisely how much protein is dissolved in cytosol in intact cells is difficult, since some proteins appear to be weakly associated with membranes or organelles in whole cells and are released into solution upon cell lysis. Indeed, in experiments where the plasma membrane of cells were carefully disrupted using saponin, without damaging the other cell membranes, only about one quarter of cell protein was released. These cells were also able to synthesize proteins if given ATP and amino acids, implying that many of the enzymes in cytosol are bound to the cytoskeleton. However, the idea that the majority of the proteins in cells are tightly bound in a network called the microtrabecular lattice is now seen as unlikely.

In prokaryotes the cytosol contains the cell's genome, within a structure known as a nucleoid. This is an irregular mass of DNA and associated proteins that control the transcription and replication of the bacterial chromosome and plasmids. In eukaryotes the genome is held within the cell nucleus, which is separated from the cytosol by nuclear pores that block the free diffusion of any molecule larger than about 10 nanometres in diameter.

This high concentration of macromolecules in cytosol causes an effect called macromolecular crowding, which is when the effective concentration of other macromolecules is increased, since they have less volume to move in. This crowding effect can produce large changes in both the rates and the position of chemical equilibrium of reactions in the cytosol. It is particularly important in its ability to alter dissociation constants by favoring the association of macromolecules, such as when multiple proteins come together to form protein complexes, or when DNA-binding proteins bind to their targets in the genome.

Organization

Although the components of the cytosol are not separated into regions by cell membranes, these components do not always mix randomly and several levels of organization can localize specific molecules to defined sites within the cytosol.

Concentration Gradients

Although small molecules diffuse rapidly in the cytosol, concentration gradients can still be produced within this compartment. A well-studied example of these are the "calcium sparks" that are produced for a short period in the region around an open calcium channel. These are about 2 micrometres in diameter and last for only a few milliseconds, although several sparks can merge to form larger gradients, called "calcium waves". Concentration gradients of other small molecules, such as oxygen and adenosine triphosphate may be produced in cells around clusters of mitochondria, although these are less well understood.

Protein complexes

Proteins can associate to form protein complexes, these often contain a set of proteins with similar functions, such as enzymes that carry out several steps in the same metabolic pathway. This organization can allow substrate channeling, which is when the product of one enzyme is passed directly to the next enzyme in a pathway without being released into solution. Channeling can make a pathway more rapid and efficient than it would be if the enzymes were randomly distributed in the cytosol, and can also prevent the release of unstable reaction intermediates. Although a wide variety of metabolic pathways involve enzymes that are tightly bound to each other, others may involve more loosely associated complexes that are very difficult to study outside the cell. Consequently, the importance of these complexes for metabolism in general remains unclear.

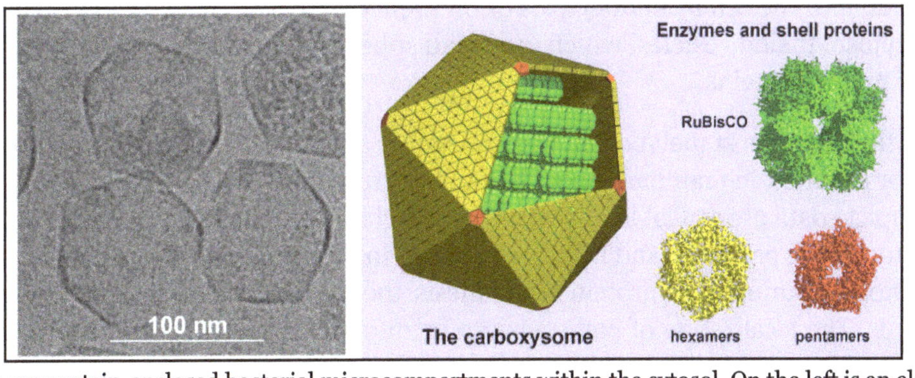

Carboxysomes are protein-enclosed bacterial microcompartments within the cytosol. On the left is an electron microscope image of carboxysomes, and on the right a model of their structure.

Protein Compartments

Some protein complexes contain a large central cavity that is isolated from the remainder of the cytosol. One example of such an enclosed compartment is the proteasome. Here, a set of subunits form a hollow barrel containing proteases that degrade cytosolic proteins. Since these would be damaging if they mixed freely with the remainder of the cytosol, the barrel is capped by a set of regulatory proteins that recognize proteins with a signal directing them for degradation (a ubiquitin tag) and feed them into the proteolytic cavity.

Another large class of protein compartments are bacterial microcompartments, which are made of a protein shell that encapsulates various enzymes. These compartments are typically about 100-200 nanometres across and made of interlocking proteins. A well-understood example is the carboxysome, which contains enzymes involved in carbon fixation such as RuBisCO.

Cytoskeletal Sieving

Although the cytoskeleton is not part of the cytosol, the presence of this network of filaments restricts the diffusion of large particles in the cell. For example, in several studies tracer particles larger than about 25 nanometres (about the size of a ribosome) were excluded from parts of the cytosol around the edges of the cell and next to the nucleus. These "excluding compartments" may contain a much denser meshwork of actin fibres than the remainder of the cytosol. These microdomains could influence the distribution of large structures such as ribosomes and organelles within the cytosol by excluding them from some areas and concentrating them in others.

Function

The cytosol has no single function and is instead the site of multiple cell processes. Examples of these processes include signal transduction from the cell membrane to sites within the cell, such as the cell nucleus, or organelles. This compartment is also the site of many of the processes of cytokinesis, after the breakdown of the nuclear membrane in mitosis. Another major function of cytosol is to transport metabolites from their site of production to where they are used. This is relatively simple for water-soluble molecules, such as amino acids, which can diffuse rapidly through the cytosol. However, hydrophobic molecules, such as fatty acids or sterols, can be transported through the cytosol by specific binding proteins, which shuttle these molecules between cell membranes. Molecules taken into the cell by endocytosis or on their way to be secreted can also be transported through the cytosol inside vesicles, which are small spheres of lipids that are moved along the cytoskeleton by motor proteins.

The cytosol is the site of most metabolism in prokaryotes, and a large proportion of the metabolism of eukaryotes. For instance, in mammals about half of the proteins in the cell are localized to the cytosol. The most complete data are available in yeast, where metabolic reconstructions indicate that the majority of both metabolic processes and metabolites occur in the cytosol. Major metabolic pathways that occur in the cytosol in animals are protein biosynthesis, the pentose phosphate pathway, glycolysis and gluconeogenesis. The localization of pathways can be different in other organisms, for instance fatty acid synthesis occurs in chloroplasts in plants and in apicoplasts in apicomplexa.

Structure and Function of Cytoplasm

Cytoplasm was discovered in 1835 and no single scientist can be credited for discovering cytoplasm the discovery was possible due to contribution of several scientists. It is worth mentioning that the discovery of different organelles in the cytoplasm was attributed to different scientist. The cytoplasm is the part of the cell outside the largest organelle, the nucleus. Cytoplasm appears as thick, gel-like semitransparent fluid that is found in both plant and animal cell. It is bounded by the plasma membrane, and contains many organelles in a eukaryotic cell (cell containing membrane bounded nucleus). The constituent parts of cytoplasm are cytosol, organelles and cytoplasmic inclusions. The cytosol, the aqueous part of the cytoplasm outside all of the organelles, also contains its own distinctive proteins.

Cytosol

Cytosol is the part of the cytoplasm that is not occupied by any organelle. It accounts for almost 70% of the total cell volume. Cytosol (cytoplasmic matrix) like many colloidal systems, shows the property of phase reversal . Under the natural conditions, the phase reversal of the cytosol (cytoplasmic matrix) depends on various physiological, mechanical and biochemical activities of the cell. It is a gelatinous substance consisting mainly of cytoskeleton filaments, organic molecules, salt and water. Chemically, the cytoplasmic matrix is composed of many chemical elements in the form of atoms, ions and molecules. Of the 92 naturally occurring elements, approximately 46 are found in the cytosol (cytoplasmic matrix). Twenty four of these are essential elements , while others are present in cytosol only because they exist in the environment with which the organism interacts. Of the 24 essential elements, six play especially important roles in living systems. These major elements are carbon (C, 20 per cent), hydrogen (H, 10 per cent), nitrogen (N, 3 per cent), oxygen (O, 62 per cent), phosphorus (P, 1.14 per cent) and sulphur (S,0.14 per cent). Most organic molecules are built with these six elements. Another five essential elements found in less abundance in living systems are calcium (Ca, 2.5 per cent), potassium (K, 0.11 per cent), sodium (Na, 0.10 per cent), chlorine (Cl, 0.16 per cent) and magnesium (Mg, 0.07 per cent).Several other elements, called trace elements , are also found in minute amounts in animal and plant cell cytosol. These are iron (Fe, 0.10 per cent), iodine (I, 0.014 per cent), molybdenum (Mo), manganese (Mn), Cobalt (Co), zinc (Zn), selenium (Se), copper (Cu), chromium (Cr), tin (Sn), vanadium (V), silicon (Si), nickel (Ni), fluorine (F) and boron (B).

The cytoplasmic matrix consists of various kinds of ions. The ions are important in maintaining osmotic pressure and acid-base balance in the cells. Retention of ions in the matrix produces an increase in osmotic pressure and, thus, the entrance of water in the cell. The concentration of various ions in the intracellular fluid (matrix) differs from that in the interstitial fluid. For example, in the cell K^+ and Mg^{++} can be high, and Na^+ and Cl-high outside the cell. In muscle and nerve cells a high order of difference exists between intracellular K^+ and extracellular Na^+. Free calcium ions (Ca^{++}) may occur in cells or circulating blood. Silicon ions occur in the epithelium cells of grasses.

Chemical compounds present in cytosol are conventionally divided into two groups: organic and inorganic. Organic compounds form 30 per cent of a cell, rest are the inorganic substances such as water and other substances. The inorganic compounds are those compounds which normally found in the bulk of the physical, non-living universe, such as elements, metals, non-metals, and their compounds such as water, salts and variety of electrolytes and non-electrolytes. We have discussed a lot about the inorganic substances except the water which will be discussed in the following paragraph. The main organic compounds of the matrix are the carbohydrates, lipids, proteins, vitamins, hormones and nucleotides.

Cell Nucleus

In cell biology, the nucleus is a membrane-enclosed organelle found in eukaryotic cells. Eukaryotes usually have a single nucleus, but a few cell types, such as mammalian red blood cells, have no nuclei, and a few others have many.

Cell nuclei contain most of the cell's genetic material, organized as multiple long linear DNA molecules in complex with a large variety of proteins, such as histones, to form chromosomes. The genes within these chromosomes are the cell's nuclear genome and are structured in such a way to promote cell function. The nucleus maintains the integrity of genes and controls the activities of the cell by regulating gene expression—the nucleus is, therefore, the control center of the cell. The main structures making up the nucleus are the nuclear envelope, a double membrane that encloses the entire organelle and isolates its contents from the cellular cytoplasm, and the nuclear matrix (which includes the nuclear lamina), a network within the nucleus that adds mechanical support, much like the cytoskeleton, which supports the cell as a whole.

Because the nuclear membrane is impermeable to large molecules, nuclear pores are required to regulate nuclear transport of molecules across the envelope. The pores cross both nuclear membranes, providing a channel through which larger molecules must be actively transported by carrier proteins while allowing free movement of small molecules and ions. Movement of large molecules such as proteins and RNA through the pores is required for both gene expression and the maintenance of chromosomes. Although the interior of the nucleus does not contain any membrane-bound sub compartments, its contents are not uniform, and a number of *sub-nuclear bodies* exist, made up of unique proteins, RNA molecules, and particular parts of the chromosomes. The best-known of these is the nucleolus, which is mainly involved in the assembly of ribosomes. After being produced in the nucleolus, ribosomes are exported to the cytoplasm where they translate mRNA.

History

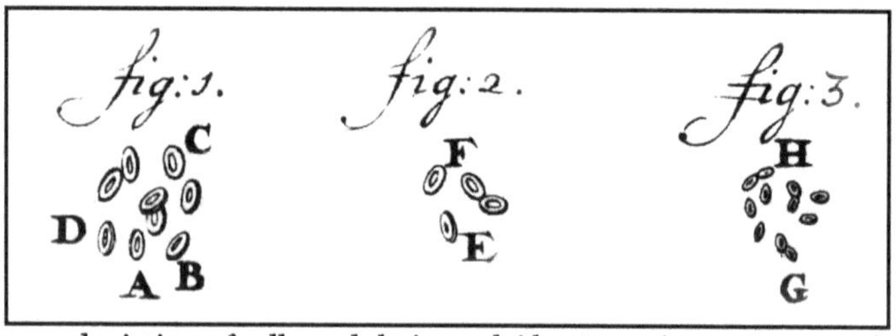

Oldest known depiction of cells and their nuclei by Antonie van Leeuwenhoek, 1719.

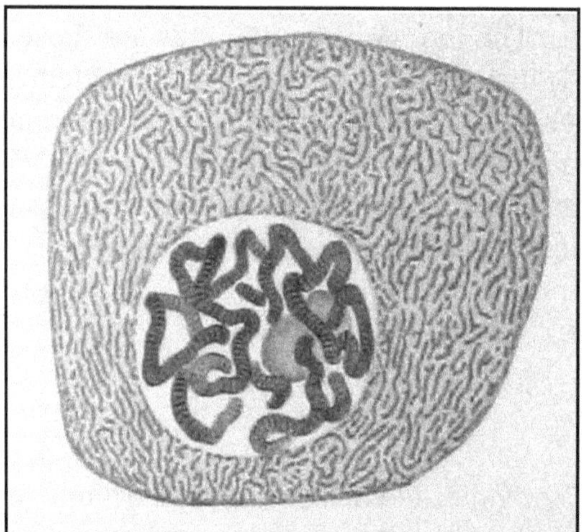

Drawing of a *Chironomus* salivary gland cell published by Walther Flemming in 1882. The nucleus contains Polytene chromosomes.

The nucleus was the first organelle to be discovered. What is most likely the oldest preserved drawing dates back to the early microscopist Antonie van Leeuwenhoek (1632–1723). He observed a "lumen", the nucleus, in the red blood cells of salmon. Unlike mammalian red blood cells, those of other vertebrates still contain nuclei.

The nucleus was also described by Franz Bauer in 1804 and in more detail in 1831 by Scottish botanist Robert Brown in a talk at the Linnean Society of London. Brown was studying orchids under microscope when he observed an opaque area, which he called the "areola" or "nucleus", in the cells of the flower's outer layer.

He did not suggest a potential function. In 1838, Matthias Schleiden proposed that the nucleus plays a role in generating cells, thus he introduced the name "cytoblast" (cell builder). He believed that he had observed new cells assembling around "cytoblasts". Franz Meyen was a strong opponent of this view, having already described cells multiplying by division and believing that many cells would have no nuclei. The idea that cells can be generated de novo, by the "cytoblast" or otherwise, contradicted work by Robert Remak (1852) and Rudolf Virchow (1855) who decisively propagated the new paradigm that cells are generated solely by cells ("Omnis cellula e cellula"). The function of the nucleus remained unclear.

Between 1877 and 1878, Oscar Hertwig published several studies on the fertilization of sea urchin eggs, showing that the nucleus of the sperm enters the oocyte and fuses with its nucleus. This was the first time it was suggested that an individual develops from a (single) nucleated cell. This was in contradiction to Ernst Haeckel's theory that the complete phylogeny of a species would be repeated during embryonic development, including generation of the first nucleated cell from a "monerula", a structureless mass of primordial mucus ("Urschleim"). Therefore, the necessity of the sperm nucleus for fertilization was discussed for quite some time. However, Hertwig confirmed his observation in other animal groups, including amphibians and molluscs. Eduard Strasburger produced the same results for plants in 1884. This paved the way to assign the nucleus an important role in heredity. In 1873, August Weismann postulated the equivalence of the maternal and paternal germ *cells* for heredity. The function of the nucleus as carrier of genetic information became clear only later, after mitosis was discovered and the Mendelian rules were rediscovered at the beginning of the 20th century; the chromosome theory of heredity was therefore developed.

Structures

The nucleus is the largest cellular organelle in animal cells. In mammalian cells, the average diameter of the nucleus is approximately 6 micrometres (μm), which occupies about 10% of the total cell volume. The viscous liquid within it is called nucleoplasm (or karyolymph), and is similar in composition to the cytosol found outside the nucleus. It appears as a dense, roughly spherical or irregular organelle. The composition by dry weight of the nucleus is approximately: DNA 9%, RNA 1%, Histone Protein 11%, Residual Protein 14%, Acidic Proteins 65%.

Nuclear Envelope and Pores

The nuclear envelope, otherwise known as nuclear membrane, consists of two cellular membranes, an inner and an outer membrane, arranged parallel to one another and separated by 10 to 50 nanometres (nm). The nuclear envelope completely encloses the nucleus and separates the cell's genetic material from the surrounding cytoplasm, serving as a barrier to prevent macromolecules from diffusing freely between the nucleoplasm and the cytoplasm. The outer nuclear membrane is continuous with the membrane of the rough endoplasmic reticulum (RER), and is similarly studded with ribosomes. The space between the membranes is called the perinuclear space and is continuous with the RER lumen.

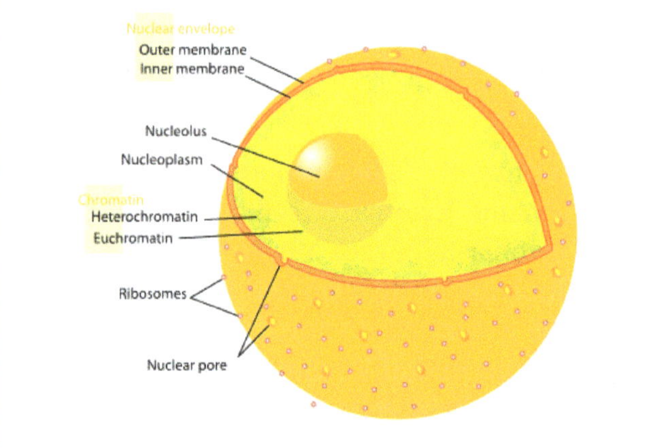

The eukaryotic cell nucleus. Visible in this diagram are the ribosome-studded double membranes of the nuclear envelope, the DNA (complexed as chromatin), and the nucleolus. Within the cell nucleus is a viscous liquid called nucleoplasm, similar to the cytoplasm found outside the nucleus.

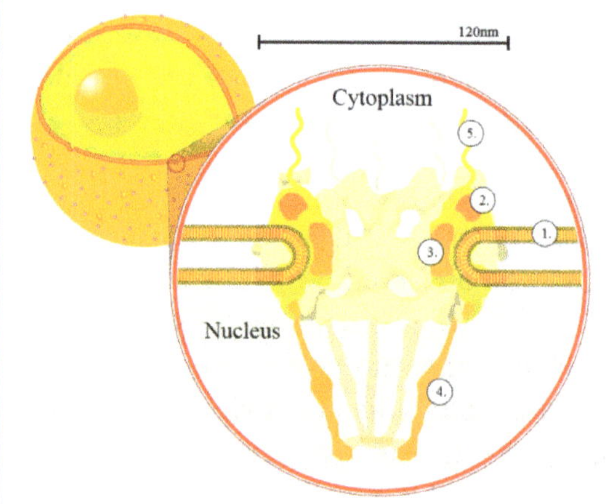

A cross section of a nuclear pore on the surface of the nuclear envelope (1). Other diagram labels show (2) the outer ring, (3) spokes, (4) basket, and (5) filaments.

Nuclear pores, which provide aqueous channels through the envelope, are composed of multiple proteins, collectively referred to as nucleoporins. The pores are about 125 million daltons in molecular weight and consist of around 50 (in yeast) to several hundred proteins (in vertebrates). The pores are 100 nm in total diameter; however, the gap through which molecules freely diffuse is only about 9 nm wide, due to the presence of regulatory systems within the center of the pore. This size selectively allows the passage of small water-soluble molecules while preventing larger molecules, such as nucleic acids and larger proteins, from inappropriately entering or exiting the nucleus. These large molecules must be actively transported into the nucleus instead. The nucleus of a typical mammalian cell will have about 3000 to 4000 pores throughout its envelope, each of which contains an eightfold-symmetric ring-shaped structure at a position where the inner and outer membranes fuse. Attached to the ring is a structure called the *nuclear basket* that extends into the nucleoplasm, and a series of filamentous extensions that reach into the cytoplasm. Both structures serve to mediate binding to nuclear transport proteins.

Most proteins, ribosomal subunits, and some DNAs are transported through the pore complexes in a process mediated by a family of transport factors known as karyopherins. Those karyopherins that mediate movement into the nucleus are also called importins, whereas those that mediate movement out of the nucleus are called exportins. Most karyopherins interact directly with their cargo, although some use adaptor proteins. Steroid hormones such as cortisol and aldosterone, as well as other small lipid-soluble molecules involved in intercellular signaling, can diffuse through the cell membrane and into the cytoplasm, where they bind nuclear receptor proteins that are trafficked into the nucleus. There they serve as transcription factors when bound to their ligand; in the absence of a ligand, many such receptors function as histone deacetylases that repress gene expression.

Nuclear Lamina

In animal cells, two networks of intermediate filaments provide the nucleus with mechanical support: The nuclear lamina forms an organized meshwork on the internal face of the envelope, while less organized support is provided on the cytosolic face of the envelope. Both systems provide structural support for the nuclear envelope and anchoring sites for chromosomes and nuclear pores.

The nuclear lamina is composed mostly of lamin proteins. Like all proteins, lamins are synthesized in the cytoplasm and later transported to the nucleus interior, where they are assembled before being incorporated into the existing network of nuclear lamina. Lamins found on the cytosolic face of the membrane, such as emerin and nesprin, bind to the cytoskeleton to provide structural support. Lamins are also found inside the nucleoplasm where they form another regular structure, known as the *nucleoplasmic veil*, that is visible using fluorescence microscopy. The actual function of the veil is not clear, although it is excluded from the nucleolus and is present during interphase. Lamin structures that make up the veil, such as LEM3, bind chromatin and disrupting their structure inhibits transcription of protein-coding genes.

Like the components of other intermediate filaments, the lamin monomer contains an alpha-helical domain used by two monomers to coil around each other, forming a dimer structure called a coiled coil. Two of these dimer structures then join side by side, in an antiparallel arrangement, to form a tetramer called a *protofilament*. Eight of these protofilaments form a lateral arrangement that is twisted to form a ropelike *filament*. These filaments can be assembled or disassembled in a dynamic manner, meaning that changes in the length of the filament depend on the competing rates of filament addition and removal.

Mutations in lamin genes leading to defects in filament assembly cause a group of rare genetic disorders known as *laminopathies*. The most notable laminopathy is the family of diseases known as progeria, which causes the appearance of premature aging in its sufferers. The exact mechanism by which the associated biochemical changes give rise to the aged phenotype is not well understood.

Chromosomes

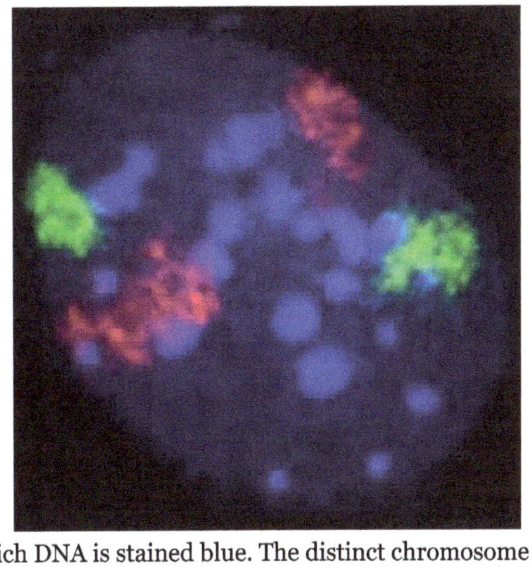

A mouse fibroblast nucleus in which DNA is stained blue. The distinct chromosome territories of chromosome 2 (red) and chromosome 9 (green) are stained with fluorescent in situ hybridization.

The cell nucleus contains the majority of the cell's genetic material in the form of multiple linear DNA molecules organized into structures called chromosomes. Each human cell contains roughly two meters of DNA. During most of the cell cycle these are organized in a DNA-protein complex known as chromatin, and during cell division the chromatin can be seen to form the well-defined chromosomes familiar from a karyotype. A small fraction of the cell's genes are located instead in the mitochondria.

There are two types of chromatin. Euchromatin is the less compact DNA form, and contains genes that are frequently expressed by the cell. The other type, heterochromatin, is the more compact form, and contains DNA that is infrequently transcribed. This structure is further categorized into *facultative* heterochromatin, consisting of genes that are organized as heterochromatin only in certain cell types or at certain stages of development, and *constitutive* heterochromatin that consists of chromosome structural components such as telomeres and centromeres. During interphase the chromatin organizes itself into discrete individual patches, called *chromosome territories*. Active genes, which are generally found in the euchromatic region of the chromosome, tend to be located towards the chromosome's territory boundary.

Antibodies to certain types of chromatin organization, in particular, nucleosomes, have been associated with a number of autoimmune diseases, such as systemic lupus erythematosus. These are known as anti-nuclear antibodies (ANA) and have also been observed in concert with multiple sclerosis as part of general immune system dysfunction. As in the case of progeria, the role played by the antibodies in inducing the symptoms of autoimmune diseases is not obvious.

Nucleolus

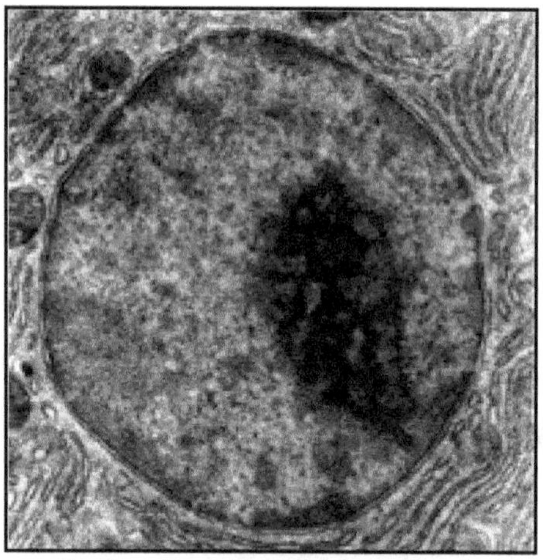

An electron micrograph of a cell nucleus, showing the darkly stained nucleolus.

The nucleolus is a discrete densely stained structure found in the nucleus. It is not surrounded by a membrane, and is sometimes called a *suborganelle*. It forms around tandem repeats of rDNA, DNA coding for ribosomal RNA (rRNA). These regions are called nucleolar organizer regions (NOR). The main roles of the nucleolus are to synthesize rRNA and assemble ribosomes. The structural cohesion of the nucleolus depends on its activity, as ribosomal assembly in the nucleolus results in the transient association of nucleolar components, facilitating further ribosomal assembly, and hence further association. This model is supported by observations that inactivation of rDNA results in intermingling of nucleolar structures.

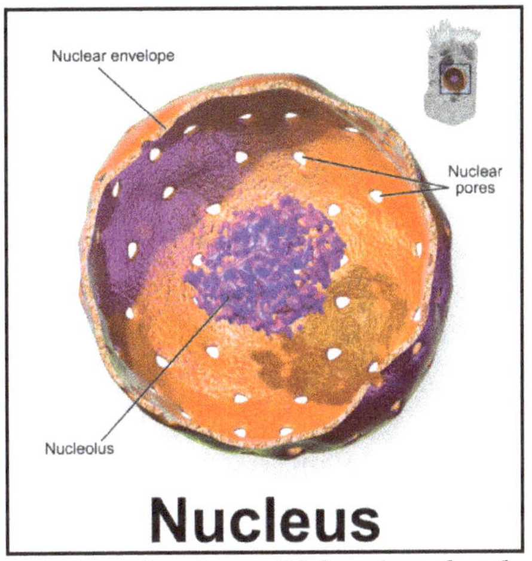

3D rendering of nucleus with location of nucleolus.

In the first step of ribosome assembly, a protein called RNA polymerase I transcribes rDNA, which forms a large pre-rRNA precursor. This is cleaved into the subunits 5.8S, 18S, and 28S rRNA. The transcription, post-transcriptional processing, and assembly of rRNA occurs in the nucleolus, aided by small nucleolar RNA (snoRNA) molecules, some of which are derived from spliced introns from messenger RNAs encoding genes related to ribosomal function. The assembled ribosomal subunits are the largest structures passed through the nuclear pores.

When observed under the electron microscope, the nucleolus can be seen to consist of three distinguishable regions: the innermost *fibrillar centers* (FCs), surrounded by the *dense fibrillar component* (DFC), which in turn is bordered by the *granular component* (GC). Transcription of the rDNA occurs either in the FC or at the FC-DFC boundary, and, therefore, when rDNA transcription in the cell is increased, more FCs are detected. Most of the cleavage and modification of rRNAs occurs in the DFC, while the latter steps involving protein assembly onto the ribosomal subunits occur in the GC.

Other Subnuclear Bodies

Subnuclear structure sizes	
Structure name	**Structure diameter**
Cajal bodies	0.2–2.0 μm
PIKA	5 μm
PML bodies	0.2–1.0 μm
Paraspeckles	0.2–1.0 μm
Speckles	20–25 nm

Besides the nucleolus, the nucleus contains a number of other non-membrane-delineated bodies. These include Cajal bodies, Gemini of coiled bodies, polymorphic interphase karyosomal association (PIKA), promyelocytic leukaemia (PML) bodies, paraspeckles, and splicing speckles. Although little is known about a number of these domains, they are significant in that they show that the nucleoplasm is not a uniform mixture, but rather contains organized functional subdomains.

Other subnuclear structures appear as part of abnormal disease processes. For example, the presence of small intranuclear rods has been reported in some cases of nemaline myopathy. This condition typically results from mutations in actin, and the rods themselves consist of mutant actin as well as other cytoskeletal proteins.

Cajal Bodies and Gems

A nucleus typically contains between 1 and 10 compact structures called Cajal bodies or coiled bodies (CB), whose diameter measures between 0.2 μm and 2.0 μm depending on the cell type and species. When seen under an electron microscope, they resemble balls of tangled thread and are dense foci of distribution for the protein coilin. CBs are involved in a number of different roles relating to RNA processing, specifically small nucleolar RNA (snoRNA) and small nuclear RNA (snRNA) maturation, and histone mRNA modification.

Similar to Cajal bodies are Gemini of Cajal bodies, or gems, whose name is derived from the Gemini constellation in reference to their close "twin" relationship with CBs. Gems are similar in size and shape to CBs, and in fact are virtually indistinguishable under the microscope. Unlike CBs, gems do not contain small nuclear ribonucleoproteins (snRNPs), but do contain a protein called survival of motor neuron (SMN) whose function relates to snRNP biogenesis. Gems are believed to assist CBs in snRNP biogenesis, though it has also been suggested from microscopy evidence that CBs and gems are different manifestations of the same structure. Later ultrastructural studies have shown gems to be twins of Cajal bodies with the difference being in the coilin component; Cajal bodies are SMN positive and coilin positive, and gems are SMN positive and coilin negative.

RAFA and PTF Domains

RAFA domains, or polymorphic interphase karyosomal associations, were first described in microscopy studies in 1991. Their function remains unclear, though they were not thought to be associated with active DNA replication, transcription, or RNA processing. They have been found to often associate with discrete domains defined by dense localization of the transcription factor PTF, which promotes transcription of small nuclear RNA (snRNA).

PML Bodies

Promyelocytic leukaemia bodies (PML bodies) are spherical bodies found scattered throughout the nucleoplasm, measuring around 0.1–1.0 μm. They are known by a number of other names, including nuclear domain 10 (ND10), Kremer bodies, and PML oncogenic domains. PML bodies are named after one of their major components, the promyelocytic leukemia protein (PML). They are often seen in the nucleus in association with Cajal bodies and cleavage bodies. PML bodies belong to the nuclear matrix, an ill-defined super-structure of the nucleus proposed to anchor and regulate many nuclear functions, including DNA replication, transcription, or epigenetic silencing. The PML protein is the key organizer of these domains that recruits an ever-growing number of proteins, whose only common known feature to date is their ability to be SUMOylated. Yet, pml-/- mice (which have their PML gene deleted) cannot assemble nuclear bodies, develop normally and live well, demonstrating that PML bodies are dispensable for most basic biological functions.

Splicing Speckles

Speckles are subnuclear structures that are enriched in pre-messenger RNA splicing factors and are located in the interchromatin regions of the nucleoplasm of mammalian cells. At the fluorescence-microscope level they appear as irregular, punctate structures, which vary in size and shape, and when examined by electron microscopy they are seen as clusters of interchromatin granules. Speckles are dynamic structures, and both their protein and RNA-protein components can cycle continuously between speckles and other nuclear locations, including active transcription sites. Studies on the composition, structure and behaviour of speckles have provided a model for understanding the functional compartmentalization of the nucleus and the organization of the gene-expression machinery splicing snRNPs and other splicing proteins necessary for pre-mRNA processing. Because of a cell's changing requirements, the composition and location of these bodies changes according to mRNA transcription and regulation via phosphorylation of specific proteins. The splicing speckles are also known as nuclear speckles (nuclear specks), splicing factor compartments (SF compartments), interchromatin granule clusters (IGCs), B snurposomes. B snurposomes are found in the amphibian oocyte nuclei and in *Drosophila melanogaster* embryos. B snurposomes appear alone or attached to the Cajal bodies in the electron micrographs of the amphibian nuclei. IGCs function as storage sites for the splicing factors.

Paraspeckles

Discovered by Fox et al. in 2002, paraspeckles are irregularly shaped compartments in the nucleus' interchromatin space. First documented in HeLa cells, where there are generally 10–30 per nucleus, paraspeckles are now known to also exist in all human primary cells, transformed cell lines, and tissue sections. Their name is derived from their distribution in the nucleus; the "para" is short for parallel and the "speckles" refers to the splicing speckles to which they are always in close proximity.

Paraspeckles are dynamic structures that are altered in response to changes in cellular metabolic activity. They are transcription dependent and in the absence of RNA Pol II transcription, the paraspeckle disappears and all of its associated protein components (PSP1, p54nrb, PSP2, CFI(m)68, and PSF) form a crescent shaped perinucleolar cap in the nucleolus. This phenomenon is demonstrated during the cell cycle. In the cell cycle, paraspeckles are present during interphase and during all of mitosis except for telophase. During telophase, when the two daughter nuclei are formed, there is no RNA Pol II transcription so the protein components instead form a perinucleolar cap.

Perichromatin Fibrils

Perichromatin fibrils are visible only under electron microscope. They are located next to the transcriptionally active chromatin and are hypothesized to be the sites of active pre-mRNA processing.

Function

The nucleus provides a site for genetic transcription that is segregated from the location of translation in the cytoplasm, allowing levels of gene regulation that are not available to prokaryotes. The main function of the cell nucleus is to control gene expression and mediate the replication of DNA during the cell cycle.

The nucleus is an organelle found in eukaryotic cells. Inside its fully enclosed nuclear membrane, it contains the majority of the cell's genetic material. This material is organized as DNA molecules, along with a variety of proteins, to form chromosomes.

Cell Compartmentalization

The nuclear envelope allows the nucleus to control its contents, and separate them from the rest of the cytoplasm where necessary. This is important for controlling processes on either side of the nuclear membrane. In most cases where a cytoplasmic process needs to be restricted, a key participant is removed to the nucleus, where it interacts with transcription factors to downregulate the production of certain enzymes in the pathway. This regulatory mechanism occurs in the case of glycolysis, a cellular pathway for breaking down glucose to produce energy. Hexokinase is an enzyme responsible for the first the step of glycolysis, forming glucose-6-phosphate from glucose. At high concentrations of fructose-6-phosphate, a molecule made later from glucose-6-phosphate, a regulator protein removes hexokinase to the nucleus, where it forms a transcriptional repressor complex with nuclear proteins to reduce the expression of genes involved in glycolysis.

In order to control which genes are being transcribed, the cell separates some transcription factor proteins responsible for regulating gene expression from physical access to the DNA until they are activated by other signaling pathways. This prevents even low levels of inappropriate gene expression. For example, in the case of NF-κB-controlled genes, which are involved in most inflammatory responses, transcription is induced in response to a signal pathway such as that initiated by the signaling molecule TNF-α, binds to a cell membrane receptor, resulting in the recruitment of signalling proteins, and eventually activating the transcription factor NF-κB. A nuclear localisation signal on the NF-κB protein allows it to be transported through the nuclear pore and into the nucleus, where it stimulates the transcription of the target genes.

The compartmentalization allows the cell to prevent translation of unspliced mRNA. Eukaryotic mRNA contains introns that must be removed before being translated to produce functional proteins. The splicing is done inside the nucleus before the mRNA can be accessed by ribosomes for translation. Without the nucleus, ribosomes would translate newly transcribed (unprocessed) mRNA, resulting in malformed and nonfunctional proteins.

Gene Expression

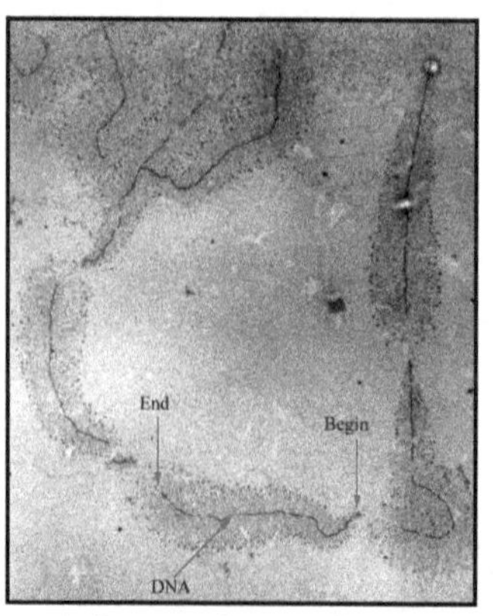

A micrograph of ongoing gene transcription of ribosomal RNA illustrating the growing primary transcripts. "Begin" indicates the 5' end of the DNA, where new RNA synthesis begins; "end" indicates the 3' end, where the primary transcripts are almost complete.

Gene expression first involves transcription, in which DNA is used as a template to produce RNA. In the case of genes encoding proteins, that RNA produced from this process is messenger RNA (mRNA), which then needs to be translated by ribosomes to form a protein. As ribosomes are located outside the nucleus, mRNA produced needs to be exported.

Since the nucleus is the site of transcription, it also contains a variety of proteins that either directly mediate transcription or are involved in regulating the process. These proteins include helicases, which unwind the double-stranded DNA molecule to facilitate access to it, RNA polymerases, which bind to the DNA promoter to synthesize the growing RNA molecule, topoisomerases, which change the amount of supercoiling in DNA, helping it wind and unwind, as well as a large variety of transcription factors that regulate expression.

Processing of Pre-mRNA

Newly synthesized mRNA molecules are known as primary transcripts or pre-mRNA. They must undergo post-transcriptional modification in the nucleus before being exported to the cytoplasm; mRNA that appears in the cytoplasm without these modifications is degraded rather than used for protein translation. The three main modifications are 5' capping, 3' polyadenylation, and RNA splicing. While in the nucleus, pre-mRNA is associated with a variety of proteins in complexes known as heterogeneous ribonucleoprotein particles (hnRNPs). Addition of the 5' cap occurs co-transcriptionally and is the first step in post-transcriptional modification. The 3' poly-adenine tail is only added after transcription is complete.

RNA splicing, carried out by a complex called the spliceosome, is the process by which introns, or regions of DNA that do not code for protein, are removed from the pre-mRNA and the remaining exons connected to re-form a single continuous molecule. This process normally occurs after 5' capping and 3' polyadenylation but can begin before synthesis is complete in transcripts with many exons. Many pre-mRNAs, including those encoding antibodies, can be spliced in multiple ways to produce different mature mRNAs that encode different protein sequences. This process is known as alternative splicing, and allows production of a large variety of proteins from a limited amount of DNA.

Dynamics and Regulation

Nuclear Transport

The entry and exit of large molecules from the nucleus is tightly controlled by the nuclear pore complexes. Although small molecules can enter the nucleus without regulation, macromolecules such as RNA and proteins require association karyopherins called importins to enter the nucleus and exportins to exit. "Cargo" proteins that must be translocated from the cytoplasm to the nucleus contain short amino acid sequences known as nuclear localization signals, which are bound by importins, while those transported from the nucleus to the cytoplasm carry nuclear export signals bound by exportins. The ability of importins and exportins to transport their cargo is regulated by GTPases, enzymes that hydrolyze the molecule guanosine triphosphate to release energy. The key GTPase in nuclear transport is Ran, which can bind either GTP or GDP (guanosine diphosphate), depending on whether it is located in the nucleus or the cytoplasm. Whereas importins depend on RanGTP to dissociate from their cargo, exportins require RanGTP in order to bind to their cargo.

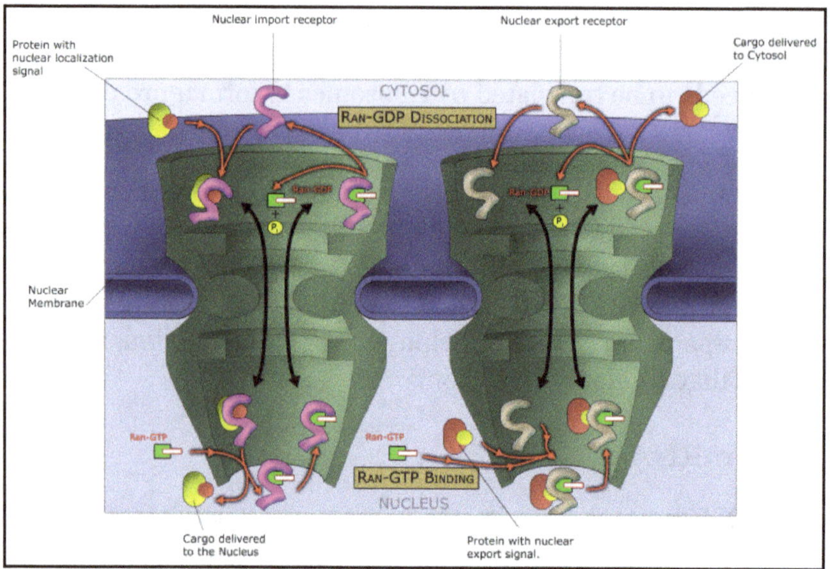

Macromolecules, such as RNA and proteins, are actively transported across the nuclear membrane in a process called the Ran-GTP nuclear transport cycle.

Nuclear import depends on the importin binding its cargo in the cytoplasm and carrying it through the nuclear pore into the nucleus. Inside the nucleus, RanGTP acts to separate the cargo from the importin, allowing the importin to exit the nucleus and be reused. Nuclear export is similar, as the exportin binds the cargo inside the nucleus in a process facilitated by RanGTP, exits through the nuclear pore, and separates from its cargo in the cytoplasm.

Specialized export proteins exist for translocation of mature mRNA and tRNA to the cytoplasm after post-transcriptional modification is complete. This quality-control mechanism is important due to these molecules' central role in protein translation. Mis-expression of a protein due to incomplete excision of exons or mis-incorporation of amino acids could have negative consequences for the cell; thus, incompletely modified RNA that reaches the cytoplasm is degraded rather than used in translation.

Assembly and Disassembly

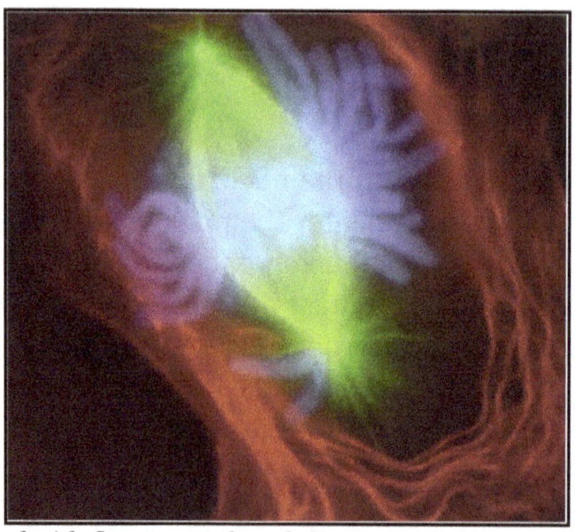

An image of a newt lung cell stained with fluorescent dyes during metaphase. The mitotic spindle can be seen, stained green, attached to the two sets of chromosomes, stained light blue. All chromosomes but one are already at the metaphase plate.

During its lifetime, a nucleus may be broken down or destroyed, either in the process of cell division or as a consequence of apoptosis (the process of programmed cell death). During these events, the structural components of the nucleus — the envelope and lamina — can be systematically degraded. In most cells, the disassembly of the nuclear envelope marks the end of the prophase of mitosis. However, this disassembly of the nucleus is not a universal feature of mitosis and does not occur in all cells. Some unicellular eukaryotes (e.g., yeasts) undergo so-called closed mitosis, in which the nuclear envelope remains intact. In closed mitosis, the daughter chromosomes migrate to opposite poles of the nucleus, which then divides in two. The cells of higher eukaryotes, however, usually undergo open mitosis, which is characterized by breakdown of the nuclear envelope. The daughter chromosomes then migrate to opposite poles of the mitotic spindle, and new nuclei reassemble around them.

At a certain point during the cell cycle in open mitosis, the cell divides to form two cells. In order for this process to be possible, each of the new daughter cells must have a full set of genes, a process requiring replication of the chromosomes as well as segregation of the separate sets. This occurs by the replicated chromosomes, the sister chromatids, attaching to microtubules, which in turn are attached to different centrosomes. The sister chromatids can then be pulled to separate locations in the cell. In many cells, the centrosome is located in the cytoplasm, outside the nucleus; the microtubules would be unable to attach to the chromatids in the presence of the nuclear envelope. Therefore, the early stages in the cell cycle, beginning in prophase and until around prometaphase, the nuclear membrane is dismantled. Likewise, during the same period, the nuclear lamina is also disassembled, a process regulated by phosphorylation of the lamins by protein kinases such as the CDC2 protein kinase. Towards the end of the cell cycle, the nuclear membrane is reformed, and around the same time, the nuclear lamina are reassembled by dephosphorylating the lamins.

However, in dinoflagellates, the nuclear envelope remains intact, the centrosomes are located in the cytoplasm, and the microtubules come in contact with chromosomes, whose centromeric regions are incorporated into the nuclear envelope (the so-called closed mitosis with extranuclear spindle). In many other protists (e.g., ciliates, sporozoans) and fungi, the centrosomes are intranuclear, and their nuclear envelope also does not disassemble during cell division.

Apoptosis is a controlled process in which the cell's structural components are destroyed, resulting in death of the cell. Changes associated with apoptosis directly affect the nucleus and its contents, for example, in the condensation of chromatin and the disintegration of the nuclear envelope and lamina. The destruction of the lamin networks is controlled by specialized apoptotic proteases called caspases, which cleave the lamin proteins and, thus, degrade the nucleus' structural integrity. Lamin cleavage is sometimes used as a laboratory indicator of caspase activity in assays for early apoptotic activity. Cells that express mutant caspase-resistant lamins are deficient in nuclear changes related to apoptosis, suggesting that lamins play a role in initiating the events that lead to apoptotic degradation of the nucleus. Inhibition of lamin assembly itself is an inducer of apoptosis.

The nuclear envelope acts as a barrier that prevents both DNA and RNA viruses from entering the nucleus. Some viruses require access to proteins inside the nucleus in order to replicate and/or assemble. DNA viruses, such as herpesvirus replicate and assemble in the cell nucleus, and exit by budding through the inner nuclear membrane. This process is accompanied by disassembly of the lamina on the nuclear face of the inner membrane.

Disease-related Dynamics

Initially, it has been suspected that immunoglobulins in general and autoantibodies in particular do not enter the nucleus. Now there is a body of evidence that under pathological conditions (e.g. lupus erythematosus) IgG can enter the nucleus.

Nuclei Per Cell

Most eukaryotic cell types usually have a single nucleus, but some have no nuclei, while others have several. This can result from normal development, as in the maturation of mammalian red blood cells, or from faulty cell division.

Anucleated Cells

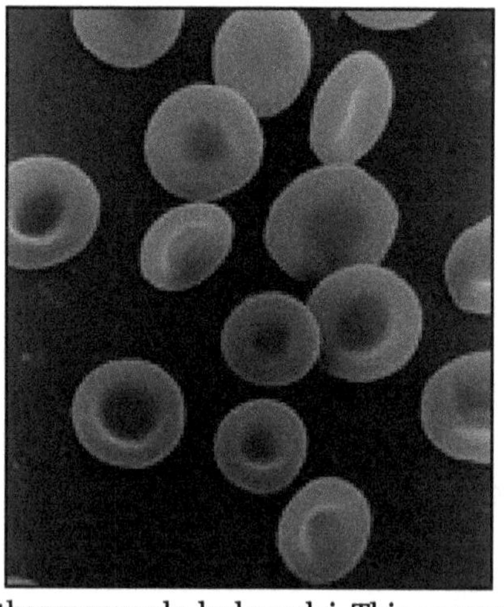

Human red blood cells, like those of other mammals, lack nuclei. This occurs as a normal part of the cells' development.

An anucleated cell contains no nucleus and is, therefore, incapable of dividing to produce daughter cells. The best-known anucleated cell is the mammalian red blood cell, or erythrocyte, which also lacks other organelles such as mitochondria, and serves primarily as a transport vessel to ferry oxygen from the lungs to the body's tissues. Erythrocytes mature through erythropoiesis in the bone marrow, where they lose their nuclei, organelles, and ribosomes. The nucleus is expelled during the process of differentiation from an erythroblast to a reticulocyte, which is the immediate precursor of the mature erythrocyte. The presence of mutagens may induce the release of some immature "micronucleated" erythrocytes into the bloodstream. Anucleated cells can also arise from flawed cell division in which one daughter lacks a nucleus and the other has two nuclei.

In flowering plants, this condition occurs in sieve tube elements.

Multinucleated Cells

Multinucleated cells contain multiple nuclei. Most acantharean species of protozoa and some fungi in mycorrhizae have naturally multinucleated cells. Other examples include the intestinal parasites in the genus *Giardia*, which have two nuclei per cell. In humans, skeletal muscle cells, called myocytes and syncytium, become multinucleated during development; the resulting arrangement of nuclei near the periphery of the cells allows maximal intracellular space for myofibrils. Multinucleated and binucleated cells can also be abnormal in humans; for example, cells arising from the fusion of monocytes and macrophages, known as giant multinucleated cells, sometimes accompany inflammation and are also implicated in tumor formation.

A number of dinoflagellates are known to have two nuclei. Unlike other multinucleated cells these nuclei contain two distinct lineages of DNA: one from the dinoflagellate and the other from a symbiotic diatom. The mitochondria and the plastids of the diatom somehow remain functional.

Evolution

As the major defining characteristic of the eukaryotic cell, the nucleus' evolutionary origin has been the subject of much speculation. Four major hypotheses have been proposed to explain the existence of the nucleus, although none have yet earned widespread support.

The first model known as the "syntrophic model" proposes that a symbiotic relationship between the archaea and bacteria created the nucleus-containing eukaryotic cell. (Organisms of the Archaea and Bacteria domain have no cell nucleus.) It is hypothesized that the symbiosis originated when ancient archaea, similar to modern methanogenic archaea, invaded and lived within bacteria similar to modern myxobacteria, eventually forming the early nucleus. This theory is analogous to the accepted theory for the origin of eukaryotic mitochondria and chloroplasts, which are thought to have developed from a similar endosymbiotic relationship between proto-eukaryotes and aerobic bacteria. The archaeal origin of the nucleus is supported by observations that archaea and eukarya have similar genes for certain proteins, including histones. Observations that myxobacteria are motile, can form multicellular complexes, and possess kinases and G proteins similar to eukarya, support a bacterial origin for the eukaryotic cell.

A second model proposes that proto-eukaryotic cells evolved from bacteria without an endosymbiotic stage. This model is based on the existence of modern planctomycetes bacteria that possess a nuclear structure with primitive pores and other compartmentalized membrane structures. A similar proposal states that a eukaryote-like cell, the chronocyte, evolved first and phagocytosed archaea and bacteria to generate the nucleus and the eukaryotic cell.

The most controversial model, known as *viral eukaryogenesis*, posits that the membrane-bound nucleus, along with other eukaryotic features, originated from the infection of a prokaryote by a virus. The suggestion is based on similarities between eukaryotes and viruses such as linear DNA strands, mRNA capping, and tight binding to proteins (analogizing histones to viral envelopes). One version of the proposal suggests that the nucleus evolved in concert with phagocytosis to form an early cellular "predator". Another variant proposes that eukaryotes originated from early archaea infected by poxviruses, on the basis of observed similarity between the DNA polymerases in modern poxviruses and eukaryotes. It has been suggested that the unresolved question of the evolution of sex could be related to the viral eukaryogenesis hypothesis.

A more recent proposal, the *exomembrane hypothesis*, suggests that the nucleus instead originated from a single ancestral cell that evolved a second exterior cell membrane; the interior membrane enclosing the original cell then became the nuclear membrane and evolved increasingly elaborate pore structures for passage of internally synthesized cellular components such as ribosomal subunits.

Nuclear Lamina

The nuclear lamina is a dense (~30 to 100 nm thick) fibrillar network inside the nucleus of most cells. It is composed of intermediate filaments and membrane associated proteins. Besides providing mechanical support, the nuclear lamina regulates important cellular events such as DNA replication and cell division. Additionally, it participates in chromatin organization and it anchors the nuclear pore complexes embedded in the nuclear envelope.

The nuclear lamina is associated with the inner face of the double bilayer nuclear envelope, whereas the outer face is continuous with the endoplasmic reticulum. The nuclear lamina is similar in structure to the nuclear matrix, but the latter extends throughout the nucleoplasm.

Structure and Composition

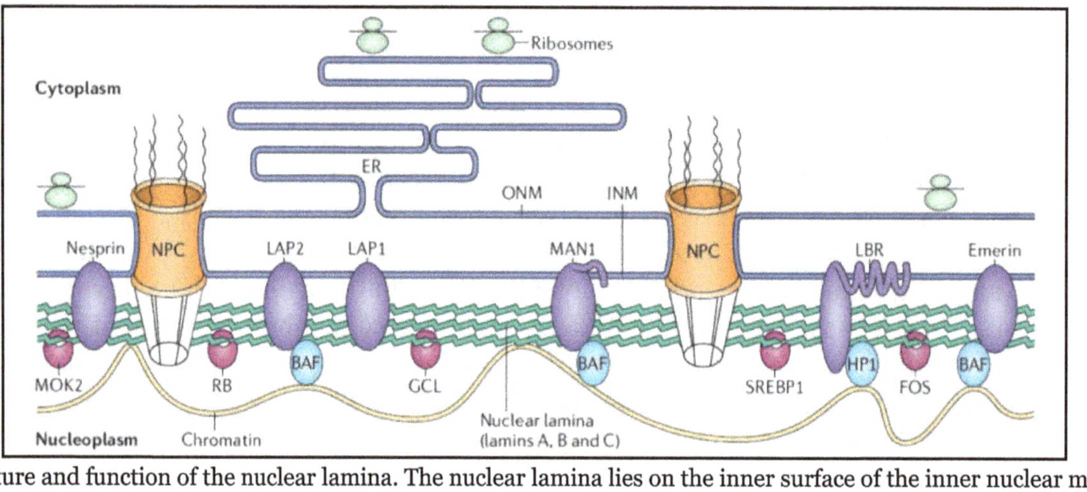

Structure and function of the nuclear lamina. The nuclear lamina lies on the inner surface of the inner nuclear membrane (INM), where it serves to maintain nuclear stability, organize chromatin and bind nuclear pore complexes (NPCs) and a steadily growing list of nuclear envelope proteins (purple) and transcription factors (pink). Nuclear envelope proteins that are bound to the lamina include nesprin, emerin, lamina-associated proteins 1 and 2 (LAP1 and LAP2), the lamin B receptor (LBR) and MAN1. Transcription factors that bind to the lamina include the retinoblastoma transcriptional regulator (RB), germ cell-less (GCL), sterol response element binding protein (SREBP1), FOS and MOK2. Barrier to autointegration factor (BAF) is a chromatin-associated protein that also binds to the nuclear lamina and several of the aforementioned nuclear envelope proteins. Heterochromatin protein 1 (HP1) binds both chromatin and the LBR. ONM, outer nuclear membrane. Coutinho et al. Immunity & Ageing 2009.

The nuclear lamina consists of two components, lamins and nuclear lamin-associated membrane proteins. The lamins are type V intermediate filaments which can be categorized as either A-type (lamin A, C) or B-type(lamin B_1, B_2) according to homology of their DNA sequences, biochemical properties and cellular localization during the cell cycle. Type V intermediate filaments differ from cytoplasmic intermediate filaments in the way that they have an extended rod domain (42 amino acid longer), that they all carry a nuclear localization signal (NLS) at their C-terminus and that they display typical tertiary structures. Lamin polypeptides have an almost complete α-helical conformation with multiple α-helical domains separated by non-α-helical linkers that are highly conserved in length and amino acid sequence. Both the C-terminus and the N-terminus are non α-helical, with the C-terminus displaying a globular structure. Their molecular weight ranges from 60 to 80 kilodaltons (kDa). In the amino acid sequence of a nuclear lamin, there are also two phosphoacceptor sites present, flanking the central rod domain. A phosphorylation event at the onset of mitosis leads to a conformational change which causes the disassembly of the nuclear lamina.

In the vertebrate genome, lamins are encoded by three genes. By alternative splicing, at least seven different polypeptides (splice variants) are obtained, some of which are specific for germ cells and play an important role in the chromatin reorganisation during meiosis. Not all organisms have the same number of lamin encoding genes; *Drosophila melanogaster* for example has only 2 genes, whereas *Caenorhabditis elegans* has only one.

The presence of lamin polypeptides is an exclusive property of Metazoan organisms.

The nuclear lamin-associated membrane proteins are either integral or peripheral membrane proteins. The most important are lamin associated polypeptide 1 and 2 (LAP_1, LAP2), emerin, lamin B-receptor (LBR), otefin and MAN_1. Due to their positioning within or their association with the inner membrane, they mediate the attachment of the nuclear lamina to the nuclear envelope.

Role and Interaction Aspects

The nuclear lamina is assembled by interactions of two lamin polypeptides in which the α-helical regions are wound around each other to form a two stranded α-helical coiled-coil structure, followed by a head-to-tail association of the multiple dimers. The linearly elongated polymer is extended laterally by a side-by-side association of polymers, resulting in a 2D structure underlying the nuclear envelope. Next to providing mechanical support to the nucleus, the nuclear lamina plays an essential role in chromatin organization, cell cycle regulation, DNA replication, DNA repair, cell differentiation and apoptosis.

Chromatin Organization

The non-random organization of the genome strongly suggests that the nuclear lamina plays a role in chromatin organization. It has been shown that lamin polypeptides have an affinity for binding chromatin through their α-helical (rod like) domains at specific DNA sequences called *matrix attachment regions* (MAR). A MAR has a length of approximately 300–1000 bp and has a high A/T content. Lamin A and B can also bind core histones through a sequence element in their tail domain.

Cell Cycle Regulation

At the onset of mitosis (prophase, prometaphase), the cellular machinery is engaged in the disassembly of various cellular components including structures such as the nuclear envelope, the nuclear lamina and the nuclear pore complexes. This nuclear breakdown is necessary to allow the mitotic spindle to interact with the (condensed) chromosomes and to bind them at their kinetochores.

These different disassembly events are initiated by the cyclin B/Cdk1 protein kinase complex (MPF). Once this complex is activated, the cell is forced into mitosis, by the subsequent activation and regulation of other protein kinases or by direct phosphorylation of structural proteins involved in this cellular reorganisation. After phosphorylation by cyclin B/Cdk_1, the nuclear lamina depolymerises and B-type lamins stay associated with the fragments of the nuclear envelope whereas A-type lamins remain completely soluble throughout the remainder of the mitotic phase.

The importance of the nuclear lamina breakdown at this stage is underlined by experiments where inhibition of the disassembly event leads to a complete cell cycle arrest.

At the end of mitosis, (anaphase, telophase) there is a nuclear reassembly which is highly regulated in time, starting with the association of 'skeletal' proteins on the surface of the still partially condensed chromosomes, followed by nuclear envelope assembly. Novel nuclear pore complexes are formed through which nuclear lamins are actively imported by use of their NLS. This typical hierarchy raises the question whether the nuclear lamina at this stage has a stabilizing role or some regulative function, for it is clear that it plays no essential part in the nuclear membrane assembly around chromatin.

Embryonic Development and Cell Differentiation

The presence of lamins in embryonic development is readily observed in various model organisms such as *Xenopus laevis*, the chick and mammals. In *Xenopus laevis*, five different types were identified which are present in different expression patterns during the different stages of the embryonic development. The major types are LI and LII, which are considered homologs of lamin B_1 and B2. LA are considered homologous to lamin A and LIII as a B-type lamin. A fourth type exists and is germ cell specific.

In the early embryonic stages of the chick, the only lamins present are B-type lamins. In further stages, the expression pattern of lamin B_1 decreases and there is a gradual increase in the expression of lamin A. Mammalian development seems to progress in a similar way. In the latter case as well it is the B-type lamins that are expressed in the early stages. Lamin B1 reaches the highest expression level, whereas the expression of B2 is relatively constant in the early stages and starts to increase after cell differentiation. With the development of the different kinds of tissue in a relatively advanced developmental stage, there is an increase in the levels of lamin A and lamin C.

These findings would indicate that in its most basic form, a functional nuclear lamina requires only B-type lamins.

DNA Replication

Various experiments show that the nuclear lamina plays a part in the elongation phase of DNA replication. It has been suggested that lamins provide a scaffold, essential for the assembly of the elongation complexes, or that it provides an initiation point for the assembly of this nuclear scaffold.

Not only nuclear lamina associated lamins are present during replication, but free lamin polypeptides are present as well and seem to have some regulative part in the replication process.

DNA Repair

Repair of DNA double-strand breaks can occur by either of two processes, non-homologous end joining (NHEJ) or homologous recombination (HR). A-type lamins promote genetic stability by maintaining levels of proteins that have key roles in NHEJ and HR. Mouse cells deficient for maturation of prelamin A show increased DNA damage and chromosome aberrations and are more sensitive to DNA damaging agents.

Apoptosis

Apoptosis (cellular suicide) is of the highest importance in homeostasis of tissue and in defending the organism against invasive entry of viruses or other pathogens. Apoptosis is a highly regulated process in which the nuclear lamina is disassembled in an early stage.

In contrast to the phosphorylation-induced disassembly during mitosis, the nuclear lamina is degraded by proteolytic cleavage, and both the lamins and the nuclear lamin-associated membrane proteins are targeted. This proteolytic activity is performed by members of the caspase-protein family who cleave the lamins after aspartic acid (Asp) residues.

Laminopathies

Defects in the genes encoding for nuclear lamin (such as lamin A and lamin B_1) have been implicated in a variety of diseases (laminopathies) such as:

- Emery–Dreifuss muscular dystrophy - A muscle wasting disease.
- Progeria - Premature aging.
- Restrictive dermopathy - A disease associated with extremely tight skin and other severe neonatal abnormalities.

Nuclear Transport

The entry and exit of large molecules from the cell nucleus is tightly controlled by the nuclear pore complexes (NPCs). Although small molecules can enter the nucleus without regulation, macromolecules such as RNA and proteins require association with transport factors like karyopherins called importins to enter the nucleus and exportins to exit.

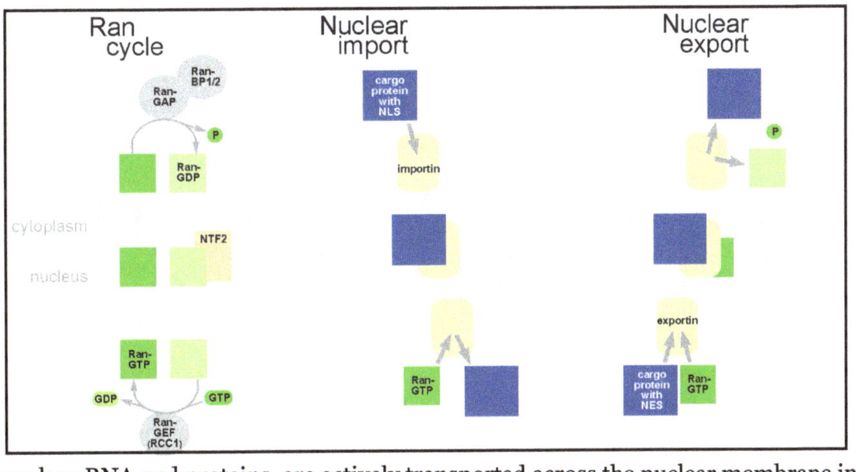

Macromolecules, such as RNA and proteins, are actively transported across the nuclear membrane in a process called the Ran-GTP nuclear transport cycle.

Protein that must be imported to the nucleus from the cytoplasm carry nuclear localization signals (NLS) that are bound by importins. A NLS is a sequence of amino acids that acts as a tag. They are diverse in their composition and most commonly hydrophilic, although hydrophobic sequences have also been documented. Proteins, transfer RNA, and assembled ribosomal subunits are exported from the nucleus due to association with exportins, which bind signaling sequences called nuclear export signals (NES). The ability of both importins and exportins to transport their cargo is regulated by the small Ras related GTPase, Ran.

GTPases are enzymes that bind to a molecule called guanosine triphosphate (GTP) which they then hydrolyze to create guanosine diphosphate (GDP) and release energy. Ran is in a different conformation depending on whether it is bound to GTP or GDP. In its GDP bound state, Ran is capable of binding karyopherins (importins and exportins). Importins release cargo upon binding to RanGTP, while exportins must bind RanGTP to form a ternary complex with their export cargo. The dominant nucleotide binding state of Ran depends on whether it is located in the nucleus (RanGTP) or the cytoplasm (RanGDP).

Nuclear Import

Importin proteins bind their cargo in the cytoplasm, after which they are able to interact with the nuclear pore complex and pass through its channel. Once inside the nucleus, interaction with Ran-GTP causes a conformational change in the importin that causes it to dissociate from its cargo. The resulting complex of importin and Ran-GTP then translocates to the cytoplasm, where a protein called Ran Binding Protein (RanBP) separates Ran-GTP from importin. Separation allows access to a GTPase activating protein (GAP) that binds Ran-GTP and induces the hydrolysis of GTP to GDP. The Ran-GDP produced from this process now binds the nuclear transport factor NUTF2 which returns it to the nucleoplasm. Now in the nucleus, the Ran-GDP interacts with a guanine nucleotide exchange factor (GEF) which replaces the GDP with GTP, resulting again in Ran-GTP, and beginning the cycle anew.

Nuclear Export

Nuclear export roughly reverses the import process; in the nucleus, the exportin binds the cargo and Ran-GTP and diffuses through the pore to the cytoplasm, where the complex dissociates. Ran-GTP binds GAP and hydrolyzes GTP, and the resulting Ran-GDP complex is restored to the nucleus where it exchanges its bound ligand for GTP. Hence, whereas importins depend on RanGTP to dissociate from their cargo, exportins require RanGTP in order to bind to their cargo.

A specialized mRNA exporter protein moves mature mRNA to the cytoplasm after post-transcriptional modification is complete. This translocation process is actively dependent on the Ran protein, although the specific mechanism is not yet well understood. Some particularly commonly transcribed genes are physically located near nuclear pores to facilitate the translocation process.

tRNA export is also dependent on the various modifications it undergoes, thus preventing export of improperly functioning tRNA. This quality control mechanism is important due to tRNA's central role in translation, where it is involved in adding amino acids to a growing peptide chain. The tRNA exporter in vertebrates is called *exportin-t*. Exportin-t binds directly to its tRNA cargo in the nucleus, a process promoted by the presence of RanGTP. Mutations that affect tRNA's structure inhibit its ability to bind to exportin-t, and consequentially, to be exported, providing the cell with another quality control step. As described above, once the complex has crossed the envelope it dissociates and releases the tRNA cargo into the cytosol.

Protein Shuttling

Many proteins are known to have both NESs and NLSs and thus shuttle constantly between the nucleus and the cytosol. In certain cases one of these steps (i.e., nuclear import or nuclear export) is regulated, often by post-translational modifications.

Protein shuttling can be assessed using a *heterokaryon fusion assay.*

Properties of Cytoplasmic Matrix

The most of the physical properties of the matrix are due to its colloidal nature. The cytosol shows Tyndal effect (light scattering by particle in colloidal solution) and Brownian motion (random moving of particles). Due to the phase reversal property of the cytoplasmic matrix, the intracellular streaming or movement of the matrix takes place and is known as the cyclosis. The cyclosis usually occurs in the sol-phase of the matrix and is effected by the hydrostatic pressure, temperature, pH, viscosity, etc. Cyclosis has been observed in most animal and plant cells. The amoeboid movement depends directly on the cyclosis. The amoeboid movement occurs in the protozoans, leucocytes, epithelia, mesenchymal and other cells. Due to cyclosis matrix moves these pseudopodia and this causes forward motion of the cell. The cytoplasmic matrix being a liquid possesses the property of surface tension. The proteins and lipids of matrix have less surface tension, therefore, occur at the surface and form the membrane, while the chemical substances such as NaCl have high surface tension, therefore, occur in deeper part of the matrix. Besides surface tension and adsorption, the matrix possesses other mechanical properties, *e.g.*, elasticity, contractility, rigidity and viscosity which provide to the matrix many physiological utilities. The colloidal system due to its stable phase gives polarity of the cell matrix which cannot be altered by centrifugation of other mechanical means. The matrix has a definite pH value and it does not tolerate significant variations in its pH. Yet various metabolic activities produce small amount of excess acids or bases which is maintained by certain chemical compounds as carbonate-bicarbonate buffers. The matrix is a living substance and possesses various biological properties as irritability, conductivity, movement, metabolic activity, growth and reproduction.

Organelles

Cytoplasm contains all the organelles like nucleus, mitochondria, endoplasmic reticulum, lysosomes and Golgi apparatus. Besides, it also contains chloroplast in plant cells. Each organelle is bounded by a lipid membrane, and has specific functions.

Cytoplasmic Inclusions

Some insoluble suspended substances found in cytosol. They are basically granules of starch and glycogen, and they can store energy. Besides, crystals of some minerals and lipid droplets can also be found in cytoplasm. Lipid droplets act as storage site of fatty acid and steroids.

Functions of Cytoplasm

Cytoplasm is the site of many vital biochemical reactions crucial for maintaining life:

1. It is the place where cell expansion and growth take place.
2. It provides a medium in which the organelles can remain suspended.
3. Besides, cytoskeleton found in cytoplasm gives the shape to the cell, and facilitates its movement.
4. It also assists the movement of different elements found within the cell. The enzymes found in the cytoplasm breaks down the macromolecules into small parts so that it can be easily used by the other organelles like mitochondria. For example, mitochondria cannot use glucose present in the cell, unless it is broken down by the enzymes into pyruvate. They act as catalysts in glycolysis, as well as in the synthesis of fatty acid, sugar and amino acid.
5. Cell reproduction, protein synthesis, anaerobic glycolysis, cytokinesis are some other vital functions that are carried out in cytoplasm.

However, the smooth operation of all these functions depend on the existence of cytoplasm, as it provides the medium for carrying out these vital processes.

Cell Organelles

Nucleus

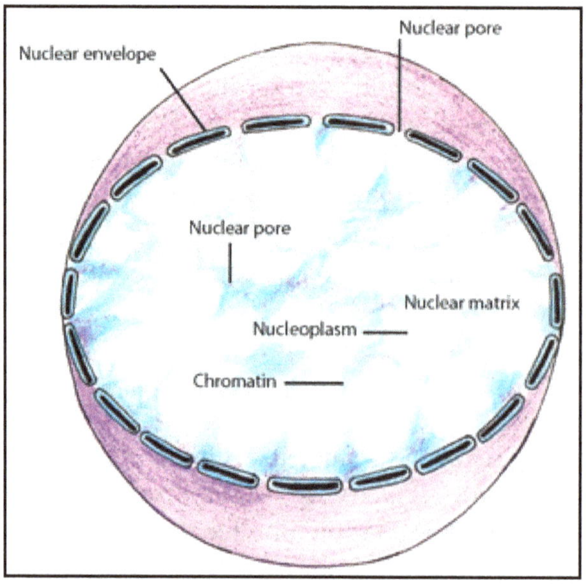

The schematic representation of nucleus.

Nucleus means kernel and was the first organelle to be discovered. It was discovered and named by Robert Brown in 1833 in the plant cells and is recognized as a constant feature of all animal and plant cells. Certain eukaryotic cells such as the mature sieve tubes of higher plants and mammalian erythrocytes contain no nucleus. It is the largest cellular organelle in eukaryotes. Prokaryotic cells lack nucleus and is complemented by nucleoid. In mammalian cells, the average diameter of the nucleus is approximately 6 micrometers (μm), occupying about 10% of the total cell volume. The contents of the nucleus are DNA genome, RNA synthetic apparatus, and a fibrous matrix. It is surrounded by two membranes, each one a phospholipid bilayer containing many different types of proteins. The inner nuclear membrane defines the nucleus itself. In most cells, the outer nuclear membrane is continuous with the rough endoplasmic reticulum, and the space between the inner and outer nuclear membranes is continuous with the lumen of the rough endoplasmic reticulum. The two nuclear membranes appear to fuse at nuclear pores, the ringlike complexes composed of specific membrane proteins through which material moves between the nucleus and the cytosol. It contains cell's genetic material, organized as multiple long linear DNA molecules in complex with histones, to form chromosomes. The genes within these chromosomes are the cell's nuclear genome. The function is to maintain the integrity of the genes that controls the activities of the cell by regulating gene expression. The schematic presentation of nucleus is in Figure.

In a growing or differentiating cell, the nucleus is metabolically active, replicating DNA and synthesizing rRNA, tRNA, and mRNA. Within the nucleus mRNA binds to specific proteins, forming ribonucleoprotein particles. Most of the cell's ribosomal RNA is synthesized in the nucleolus, a subcompartment of the nucleus that is not bounded by a phospholipid membrane. Some ribosomal proteins are added to ribosomal RNAs within the nucleolus as well. The finished or partly finished ribosomal subunits, as well as tRNAs and mRNA-containing particles, pass through a nuclear pore into the cytosol for use in protein synthesis. In a nucleus that is not dividing, the chromosomes are dispersed and not dense enough to be observed in the light microscope. Only during cell division are individual chromosomes visible by light microscopy. In the electron microscope, the nonnucleolar regions of the nucleus, called the nucleoplasm,can be seen to have dark and light staining areas. The dark areas, which are often closely associated with the nuclear membrane, contain condensed concentrated DNA, called heterochromatin. Fibrous proteins called lamins form a two-dimensional network along the inner surface of the inner membrane, giving it shape and apparently binding DNA to it. The breakdown of this network occurs early in cell division.

Cell Nucleus: Ultrastructure

The structure of a cell nucleus consists of a nuclear membrane (nuclear envelope), nucleoplasm, nucleolus, and chromosomes. Nucleoplasm, also known as karyoplasm, is the matrix present inside the nucleus.

- Nuclear Membrane

It is a double-membrane structure each 5–10 nm thick . Numerous pores occur in the envelope, allowing RNA and other chemicals to pass, but not the DNA. Because the nuclear membrane is impermeable to most molecules, nuclear pores are required to allow movement of molecules across the envelope. These pores cross both of the membranes, providing a channel that allows free movement of small molecules and ions. The movement of larger molecules such as protein requires active transport regulated by carrier proteins. Figure illustrates the nuclear membrane. The nuclear envelope (or perinuclear cisterna) encloses the DNA and defines the nuclear compartment of interphase and prophase nuclei. The spherical inner nuclearmembrane contains specific proteins that act as binding sites for the supporting fibrous sheath ofintermediate filaments (IF), called nuclear lamina. Nuclear lamina has contact with the chromatin (or chromosomes) and nuclear RNAs. The inner nuclear membrane is surrounded by the outer nuclear membrane, which closely resembles the membrane of the endoplasmic reticulum, that is continuous with it. Like the membrane of the rough ER, the outer surface of outer nuclear membrane is generally studded with ribosomes engaged in protein synthesis. The proteins made on these ribosomes are transported into space between the inner and outer nuclear membrane, called perinuclear space. The perinuclear space is a 10 to 50 nm wide fluid-filled compartment which is continuous with the ER lumen and may contain fibres, crystalline deposits, lipid droplets or electrondense material. Nuclear pores and nucleocytoplasmic traffic. The nuclear envelope in all eukaryotic forms, from yeasts to humans, is perforated by nuclear pores which have the following structure and function:

Structure of nuclear pores: Nuclear pores appear circular in surface view and have a diameter between 10nm to 100 nm. Previously it was believed that a diaphragm made of amorphous to fibrillar material extends across each pore limiting free transfer of material. Such a diaphragm called annulus has been observed in animal cells, but lack in plant cells. Recent electron microscopic studies have revealed that a nuclear pore has far more complex structure, so it is called nuclear pore complex with an estimated molecular weight of 50 to 100 million daltons. Negative staining techniques have demonstrated that pore complexes have an eight-fold or octagonal symmetry.

Nuclear Pore density: In nuclei of mammals it has been calculated that nuclear pores account for 5 to 15 per cent of the surface area of the nuclear membrane. In amphibian oocytes, certain plant cells and protozoa, the surface occupied by the nuclear pores may be as high as 20 to 36 per cent.

Arrangement of nuclear pores on nuclear envelope: In somatic cells, the nuclear pores are evenly or randomly distributed over the surface of nuclear envelope. However, pore arrangement in other cell types is not random but rather range from rows (spores of *Eqisetum*) to Clusters (oocytes of *Xenopus laevis*) to hexagonal (Malpighian tubules of leaf hoppers) packing order.

Nucleo-cytoplasmic traffic: Quite evidently there is considerable trafficking across the nuclear envelope during interphase. Ions, nucleotides and structural, catalytic and regulatory proteins are imported from the cytosol (cytoplasmic matrix); mRNA, tRNA are exported to the cytosol (cytoplasmic matrix). However, one of the main functions of the nuclear envelope is to prevent the entrance of active ribosomes into the nucleus.

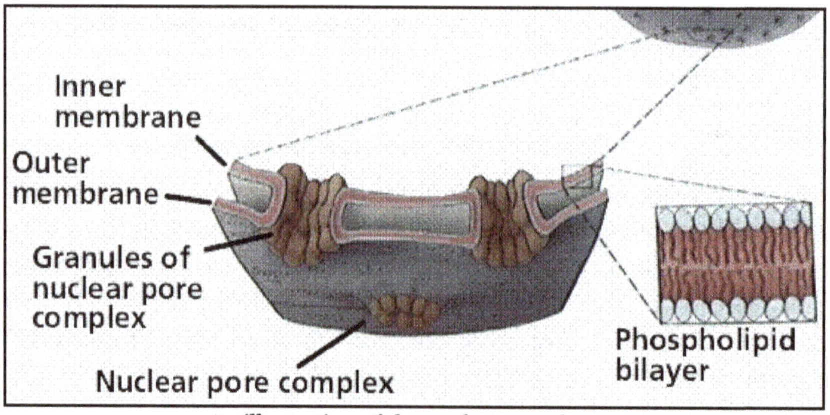

An illustration of the nuclear membrane

Nucleoplasm: The space between the nuclear envelope and the nucleolus is filled by a transparent, semi-solid, granular and slightly acidophilic ground substance or the matrix known as the nuclear sap or nucleoplasm or karyolymph. The nuclear components such as the chromatin threads and the nucleolus remain suspended in the nucleoplasm which is composed mainly of nucleoproteins but it also contains other inorganic and organic substances, namely nucleic acids, proteins, enzymes and minerals. The most common nucleic acids of the nucleoplasm are the DNA and RNA. The nucleoplasm contains many types of complex proteins categorized into: (i) Basic proteins. The proteins which take basic stain are known as the basic proteins. The most important basic proteins of the nucleus are nucleoprotamines and the nucleohistones. (ii) Non-histone or Acidic proteins. The acidic proteins either occur in the nucleoplasm or in the chromatin. The most abundant acidic proteins of the euchromatin (a type of chromatin) are the phosphoproteins. The nucleoplasm contains many enzymes which are necessary for the synthesis of the DNA and RNA. Most of the nuclear enzymes are composed of non-histone (acidic) proteins. The most important nuclear enzymes are the DNA polymerase, RNA polymerase, NAD synthetase, nucleoside triphosphatase, adenosine diaminase, nucleoside phosphorylase, guanase, aldolase, enolase, 3-phosphoglyceraldehyde dehydrogenase and pyruvate kinase. The nucleoplasm also contains certain cofactors and coenzymes such as ATP and acetyl CoA. The nucleoplasm has small lipid content. The nucleoplasm also contains several inorganic compounds such as phosphorus, potassium, sodium, calcium and magnesium. The chromatin comparatively contains large amount of these minerals than the nucleoplasm.

The nucleoplasm contains many thread-like, coiled and much elongated structures which take readily the basic stains such as the basic fuchsin. These thread-like structures are known as the chromatin (*chrome*=colour) substance or chromatin fibres.

Nucleolus: Most cells contain in their nuclei one or more prominent spherical colloidal acidophilic bodies, called nucleoli. However, cells of bacteria and yeast lack nucleolus. The nucleolus is mainly involved in the assembly of ribosomes. After being produced in the nucleolus, ribosomes are exported to the cytoplasm where they translate mRNA. Some of the eukaryotic organisms have nucleus that contains up to four nucleoli. The nucleolus plays an indirect role in protein synthesis by producing ribosomes. Nucleolus disappears when a cell undergoes division and is reformed after the completion of cell-division. The size of the nucleolus is found to be related with the synthetic activity of the cell. Therefore, the cells with little or no synthetic activities, sperm cells, blastomeres, muscle cell, etc., are found to contain smaller or no nucleoli, while the oocytes, neurons and secretory cells which synthesize the proteins or other substances contain comparatively large-sized nucleoli. The number of the nucleoli in the nucleus depends on the species and the number of the chromosomes. The number of the nucleoli in the cells may be one, two or four. A nucleolus is often associated with the nucleolar organizer (NO) which represents the secondary constriction of the nucleolar organizing chromosomes, and are 10 in number in human beings. Nucleolar organizer consists of the genes for 18S, 5.8S and 28S rRNAs. The genes for fourth type of rRNA, *i.e.,* 5S rRNA occur outside the nucleolar organizer. Nucleolus is not bounded by any limiting membrane; calcium ions are supposed to maintain its intact organization. Nucleolus also contains some enzymes such as acid phosphatase, nucleoside phosphorylase and NAD^+ synthesizing enzymes for the synthesis of some coenzymes, nucleotides and ribosomal RNA. RNA methylase enzyme which transfers methyl groups to the nitrogen bases occurs in the nucleolus of some cells. Functionally nucleolus is the site where biogenesis of ribosomal subunits (40S and 60S) takes place. In it three types of rRNAs, namely 18S, 5.8S and 28S rRNAs, are transcribed as parts of a much longer precursor molecule (45S transcript) which undergoes processing (RNA splicing) by the help of two types of proteins such as nucleolin and

U3 sn RNP (U3 is a 250 nucleotide containing RNA, sn RNP represents small nuclear ribonucleoprotein). The 5S r RNA is transcribed on the chromosome existing outside the nucleolus and the 70S types of ribosomal proteins are synthesized in the cytoplasm. All of these components of the ribosomes migrate to the nucleolus, where they are assembled into two types of ribosomal subunits which are transported back to the cytoplasm. The smaller (40S) ribosomal subunits are formed and migrate to the cytoplasm much earlier than larger (60S) ribosomal subunits; therefore, nucleolus contains many more incomplete 60S ribosomal subunits than the 40S ribosomal subunits. Such a time lag in the migration of 60S and 40S ribosomal subunits, prevents functional ribosomes from gaining access to the incompletely processed heterogeneous RNA (hn RNA; the precursor of m RNA) molecule inside the nucleus.

Functions of the Nucleus

Speaking about the functions of a cell nucleus, it controls the hereditary characteristics of an organism. This organelle is also responsible for the protein synthesis, cell division, growth, and differentiation. Some important functions carried out by a cell nucleus are:

1. Storage of hereditary material, the genes in the form of long and thin DNA (deoxyribonucleic acid) strands, referred to as chromatins.
2. Storage of proteins and RNA (ribonucleic acid) in the nucleolus.
3. Nucleus is a site for transcription in which messenger RNA (mRNA) are produced for the protein synthesis.
4. Exchange of hereditary molecules (DNA and RNA) between the nucleus and rest of the cell.
5. During the cell division, chromatins are arranged into chromosomes in the nucleus.
6. Production of ribosomes (protein factories) in the nucleolus.
7. Selective transportation of regulatory factors and energy molecules through nuclear pores.

As the nucleus regulates the integrity of genes and gene expression, it is also referred to as the control center of a cell. Overall, the cell nucleus stores all the chromosomal DNA of an organism.

Mitochondrion

The mitochondrion (plural mitochondria) is a double membrane-bound organelle found in all eukaryotic organisms. Some cells in some multicellular organisms may however lack them (for example, mature mammalian red blood cells). A number of unicellular organisms, such as microsporidia, parabasalids, and diplomonads, have also reduced or transformed their mitochondria into other structures. To date, only one eukaryote, *Monocercomonoides*, is known to have completely lost its mitochondria. The word mitochondrion comes from the Greek *mitos*, "thread", and *chondrion*, "granule" or "grain-like". Mitochondria generate most of the cell's supply of adenosine triphosphate (ATP), used as a source of chemical energy.

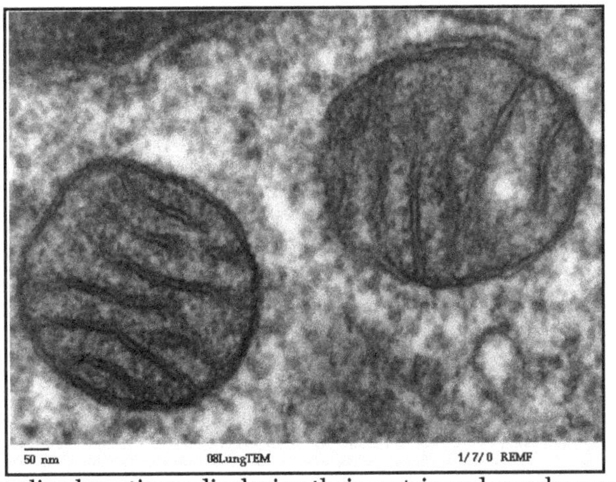

Two mitochondria from mammalian lung tissue displaying their matrix and membranes as shown by electron microscopy.

Mitochondria are commonly between 0.75 and 3 μm in diameter but vary considerably in size and structure. Unless specifically stained, they are not visible. In addition to supplying cellular energy, mitochondria are involved in other tasks, such as signaling, cellular differentiation, and cell death, as well as maintaining control of the cell cycle and cell growth. Mitochondrial biogenesis is in turn temporally coordinated with these cellular processes. Mitochondria have been implicated in several human diseases, including mitochondrial disorders, cardiac dysfunction, heart failure and autism.

The number of mitochondria in a cell can vary widely by organism, tissue, and cell type. For instance, red blood cells have no mitochondria, whereas liver cells can have more than 2000; The organelle is composed of compartments that carry out specialized functions. These compartments or regions include the outer membrane, the intermembrane space, the inner membrane, and the cristae and matrix.

Although most of a cell's DNA is contained in the cell nucleus, the mitochondrion has its own independent genome that shows substantial similarity to bacterial genomes. Mitochondrial proteins (proteins transcribed from mitochondrial DNA) vary depending on the tissue and the species. In humans, 615 distinct types of protein have been identified from cardiac mitochondria, whereas in rats, 940 proteins have been reported. The mitochondrial proteome is thought to be dynamically regulated.

History

The first observations of intracellular structures that probably represented mitochondria were published in the 1840s. Richard Altmann, in 1890, established them as cell organelles and called them "bioblasts". The term "mitochondria" was coined by Carl Benda in 1898. Leonor Michaelis discovered that Janus green can be used as a supravital stain for mitochondria in 1900. In 1904, Friedrich Meves, made the first recorded observation of mitochondria in plants in cells of the white waterlily, *Nymphaea alba* and in 1908, along with Claudius Regaud, suggested that they contain proteins and lipids. Benjamin F. Kingsbury, in 1912, first related them with cell respiration, but almost exclusively based on morphological observations. In 1913, particles from extracts of guinea-pig liver were linked to respiration by Otto Heinrich Warburg, which he called "grana".

Warburg and Heinrich Otto Wieland, who had also postulated a similar particle mechanism, disagreed on the chemical nature of the respiration. It was not until 1925, when David Keilin discovered cytochromes, that the respiratory chain was described.

In 1939, experiments using minced muscle cells demonstrated that cellular respiration using one oxygen atom can form two adenosine triphosphate (ATP) molecules, and, in 1941, the concept of the phosphate bonds of ATP being a form of energy in cellular metabolism was developed by Fritz Albert Lipmann. In the following years, the mechanism behind cellular respiration was further elaborated, although its link to the mitochondria was not known. The introduction of tissue fractionation by Albert Claude allowed mitochondria to be isolated from other cell fractions and biochemical analysis to be conducted on them alone. In 1946, he concluded that cytochrome oxidase and other enzymes responsible for the respiratory chain were isolated to the mitchondria. Eugene Kennedy and Albert Lehninger discovered in 1948 that mitochondria are the site of oxidative phosphorylation in eukaryotes. Over time, the fractionation method was further developed, improving the quality of the mitochondria isolated, and other elements of cell respiration were determined to occur in the mitochondria.

The first high-resolution electron micrographs appeared in 1952, replacing the Janus Green stains as the preferred way of visualising the mitochondria. This led to a more detailed analysis of the structure of the mitochondria, including confirmation that they were surrounded by a membrane. It also showed a second membrane inside the mitochondria that folded up in ridges dividing up the inner chamber and that the size and shape of the mitochondria varied from cell to cell.

The popular term "powerhouse of the cell" was coined by Philip Siekevitz in 1957.

In 1967, it was discovered that mitochondria contained ribosomes. In 1968, methods were developed for mapping the mitochondrial genes, with the genetic and physical map of yeast mitochondrial DNA being completed in 1976.

Origin and Evolution

There are two hypotheses about the origin of mitochondria: endosymbiotic and autogenous. The endosymbiotic hypothesis suggests that mitochondria were originally prokaryotic cells, capable of implementing oxidative mechanisms that were not possible for eukaryotic cells; they became endosymbionts living inside the eukaryote. In the autogenous hypothesis, mitochondria were born by splitting off a portion of DNA from the nucleus of the eukaryotic cell at the time of divergence with the prokaryotes; this DNA portion would have been enclosed by membranes, which could not be crossed by proteins. Since mitochondria have many features in common with bacteria, the endosymbiotic hypothesis is more widely accepted.

A mitochondrion contains DNA, which is organized as several copies of a single, circular chromosome. This mitochondrial chromosome contains genes for redox proteins, such as those of the respiratory chain. The CoRR hypothesis proposes that this co-location is required for redox regulation. The mitochondrial genome codes for some RNAs of ribosomes, and the 22 tRNAs necessary for the translation of messenger RNAs into protein. The circular structure is also found in prokaryotes. The proto-mitochondrion was probably closely related to the *Rickettsia*. However, the exact relationship of the ancestor of mitochondria to the alphaproteobacteria and whether the mitochondrion was formed at the same time or after the nucleus, remains controversial.

A recent study by researchers of the University of Hawaii at Manoa and the Oregon State University indicates that the SAR11 clade of bacteria shares a relatively recent common ancestor with the mitochondria existing in most eukaryotic cells.

The ribosomes coded for by the mitochondrial DNA are similar to those from bacteria in size and structure. They closely resemble the bacterial 70S ribosome and not the 80S cytoplasmic ribosomes, which are coded for by nuclear DNA.

The endosymbiotic relationship of mitochondria with their host cells was popularized by Lynn Margulis. The endosymbiotic hypothesis suggests that mitochondria descended from bacteria that somehow survived endocytosis by another cell, and became incorporated into the cytoplasm. The ability of these bacteria to conduct respiration in host cells that had relied on glycolysis and fermentation would have provided a considerable evolutionary advantage. This symbiotic relationship probably developed 1.7 to 2 billion years ago.

A few groups of unicellular eukaryotes have only vestigial mitochondria or derived structures: the microsporidians, metamonads, and archamoebae. These groups appear as the most primitive eukaryotes on phylogenetic trees constructed using rRNA information, which once suggested that they appeared before the origin of mitochondria. However, this is now known to be an artifact of long-branch attraction—they are derived groups and retain genes or organelles derived from mitochondria (e.g., mitosomes and hydrogenosomes).

Monocercomonoides appear to have lost their mitochondria completely and at least some of the mitochondrial functions seem to be carried out by cytoplasmic proteins now.

Structure

Mitochondrion ultrastructure *(interactive diagram)* A mitochondrion has a double membrane; the inner one contains its chemiosmotic apparatus and has deep grooves which increase its surface area. While commonly depicted as an "orange sausage with a blob inside of it" (like it is here), mitochondria can take many shapes and their intermembrane space is quite thin.

A mitochondrion contains outer and inner membranes composed of phospholipid bilayers and proteins. The two membranes have different properties. Because of this double-membraned organization, there are five distinct parts to a mitochondrion. They are:

1. the outer mitochondrial membrane,
2. the intermembrane space (the space between the outer and inner membranes),
3. the inner mitochondrial membrane,
4. the cristae space (formed by infoldings of the inner membrane),
5. the matrix (space within the inner membrane),

Mitochondria stripped of their outer membrane are called mitoplasts.

Outer Membrane

The outer mitochondrial membrane, which encloses the entire organelle, is 60 to 75 angstroms (Å) thick. It has a protein-to-phospholipid ratio similar to that of the eukaryotic plasma membrane (about 1:1 by weight). It contains large numbers of integral membrane proteins called porins. These porins form channels that allow molecules of 5000 daltons or less in molecular weight to freely diffuse from one side of the membrane to the other. Larger proteins can enter the mitochondrion if a signaling sequence at their N-terminus binds to a large multisubunit protein called translocase of the outer membrane, which then actively moves them across the membrane. Mitochondrial pro-proteins are imported through specialised translocation complexes. The outer membrane also contains enzymes involved in such diverse activities as the elongation of fatty acids, oxidation of epinephrine, and the degradation of tryptophan. These enzymes include monoamine oxidase, rotenone-insensitive NADH-cytochrome c-reductase, kynurenine hydroxylase and fatty acid Co-A ligase. Disruption of the outer membrane permits proteins in the intermembrane space to leak into the cytosol, leading to certain cell death. The mitochondrial outer membrane can associate with the endoplasmic reticulum (ER) membrane, in a structure called MAM (mitochondria-associated ER-membrane). This is important in the ER-mitochondria calcium signaling and is involved in the transfer of lipids between the ER and mitochondria. Outside the outer membrane there are small (diameter: 60Å) particles named sub-units of Parson.

Intermembrane Space

The intermembrane space is the space between the outer membrane and the inner membrane. It is also known as perimitochondrial space. Because the outer membrane is freely permeable to small molecules, the concentrations of small molecules, such as ions and sugars, in the intermembrane space is the same as in the cytosol. However, large proteins must have a specific signaling sequence to be transported across the outer membrane, so the protein composition of this space is different from the protein composition of the cytosol. One protein that is localized to the intermembrane space in this way is cytochrome.

Inner Membrane

The inner mitochondrial membrane contains proteins with five types of functions:

1. Those that perform the redox reactions of oxidative phosphorylation.
2. ATP synthase, which generates ATP in the matrix.
3. Specific transport proteins that regulate metabolite passage into and out of the matrix.
4. Protein import machinery.
5. Mitochondrial fusion and fission protein.

It contains more than 151 different polypeptides, and has a very high protein-to-phospholipid ratio (more than 3:1 by weight, which is about 1 protein for 15 phospholipids). The inner membrane is home to around 1/5 of the total protein in a mitochondrion. In addition, the inner membrane is rich in an unusual phospholipid, cardiolipin. This phospholipid was originally discovered in cow hearts in 1942, and is usually characteristic of mitochondrial and bacterial plasma membranes. Cardiolipin contains four fatty acids rather than two, and may help to make the inner membrane impermeable. Unlike the outer membrane, the inner membrane doesn't contain porins, and is highly impermeable to all molecules. Almost all ions and molecules require special membrane transporters to enter or exit the matrix. Proteins are ferried into the matrix via the translocase of the inner membrane (TIM) complex or via Oxa1. In addition, there is a membrane potential across the inner membrane, formed by the action of the enzymes of the electron transport chain.

Cristae

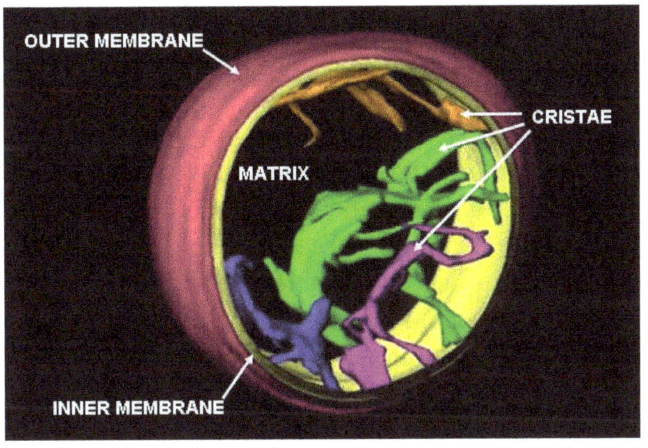

Cross-sectional image of cristae in rat liver mitochondrion to demonstrate the likely 3D structure and relationship to the inner membrane.

The inner mitochondrial membrane is compartmentalized into numerous cristae, which expand the surface area of the inner mitochondrial membrane, enhancing its ability to produce ATP. For typical liver mitochondria, the area of the inner membrane is about five times as large as the outer membrane. This ratio is variable and mitochondria from cells that have a greater demand for ATP, such as muscle cells, contain even more cristae. These folds are studded with small round bodies known as F_1 particles or oxysomes. These are not simple random folds but rather invaginations of the inner membrane, which can affect overall chemiosmotic function.

One recent mathematical modeling study has suggested that the optical properties of the cristae in filamentous mitochondria may affect the generation and propagation of light within the tissue.

Matrix

The matrix is the space enclosed by the inner membrane. It contains about 2/3 of the total protein in a mitochondrion. The matrix is important in the production of ATP with the aid of the ATP synthase contained in the inner membrane. The matrix contains a highly concentrated mixture of hundreds of enzymes, special mitochondrial ribosomes, tRNA, and several copies of the mitochondrial DNA genome. Of the enzymes, the major functions include oxidation of pyruvate and fatty acids, and the citric acid cycle.

Mitochondria have their own genetic material, and the machinery to manufacture their own RNAs and proteins. A published human mitochondrial DNA sequence revealed 16,569 base pairs encoding 37 genes: 22 tRNA, 2 rRNA, and 13 peptide genes. The 13 mitochondrial peptides in humans are integrated into the inner mitochondrial membrane, along with proteins encoded by genes that reside in the host cell's nucleus.

Mitochondria-associated ER Membrane (MAM)

The mitochondria-associated ER membrane (MAM) is another structural element that is increasingly recognized for its critical role in cellular physiology and homeostasis. Once considered a technical snag in cell fractionation techniques, the alleged ER vesicle contaminants that invariably appeared in the mitochondrial fraction have been re-identified as membranous structures derived from the MAM—the interface between mitochondria and the ER. Physical coupling between these two organelles had previously been observed in electron micrographs and has more recently been probed with fluorescence microscopy. Such studies estimate that at the MAM, which may comprise up to 20% of the mitochondrial outer membrane, the ER and mitochondria are separated by a mere 10–25 nm and held together by protein tethering complexes.

Purified MAM from subcellular fractionation has been shown to be enriched in enzymes involved in phospholipid exchange, in addition to channels associated with Ca^{2+} signaling. These hints of a prominent role for the MAM in the regulation of cellular lipid stores and signal transduction have been borne out, with significant implications for mitochondrial-associated cellular phenomena, as discussed below. Not only has the MAM provided insight into the mechanistic basis underlying such physiological processes as intrinsic apoptosis and the propagation of calcium signaling, but it also favors a more refined view of the mitochondria. Though often seen as static, isolated 'powerhouses' hijacked for cellular metabolism through an ancient endosymbiotic event, the evolution of the MAM underscores the extent to which mitochondria have been integrated into overall cellular physiology, with intimate physical and functional coupling to the endomembrane system.

Phospholipid Transfer

The MAM is enriched in enzymes involved in lipid biosynthesis, such as phosphatidylserine synthase on the ER face and phosphatidylserine decarboxylase on the mitochondrial face. Because mitochondria are dynamic organelles constantly undergoing fission and fusion events, they require a constant and well-regulated supply of phospholipids for membrane integrity. But mitochondria are not only a destination for the phospholipids they finish synthesis of; rather, this organelle also plays a role in inter-organelle trafficking of the intermediates and products of phospholipid biosynthetic pathways, ceramide and cholesterol metabolism, and glycosphingolipid anabolism.

Such trafficking capacity depends on the MAM, which has been shown to facilitate transfer of lipid intermediates between organelles. In contrast to the standard vesicular mechanism of lipid transfer, evidence indicates that the physical proximity of the ER and mitochondrial membranes at the MAM allows for lipid flipping between opposed bilayers. Despite this unusual and seemingly energetically unfavorable mechanism, such transport does not require ATP. Instead, in yeast, it has been shown to be dependent on a multiprotein tethering structure termed the ER-mitochondria encounter structure, or ERMES, although it remains unclear whether this structure directly mediates lipid transfer or is required to keep the membranes in sufficiently close proximity to lower the energy barrier for lipid flipping.

The MAM may also be part of the secretory pathway, in addition to its role in intracellular lipid trafficking. In particular, the MAM appears to be an intermediate destination between the rough ER and the Golgi in the pathway that leads to very-low-density lipoprotein, or VLDL, assembly and secretion. The MAM thus serves as a critical metabolic and trafficking hub in lipid metabolism.

Calcium Signaling

A critical role for the ER in calcium signaling was acknowledged before such a role for the mitochondria was widely accepted, in part because the low affinity of Ca^{2+} channels localized to the outer mitochondrial membrane seemed to fly in the face of this organelle's purported responsiveness to changes in intracellular Ca^{2+} flux. But the presence of the MAM resolves this apparent contradiction: the close physical association between the two organelles results in Ca^{2+} microdomains at contact points that facilitate efficient Ca^{2+} transmission from the ER to the mitochondria. Transmission occurs in response to so-called "Ca^{2+} puffs" generated by spontaneous clustering and activation of IP3R, a canonical ER membrane Ca^{2+} channel.

The fate of these puffs—in particular, whether they remain restricted to isolated locales or integrated into Ca^{2+} waves for propagation throughout the cell—is determined in large part by MAM dynamics. Although reuptake of Ca^{2+} by the ER (concomitant with its release) modulates the intensity of the puffs, thus insulating mitochondria to a certain degree from high Ca^{2+} exposure, the MAM often serves as a firewall that essentially buffers Ca^{2+} puffs by acting as a sink into which free ions released into the cytosol can be funneled. This Ca^{2+} tunneling occurs through the low-affinity Ca^{2+} receptor VDAC1, which recently has been shown to be physically tethered to the IP3R clusters on the ER membrane and enriched at the MAM. The ability of mitochondria to serve as a Ca^{2+} sink is a result of the electrochemical gradient generated during oxidative phosphorylation, which makes tunneling of the cation an exergonic process. Normally, mild calcium influx from cytosol into the mitochondrial matrix causes transient depolarization that is corrected by pumping out protons.

But transmission of Ca^{2+} is not unidirectional; rather, it is a two-way street. The properties of the Ca^{2+} pump SERCA and the channel IP3R present on the ER membrane facilitate feedback regulation coordinated by MAM function. In particular, the clearance of Ca^{2+} by the MAM allows for spatio-temporal patterning of Ca^{2+} signaling because Ca^{2+} alters IP3R activity in a biphasic manner. SERCA is likewise affected by mitochondrial feedback: uptake of Ca^{2+} by the MAM stimulates ATP production, thus providing energy that enables SERCA to reload the ER with Ca^{2+} for continued Ca^{2+} efflux at the MAM. Thus, the MAM is not a passive buffer for Ca^{2+} puffs; rather it helps modulate further Ca^{2+} signaling through feedback loops that affect ER dynamics.

Regulating ER release of Ca^{2+} at the MAM is especially critical because only a certain window of Ca^{2+} uptake sustains the mitochondria, and consequently the cell, at homeostasis. Sufficient intraorganelle Ca^{2+} signaling is required to stimulate metabolism by activating dehydrogenase enzymes critical to flux through the citric acid cycle. However, once Ca^{2+} signaling in the mitochondria passes a certain threshold, it stimulates the intrinsic pathway of apoptosis in part by collapsing the mitochondrial membrane potential required for metabolism. Studies examining the role of pro- and anti-apoptotic factors support this model; for example, the anti-apoptotic factor Bcl-2 has been shown to interact with IP3Rs to reduce Ca^{2+} filling of the ER, leading to reduced efflux at the MAM and preventing collapse of the mitochondrial membrane potential post-apoptotic stimuli. Given the need for such fine regulation of Ca^{2+} signaling, it is perhaps unsurprising that dysregulated mitochondrial Ca^{2+} has been implicated in several neurodegenerative diseases, while the catalogue of tumor suppressors includes a few that are enriched at the MAM.

Molecular Basis for Tethering

Recent advances in the identification of the tethers between the mitochondrial and ER membranes suggest that the scaffolding function of the molecular elements involved is secondary to other, non-structural functions. In yeast, ERMES, a multiprotein complex of interacting ER- and mitochondrial-resident membrane proteins, is required for lipid transfer at the MAM and exemplifies this principle. One of its components, for example, is also a constituent of the protein complex required for insertion of transmembrane beta-barrel proteins into the lipid bilayer. However, a homologue of the ERMES complex has not yet been identified in mammalian cells. Other proteins implicated in scaffolding likewise have functions independent of structural tethering at the MAM; for example, ER-resident and mitochondrial-resident mitofusins form heterocomplexes that regulate the number of inter-organelle contact sites, although mitofusins were first identified for their role in fission and fusion events between individual mitochondria. Glucose-related protein 75 (grp75) is another dual-function protein. In addition to the matrix pool of grp75, a portion serves as a chaperone that physically links the mitochondrial and ER Ca^{2+} channels VDAC and IP3R for efficient Ca^{2+} transmission at the MAM. Another potential tether is Sigma-1R, a non-opioid receptor whose stabilization of ER-resident IP3R may preserve communication at the MAM during the metabolic stress response.

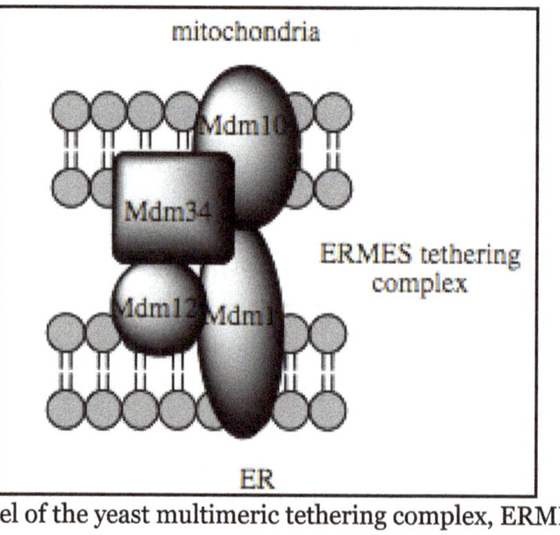

Model of the yeast multimeric tethering complex, ERMES.

Perspective

The MAM is a critical signaling, metabolic, and trafficking hub in the cell that allows for the integration of ER and mitochondrial physiology. Coupling between these organelles is not simply structural but functional as well and critical for overall cellular physiology and homeostasis. The MAM thus offers a perspective on mitochondria that diverges from the traditional view of this organelle as a static, isolated unit appropriated for its metabolic capacity by the cell. Instead, this mitochondrial-ER interface emphasizes the integration of the mitochondria, the product of an endosymbiotic event, into diverse cellular processes.

Organization and Distribution

Mitochondria (and related structures) are found in all eukaryotes (except one—the Oxymonad *Monocercomonoides* sp.). Although commonly depicted as bean-like structures they form a highly dynamic network in the majority of cells where they constantly undergo fission and fusion.

Mitochondria vary in number and location according to cell type. A single mitochondrion is often found in unicellular organisms. Conversely, numerous mitochondria are found in human liver cells, with about 1000–2000 mitochondria per cell, making up 1/5 of the cell volume. The mitochondrial content of otherwise similar cells can vary substantially in size and membrane potential, with differences arising from sources including uneven partitioning at cell divisions, leading to extrinsic differences in ATP levels and downstream cellular processes. The mitochondria can be found nestled between myofibrils of muscle or wrapped around the sperm flagellum. Often, they form a complex 3D branching network inside the cell with the cytoskeleton. The association with the cytoskeleton determines mitochondrial shape, which can affect the function as well: different structures of the mitochondrial network may afford the population a variety of physical, chemical, and signalling advantages or disadvantages. Mitochondria in cells are always distributed along microtubules and the distribution of these organelles is also correlated with the endoplasmic reticulum. Recent evidence suggests that vimentin, one of the components of the cytoskeleton, is also critical to the association with the cytoskeleton.

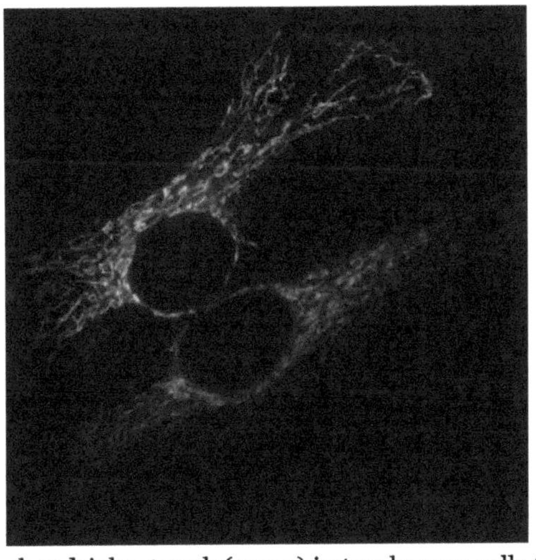

Typical mitochondrial network (green) in two human cells (HeLa cells).

Function

The most prominent roles of mitochondria are to produce the energy currency of the cell, ATP (i.e., phosphorylation of ADP), through respiration, and to regulate cellular metabolism. The central set of reactions involved in ATP production are collectively known as the citric acid cycle, or the Krebs cycle. However, the mitochondrion has many other functions in addition to the production of ATP.

Energy Conversion

A dominant role for the mitochondria is the production of ATP, as reflected by the large number of proteins in the inner membrane for this task. This is done by oxidizing the major products of glucose: pyruvate, and NADH, which are produced in the cytosol. This type of cellular respiration known as aerobic respiration, is dependent on the presence of oxygen. When oxygen is limited, the glycolytic products will be metabolized by anaerobic fermentation, a process that is independent of the mitochondria. The production of ATP from glucose has an approximately 13-times higher yield during aerobic respiration compared to fermentation. Recently it has been shown that plant mitochondria can produce a limited amount of ATP without oxygen by using the alternate substrate nitrite. ATP crosses out through the inner membrane with the help of a specific protein, and across the outer membrane via porins. ADP returns via the same route.

Pyruvate and the Citric Acid Cycle

Pyruvate molecules produced by glycolysis are actively transported across the inner mitochondrial membrane, and into the matrix where they can either be oxidized and combined with coenzyme A to form CO_2, acetyl-CoA, and NADH, or they can be carboxylated (by pyruvate carboxylase) to form oxaloacetate. This latter reaction "fills up" the amount of oxaloacetate in the citric acid cycle, and is therefore an anaplerotic reaction, increasing the cycle's capacity to metabolize acetyl-CoA when the tissue's energy needs (e.g. in muscle) are suddenly increased by activity.

In the citric acid cycle, all the intermediates (e.g. citrate, iso-citrate, alpha-ketoglutarate, succinate, fumarate, malate and oxaloacetate) are regenerated during each turn of the cycle. Adding more of any of these intermediates to the mitochondrion therefore means that the additional amount is retained within the cycle, increasing all the other intermediates as one is converted into the other. Hence, the addition of any one of them to the cycle has an anaplerotic effect, and its removal has a cataplerotic effect. These anaplerotic and cataplerotic reactions will, during the course of the cycle, increase or decrease the amount of oxaloacetate available to combine with acetyl-CoA to form citric acid. This in turn increases or decreases the rate of ATP production by the mitochondrion, and thus the availability of ATP to the cell.

Acetyl-CoA, on the other hand, derived from pyruvate oxidation, or from the beta-oxidation of fatty acids, is the only fuel to enter the citric acid cycle. With each turn of the cycle one molecule of acetyl-CoA is consumed for every molecule of oxaloacetate present in the mitochondrial matrix, and is never regenerated. It is the oxidation of the acetate portion of acetyl-CoA that produces CO_2 and water, with the energy thus released captured in the form of ATP.

In the liver, the carboxylation of cytosolic pyruvate into intra-mitochondrial oxaloacetate is an early step in the gluconeogenic pathway, which converts lactate and de-aminated alanine into glucose, under the influence of high levels of glucagon and/or epinephrine in the blood. Here, the addition of oxaloacetate to the mitochondrion does not have a net anaplerotic effect, as another citric acid cycle intermediate (malate) is immediately removed from the mitochondrion to be converted into cytosolic oxaloacetate, which is ultimately converted into glucose, in a process that is almost the reverse of glycolysis.

The enzymes of the citric acid cycle are located in the mitochondrial matrix, with the exception of succinate dehydrogenase, which is bound to the inner mitochondrial membrane as part of Complex II. The citric acid cycle oxidizes the acetyl-CoA to carbon dioxide, and, in the process, produces reduced cofactors (three molecules of NADH and one molecule of $FADH_2$) that are a source of electrons for the *electron transport chain*, and a molecule of GTP (that is readily converted to an ATP).

NADH and $FADH_2$: The Electron Transport Chain

The redox energy from NADH and $FADH_2$ is transferred to oxygen (O_2) in several steps via the electron transport chain. These energy-rich molecules are produced within the matrix via the citric acid cycle but are also produced in the cytoplasm by glycolysis. Reducing equivalents from the cytoplasm can be imported via the malate-aspartate shuttle system of antiporter proteins or feed into the electron transport chain using a glycerol phosphate shuttle. Protein complexes in the inner membrane (NADH dehydrogenase (ubiquinone), cytochrome c reductase, and cytochrome c oxidase) perform the transfer and the incremental release of energy is used to pump protons (H^+) into the intermembrane space. This process is efficient, but a small percentage of electrons may prematurely reduce oxygen, forming reactive oxygen species such as superoxide. This can cause oxidative stress in the mitochondria and may contribute to the decline in mitochondrial function associated with the aging process.

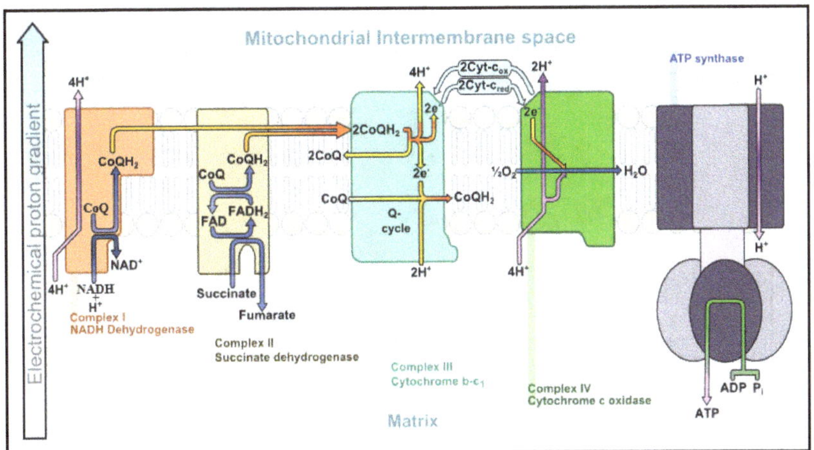

Diagram of the electron transport chain in the mitochondrial intermembrane space.

As the proton concentration increases in the intermembrane space, a strong electrochemical gradient is established across the inner membrane. The protons can return to the matrix through the ATP synthase complex, and their potential energy is used to synthesize ATP from ADP and inorganic phosphate (P_i). This process is called chemiosmosis, and was first described by Peter Mitchell who was awarded the 1978 Nobel Prize in Chemistry for his work. Later, part of the 1997 Nobel Prize in Chemistry was awarded to Paul D. Boyer and John E. Walker for their clarification of the working mechanism of ATP synthase.

Heat Production

Under certain conditions, protons can re-enter the mitochondrial matrix without contributing to ATP synthesis. This process is known as *proton leak* or *mitochondrial uncoupling* and is due to the facilitated diffusion of protons into the matrix. The process results in the unharnessed potential energy of the proton electrochemical gradient being released as heat. The process is mediated by a proton channel called thermogenin, or UCP1. Thermogenin is a 33 kDa protein first discovered in 1973. Thermogenin is primarily found in brown adipose tissue, or brown fat, and is responsible for non-shivering thermogenesis. Brown adipose tissue is found in mammals, and is at its highest levels in early life and in hibernating animals. In humans, brown adipose tissue is present at birth and decreases with age.

Storage of Calcium Ions

The concentrations of free calcium in the cell can regulate an array of reactions and is important for signal transduction in the cell. Mitochondria can transiently store calcium, a contributing process for the cell's homeostasis of calcium. In fact, their ability to rapidly take in calcium for later release makes them very good "cytosolic buffers" for calcium. The endoplasmic reticulum (ER) is the most significant storage site of calcium, and there is a significant interplay between the mitochondrion and ER with regard to calcium. The calcium is taken up into the matrix by the mitochondrial calcium uniporter on the inner mitochondrial membrane. It is primarily driven by the mitochondrial membrane potential. Release of this calcium back into the cell's interior can occur via a sodium-calcium exchange protein or via "calcium-induced-calcium-release" pathways. This can initiate calcium spikes or calcium waves with large changes in the membrane potential. These can activate a series of second messenger system proteins that can coordinate processes such as neurotransmitter release in nerve cells and release of hormones in endocrine cells.

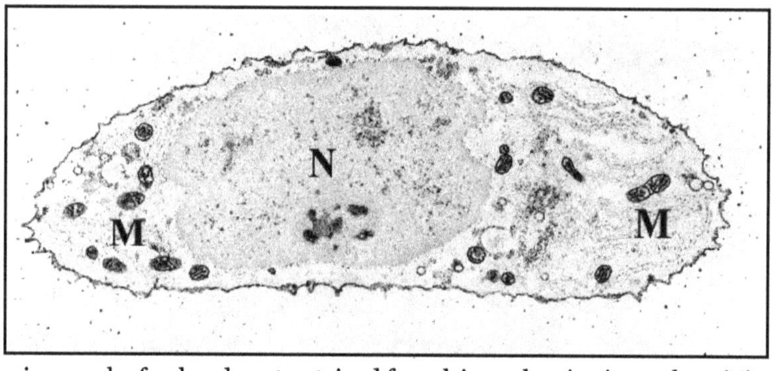

Transmission electron micrograph of a chondrocyte, stained for calcium, showing its nucleus (N) and mitochondria (M).

Ca^{2+} influx to the mitochondrial matrix has recently been implicated as a mechanism to regulate respiratory bioenergetics by allowing the electrochemical potential across the membrane to transiently "pulse" from $\Delta\Psi$-dominated to pH-dominated, facilitating a reduction of oxidative stress. In neurons, concomitant increases in cytosolic and mitochondrial calcium act to synchronize neuronal activity with mitochondrial energy metabolism. Mitochondrial matrix calcium levels can reach the tens of micromolar levels, which is necessary for the activation of isocitrate dehydrogenase, one of the key regulatory enzymes of the Kreb's cycle.

Additional Functions

Mitochondria play a central role in many other metabolic tasks, such as:

- Signaling through mitochondrial reactive oxygen species.
- Regulation of the membrane potential.
- Apoptosis-programmed cell death.
- Calcium signaling (including calcium-evoked apoptosis).
- Regulation of cellular metabolism.
- Certain heme synthesis reactions.
- Steroid synthesis.
- Hormonal signaling Mitochondria are sensitive and responsive to hormones, in part by the action of mitochondrial estrogen receptors (mtERs). These receptors have been found in various tissues and cell types, including brain and heart.

Some mitochondrial functions are performed only in specific types of cells. For example, mitochondria in liver cells contain enzymes that allow them to detoxify ammonia, a waste product of protein metabolism. A mutation in the genes regulating any of these functions can result in mitochondrial diseases.

Cellular Proliferation Regulation

The relationship between cellular proliferation and mitochondria has been investigated using cervical cancer HeLa cells. Tumor cells require an ample amount of ATP (Adenosine triphosphate) in order to synthesize bioactive compounds such as lipids, proteins, and nucleotides for rapid cell proliferation. The majority of ATP in tumor cells is generated via the oxidative phosphorylation pathway (OxPhos). Interference with OxPhos have shown to cause cell cycle arrest suggesting that mitochondria play a role in cell proliferation. Mitochondrial ATP production is also vital for cell division in addition to other basic functions in the cell including the regulation of cell volume, solute concentration, and cellular architecture. ATP levels differ at various stages of the cell cycle suggesting that there is a relationship between the abundance of ATP and the cell's ability to enter a new cell cycle. ATP's role in the basic functions of the cell make the cell cycle sensitive to changes in the availability of mitochondrial derived ATP. The variation in ATP levels at different stages of the cell cycle support the hypothesis that mitochondria play an important role in cell cycle regulation. Although the specific mechanisms between mitochondria and the cell cycle regulation is not well understood, studies have shown that low energy cell cycle checkpoints monitor the energy capability before committing to another round of cell division.

Genome

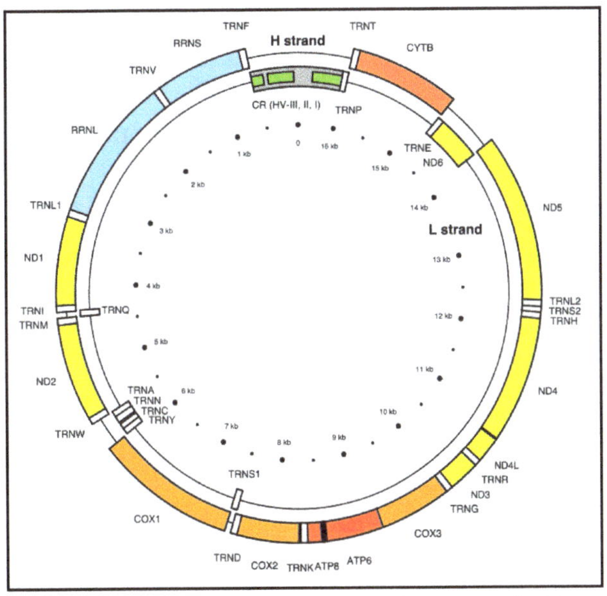

The 16,569 bp human mitochondrial genome encoding 37 genes, *i.e.*, 28 on the H-strand and 9 on the L-strand.

Mitochondria contain their own genome, an indication that they are derived from bacteria through endosymbiosis. However, the ancestral endosymbiont genome has lost most of its genes so that the mitochondrial genome is one of the most reduced genomes across organisms.

The human mitochondrial genome is a circular DNA molecule of about 16 kilobases. It encodes 37 genes: 13 for subunits of respiratory complexes I, III, IV and V, 22 for mitochondrial tRNA (for the 20 standard amino acids, plus an extra gene for leucine and serine), and 2 for rRNA. One mitochondrion can contain two to ten copies of its DNA.

As in prokaryotes, there is a very high proportion of coding DNA and an absence of repeats. Mitochondrial genes are transcribed as multigenic transcripts, which are cleaved and polyadenylated to yield mature mRNAs. Not all proteins necessary for mitochondrial function are encoded by the mitochondrial genome; most are coded by genes in the cell nucleus and the corresponding proteins are imported into the mitochondrion. The exact number of genes encoded by the nucleus and the mitochondrial genome differs between species. Most mitochondrial genomes are circular, although exceptions have been reported. In general, mitochondrial DNA lacks introns, as is the case in the human mitochondrial genome; however, introns have been observed in some eukaryotic mitochondrial DNA, such as that of yeast and protists, including *Dictyostelium discoideum*. Between protein-coding regions, tRNAs are present. During transcription, the tRNAs acquire their characteristic L-shape that gets recognized and cleaved by specific enzymes. Mitochondrial tRNA genes have different sequences from the nuclear tRNAs but lookalikes of mitochondrial tRNAs have been found in the nuclear chromosomes with high sequence similarity.

In animals, the mitochondrial genome is typically a single circular chromosome that is approximately 16 kb long and has 37 genes. The genes, while highly conserved, may vary in location. Curiously, this pattern is not found in the human body louse (*Pediculus humanus*). Instead, this mitochondrial genome is arranged in 18 minicircular chromosomes, each of which is 3–4 kb long and has one to three genes. This pattern is also found in other sucking lice, but not in chewing lice. Recombination has been shown to occur between the minichromosomes. The reason for this difference is not known.

Alternative Genetic Code

While slight variations on the standard genetic code had been predicted earlier, none was discovered until 1979, when researchers studying human mitochondrial genes determined that they used an alternative code. However, the mitochondria of many other eukaryotes, including most plants, use the standard code. Many slight variants have been discovered since, including various alternative mitochondrial codes. Further, the AUA, AUC, and AUU codons are all allowable start codons.

Exceptions to the standard genetic code in mitochondria			
Organism	**Codon**	**Standard**	**Mitochondria**
Mammals	AGA, AGG	Arginine	Stop codon
Invertebrates	AGA, AGG	Arginine	Serine
Fungi	CUA	Leucine	Threonine
All of the above	AUA	Isoleucine	Methionine
	UGA	Stop codon	Tryptophan

Some of these differences should be regarded as pseudo-changes in the genetic code due to the phenomenon of RNA editing, which is common in mitochondria. In higher plants, it was thought that CGG encoded for tryptophan and not arginine; however, the codon in the processed RNA was discovered to be the UGG codon, consistent with the standard genetic code for tryptophan. Of note, the arthropod mitochondrial genetic code has undergone parallel evolution within a phylum, with some organisms uniquely translating AGG to lysine.

Evolution and Diversity

Mitochondrial genomes have far fewer genes than the bacteria from which they are thought to be descended. Although some have been lost altogether, many have been transferred to the nucleus, such as the respiratory complex II protein subunits. This is thought to be relatively common over evolutionary time. A few organisms, such as the *Cryptosporidium*, actually have mitochondria that lack any DNA, presumably because all their genes have been lost or transferred. In *Cryptosporidium*, the mitochondria have an altered ATP generation system that renders the parasite resistant to many classical mitochondrial inhibitors such as cyanide, azide, and atovaquone.

Replication and Inheritance

Mitochondria divide by binary fission, similar to bacterial cell division. The regulation of this division differs between eukaryotes. In many single-celled eukaryotes, their growth and division is linked to the cell cycle. For example, a single mitochondrion may divide synchronously with the nucleus. This division and segregation process must be tightly controlled so that each daughter cell receives at least one mitochondrion. In other eukaryotes (in mammals for example), mitochondria may replicate their DNA and divide mainly in response to the energy needs of the cell, rather than in phase with the cell cycle. When the energy needs of a cell are high, mitochondria grow and divide. When the energy use is low, mitochondria are destroyed or become inactive. In such examples, and in contrast to the situation in many single celled eukaryotes, mitochondria are apparently randomly distributed to the daughter cells during the division of the cytoplasm. Understanding of mitochondrial dynamics, which is described as the balance between mitochondrial fusion and fission, has revealed that functional and structural alterations in mitochondrial morphology are important factors in pathologies associated with several disease conditions.

The hypothesis of mitochondrial binary fission has relied on the visualization by fluorescence microscopy and conventional transmission electron microscopy (TEM). The resolution of fluorescence microscopy(~200 nm) is insufficient to distinguish structural details, such as double mitochondrial membrane in mitochondrial division or even to distinguish individual mitochondria when several are close together. Conventional TEM has also some technical limitations in verifying mitochondrial division. Cryo-electron tomography was recently used to visualize mitochondrial division in frozen hydrated intact cells. It revealed that mitochondria divide by budding.

An individual's mitochondrial genes are not inherited by the same mechanism as nuclear genes. Typically, the mitochondria are inherited from one parent only. In humans, when an egg cell is fertilized by a sperm, the egg nucleus and sperm nucleus each contribute equally to the genetic makeup of the zygote nucleus. In contrast, the mitochondria, and therefore the mitochondrial DNA, usually come from the egg only. The sperm's mitochondria enter the egg, but do not contribute genetic information to the embryo. Instead, paternal mitochondria are marked with ubiquitin to select them for later destruction inside the embryo. The egg cell contains relatively few mitochondria, but it is these mitochondria that survive and divide to populate the cells of the adult organism. Mitochondria are, therefore, in most cases inherited only from mothers, a pattern known as maternal inheritance. This mode is seen in most organisms, including the majority of animals. However, mitochondria in some species can sometimes be inherited paternally. This is the norm among certain coniferous plants, although not in pine trees and yews. For Mytilids, paternal inheritance only occurs within males of the species. It has been suggested that it occurs at a very low level in humans. There is a recent suggestion that mitochondria that shorten male lifespan stay in the system because they are inherited only through the mother. By contrast, natural selection weeds out mitochondria that reduce female survival as such mitochondria are less likely to be passed on to the next generation. Therefore, it is suggested that human females and female animals tend to live longer than males. The authors claim that this is a partial explanation.

Uniparental inheritance leads to little opportunity for genetic recombination between different lineages of mitochondria, although a single mitochondrion can contain 2–10 copies of its DNA. For this reason, mitochondrial DNA is usually thought to reproduce by binary fission. What recombination does take place maintains genetic integrity rather than maintaining diversity. However, there are studies showing evidence of recombination in mitochondrial DNA. It is clear that the enzymes necessary for recombination are present in mammalian cells. Further, evidence suggests that animal mitochondria can undergo recombination. The data are a bit more controversial in humans, although indirect evidence of recombination exists. If recombination does not occur, the whole mitochondrial DNA sequence represents a single haplotype, which makes it useful for studying the evolutionary history of populations.

Entities undergoing uniparental inheritance and with little to no recombination may be expected to be subject to Muller's ratchet, the inexorable accumulation of deleterious mutations until functionality is lost. Animal populations of mitochondria avoid this buildup through a developmental process known as the mtDNA bottleneck. The bottleneck exploits stochastic processes in the cell to increase in the cell-to-cell variability in mutant load as an organism develops: a single egg cell with some proportion of mutant mtDNA thus produces an embryo where different cells have different mutant loads. Cell-level selection may then act to remove those cells with more mutant mtDNA, leading to a stabilisation or reduction in mutant load between generations. The mechanism underlying the bottleneck is debated, with a recent mathematical and experimental metastudy providing evidence for a combination of random partitioning of mtDNAs at cell divisions and random turnover of mtDNA molecules within the cell.

Population Genetic Studies

The near-absence of genetic recombination in mitochondrial DNA makes it a useful source of information for scientists involved in population genetics and evolutionary biology. Because all the mitochondrial DNA is inherited as a single unit, or haplotype, the relationships between mitochondrial DNA from different individuals can be represented as a gene tree. Patterns in these gene trees can be used to infer the evolutionary history of populations. The classic example of this is in human evolutionary genetics, where the molecular clock can be used to provide a recent date for mitochondrial Eve. This is often interpreted as strong support for a recent modern human expansion out of Africa. Another human example is the sequencing of mitochondrial DNA from Neanderthal bones. The relatively large evolutionary distance between the mitochondrial DNA sequences of Neanderthals and living humans has been interpreted as evidence for the lack of interbreeding between Neanderthals and anatomically modern humans.

However, mitochondrial DNA reflects only the history of the females in a population and so may not represent the history of the population as a whole. This can be partially overcome by the use of paternal genetic sequences, such as the non-recombining region of the Y-chromosome. In a broader sense, only studies that also include nuclear DNA can provide a comprehensive evolutionary history of a population.

Recent measurements of the molecular clock for mitochondrial DNA reported a value of 1 mutation every 7884 years dating back to the most recent common ancestor of humans and apes, which is consistent with estimates of mutation rates of autosomal DNA (10^{-8} per base per generation).

Dysfunction and Disease

Mitochondrial Diseases

Damage and subsequent dysfunction in mitochondria is an important factor in a range of human diseases due to their influence in cell metabolism. Mitochondrial disorders often present themselves as neurological disorders, including autism. They can also manifest as myopathy, diabetes, multiple endocrinopathy, and a variety of other systemic disorders. Diseases caused by mutation in the mtDNA include Kearns-Sayre syndrome, MELAS syndrome and Leber's hereditary optic neuropathy. In the vast majority of cases, these diseases are transmitted by a female to her children, as the zygote derives its mitochondria and hence its mtDNA from the ovum. Diseases such as Kearns-Sayre syndrome, Pearson syndrome, and progressive external ophthalmoplegia are thought to be due to large-scale mtDNA rearrangements, whereas other diseases such as MELAS syndrome, Leber's hereditary optic neuropathy, myoclonic epilepsy with ragged red fibers (MERRF), and others are due to point mutations in mtDNA.

In other diseases, defects in nuclear genes lead to dysfunction of mitochondrial proteins. This is the case in Friedreich's ataxia, hereditary spastic paraplegia, and Wilson's disease. These diseases are inherited in a dominance relationship, as applies to most other genetic diseases. A variety of disorders can be caused by nuclear mutations of oxidative phosphorylation enzymes, such as coenzyme Q10 deficiency and Barth syndrome. Environmental influences may interact with hereditary predispositions and cause mitochondrial disease. For example, there may be a link between pesticide exposure and the later onset of Parkinson's disease. Other pathologies with etiology involving mitochondrial dysfunction include schizophrenia, bipolar disorder, dementia, Alzheimer's disease, Parkinson's disease, epilepsy, stroke, cardiovascular disease, chronic fatigue syndrome, retinitis pigmentosa, and diabetes mellitus.

Mitochondria-mediated oxidative stress plays a role in cardiomyopathy in Type 2 diabetics. Increased fatty acid delivery to the heart increases fatty acid uptake by cardiomyocytes, resulting in increased fatty acid oxidation in these cells. This process increases the reducing equivalents available to the electron transport chain of the mitochondria, ultimately increasing reactive oxygen species (ROS) production. ROS increases uncoupling proteins (UCPs) and potentiate proton leakage through the adenine nucleotide translocator (ANT), the combination of which uncouples the mitochondria. Uncoupling then increases oxygen consumption by the mitochondria, compounding the increase in fatty acid oxidation. This creates a vicious cycle of uncoupling; furthermore, even though oxygen consumption increases, ATP synthesis does not increase proportionally because the mitochondria is uncoupled. Less ATP availability ultimately results in an energy deficit presenting as reduced cardiac efficiency and contractile dysfunction. To compound the problem, impaired sarcoplasmic reticulum calcium release and reduced mitochondrial reuptake limits peak cytosolic levels of the important signaling ion during muscle contraction. The decreased intra-mitochondrial calcium concentration increases dehydrogenase activation and ATP synthesis. So in addition to lower ATP synthesis due to fatty acid oxidation, ATP synthesis is impaired by poor calcium signaling as well, causing cardiac problems for diabetics.

Possible Relationships to Aging

Given the role of mitochondria as the cell's powerhouse, there may be some leakage of the high-energy electrons in the respiratory chain to form reactive oxygen species. This was thought to result in significant oxidative stress in the mitochondria with high mutation rates of mitochondrial DNA (mtDNA). Hypothesized links between aging and oxidative stress are not new and were proposed in 1956, which was later refined into the mitochondrial free radical theory of aging. A vicious cycle was thought to occur, as oxidative stress leads to mitochondrial DNA mutations, which can lead to enzymatic abnormalities and further oxidative stress.

A number of changes can occur to mitochondria during the aging process. Tissues from elderly patients show a decrease in enzymatic activity of the proteins of the respiratory chain. However, mutated mtDNA can only be found in about 0.2% of very old cells. Large deletions in the mitochondrial genome have been hypothesized to lead to high levels of oxidative stress and neuronal death in Parkinson's disease.

In Popular Culture

The science fiction novels *A Wind in the Door* and *Parasite Eve* include fictional depictions of mitochondria as major plot elements.

Madeleine L'Engle's 1973 science fantasy novel *A Wind in the Door* prominently features the mitochondria of main character Charles Wallace Murry, as being inhabited by creatures known as the farandolae. The novel also features other characters travelling inside one of Murry's mitochondria.

The 1995 horror fiction novel *Parasite Eve* by Hideaki Sena depicts mitochondria as having some consciousness and mind control abilities, attempting to use these to overtake eukaryotes as the dominant life form. This text was adapted into an eponymous film, video game, and video game sequel all involving a similar premise.

In the *Star Wars* franchise, microorganisms referred to as "midi-chlorians" give some characters the ability to sense and use the Force. George Lucas, director of the 1999 film *Star Wars: Episode I – The Phantom Menace*, in which midi-chlorians were introduced, described them as "a loose depiction of mitochondria". The non-fictional *Midichloria* genus of bacteria was later named after the midi-chlorians of *Star Wars*.

As a result of the mitochondrion's prominence in modern science education, the phrase "mitochondria is [sic] the powerhouse of the cell" became a popular Internet meme. The meme is used to imply that secondary education places an insufficient focus on life skills, compared to academic knowledge such as the role of the mitochondrion, which has been considered comparatively impractical.

References

- Nierhaus, Knud H.; Daniel N. Wilson (2004). Protein Synthesis and Ribosome Structure: Translating the Genome. Wiley-VCH. ISBN 3-527-30638-2
- "Mitochondrion – much more than an energy converter". British Society for Cell Biology. Retrieved 16, June 2020
- Matera, A. Gregory; Rebecca M. Terns; Michael P. Terns (March 2007). "Non-coding RNAs: lessons from the small nuclear and small nucleolar RNAs". Nature Reviews Molecular Cell Biology. 8 (3): 209–220. ISSN 1471-0072. PMID 17318225. doi:10.1038/nrm2124. Retrieved 11, February 2020
- Yozef Gruenbaum, Katherine L. Wilson, Amnon Harel, Michal Goldberg, Merav Cohen (2000). Nuclear Lamins – Structural Proteins with fundamental functions. Journal of Structural Biology 129, 313-323
- Fox, Archa; Wendy Bickmore (2004). "Nuclear Compartments: Paraspeckles". Nuclear Protein Database. Retrieved 05, April 2020
- Stryer, Lubert; Berg, Jeremy Mark; Tymoczko, John L. (2002). Biochemistry. San Francisco: W.H. Freeman. ISBN 0-7167-4684-0. OCLC 179705944
- Feneberg, Wolfgang; Sackmann, Erich; Westphal, Monika (2001). "Dictyostelium cells' cytoplasm as an active viscoplastic body". European Biophysics Journal. 30 (4): 284–94. doi:10.1007/s002490100135. PMID 11548131
- Lodish, H; Berk A; Matsudaira P; Kaiser CA; Krieger M; Scott MP; Zipursky SL; Darnell J. (2004). Molecular Cell Biology (5th ed.). New York: WH Freeman. ISBN 0-7167-2672-6

- Woodward, Nathaniel (January 20, 2016). "Education: Mitochondria is the powerhouse of the cell". The Eagle Online. Retrieved 06, August 2020
- Byrne, Nathan (May 13, 2016). "So, Mitochondria Aren't Actually The Powerhouse Of The Cell After All". Newsy. Retrieved 25, May 2020

Chapter 4

Concepts of Cell Biology

The pursuit of cell biology would be impossible without its various concepts and theories. Cell Theory is one such concept which states that all living organisms are made up of cells. Cell disruption and cell culture are two crucial developments in cell biology made possible by the scientific research based on this theory. These topics along with autophagy and mitophagy are discussed in this chapter.

Cell Disruption

Cell disruption is a method or process for releasing biological molecules from inside a cell.

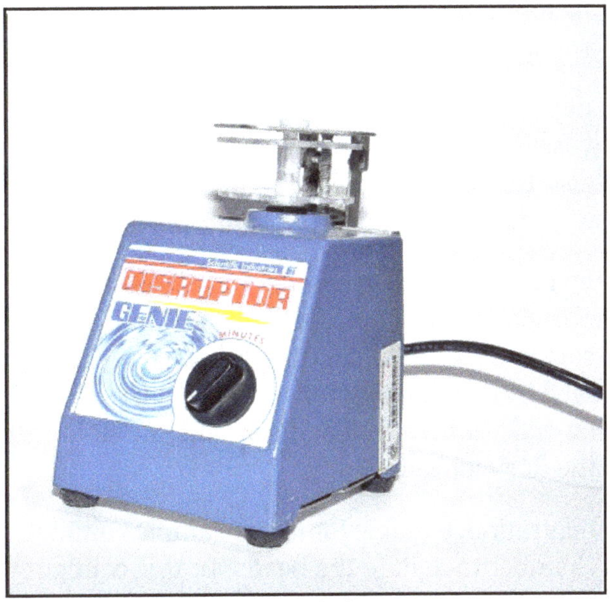

Laboratory cell disruptor.

Methods

The production of biologically interesting molecules using cloning and culturing methods allows the study and manufacture of relevant molecules. Except for excreted molecules, cells producing molecules of interest must be disrupted. This page discusses various methods.

Bead Method

A common laboratory-scale mechanical method for cell disruption uses tiny glass, ceramic or steel beads mixed with a sample suspended in aqueous media. First developed by Tim Hopkins in the late 1970s, the sample and bead mix is subjected to high level agitation by stirring or shaking. Beads collide with the cellular sample, cracking open the cell to release intercellular components. Unlike some other methods, mechanical shear is moderate during homogenization resulting in excellent membrane or subcellular preparations. The method, often called "beadbeating", works well for all types of cellular material - from spores to animal and plant tissues. It is the most widely used method of yeast lysis, and can yield breakage of over 50%. It has the advantage over other mechanical cell disruption methods of being able to disrupt very small sample sizes, process many samples at a time with no cross-contamination concerns, and does not release potentially harmful aerosols in the process.

In the simplest example of the method, an equal volume of beads are added to a cell or tissue suspension in a test tube and the sample is vigorously mixed on a common laboratory vortex mixer. While processing times are slow, taking 3-10 times longer than that in specialty shaking machines, it works well for easily disrupted cells and is inexpensive.

In most laboratories, beadbeating is done in sealed, plastic vials, centrifuge tubes, or deep-well microtiter plates. The sample and tiny beads are agitated at about 2000 oscillations per minute in specially designed vial shakers driven by high power electric motors. Cell disruption is complete in 1–3 minutes of shaking. Machines are available that can process hundreds of samples simultaneously inside deep well microplates.

Successful beadbeating is dependent not only design features of the shaking machine (which take into consideration shaking oscillations per minute, shaking throw or distance, shaking orientation and vial orientation), but also the selection of correct bead size (0.1–6 mm diameter), bead composition (glass, ceramic, steel) and bead load in the vial.

All high energy beadbeating machines warm the sample about 10 degrees/minute. This is due to frictional collisions of the beads during homogenization. Cooling of the sample during or after beadbeating may be necessary to prevent damage to heat sensitive proteins such as enzymes. Sample warming can be controlled by beadbeating for short time intervals with cooling on ice between each interval, by processing vials in pre-chilled aluminum vial holders or by circulating gaseous coolant through the machine during beadbeating.

A different beadbeater configuration, suitable for larger sample volumes, uses a fluorocarbon rotor inside a 15, 50 or 200 ml chamber to agitate the beads. In this configuration, the chamber can be surrounded by a static cooling jacket. Using the same rotor/chamber configuration, large commercial machines are available to process many liters of cell suspension. Currently, these machines are limited to processing monocellular organisms such as yeast, algae and bacteria.

Cryopulverization

Samples with a tough extracellular matrix, such as animal connective tissue, some tumor biopsy samples, venous tissue, cartilage, seeds, etc., are reduced to a fine powder by impact pulverization at liquid nitrogen temperatures. This technique, known as cryopulverization, is based on the fact that biological samples containing a significant fraction of water become brittle at extremely cold temperatures. This technique was first described by Smucker and Pfister in 1975, who referred to the technique as cryo-impacting. The authors demonstrated cells are effectively broken by this method, confirming by phase and electron microscopy that breakage planes cross cell walls and cytoplasmic membranes.

The technique can done using a mortar and pestle cooled to liquid nitrogen temperatures, but use of this classic apparatus is laborious and sample loss is often a concern. Specialised stainless steel pulverizers generically known as Tissue Pulverizers are also available for this purpose. They require less manual effort, give good sample recovery and are easy to clean between samples. Advantages of this technique are higher yields of proteins and nucleic acids from small, hard tissue samples - especially when used as a preliminary step to mechanical or chemical/solvent cell disruption methods mentioned above.

Microfluidizer Method

This method used for cell disruption strongly influences the physicochemical properties of the lysed cell suspension, such as particle size, viscosity, protein yield and enzyme activity. In recent years the Microfluidizer method has gained popularity in cell disruption due to its ease of use and efficiency at disrupting many different kinds of cells. The Microfluidizer technology was licensed from a company called Arthur D. Little and was first developed and utilized in the 1980s, initially starting as a tool for liposome creation. It has since been used in other applications such as cell disruption nanoemulsions, and solid particle size reduction, among others.

By using microchannels with fixed geometry, and an intensifier pump, high shear rates are generated that rupture the cells. This method of cell lysis can yield breakage of over 90% of E. coli cells.

Many proteins are extremely temperature-sensitive, and in many cases can start to denature at temperatures of only 4 degrees celsius. Within the microchannels, temperatures exceed 4 degrees celsius, but the machine is designed to cool quickly so that the time the cells are exposed to elevated temperatures is extremely short (residence time 25 ms-40 ms). Because of this effective temperature control, the Microfluidizer yields higher levels of active proteins and enzymes than other mechanical methods when the proteins are temperature-sensitive.

Viscosity changes are also often observed when disrupting cells. If the cell suspension viscosity is high, it can make downstream handling—such as filtration and accurate pipetting—quite difficult. The viscosity changes observed with a Microfluidizer are relatively low, and decreases with further additional passes through the machine.

In contrast to other mechanical disruption methods the Microfluidizer breaks the cell membranes efficiently but gently, resulting in relatively large cell wall fragments (450 nm), and thus making it easier to separate the cell contents. This can lead to shorter filtration times and better centrifugation separation.

Microfluidizer technology scales from one milliliter to thousands of liters.

Nitrogen Decompression

For nitrogen decompression, large quantities of nitrogen are first dissolved in the cell under high pressure within a suitable pressure vessel. Then, when the gas pressure is suddenly released, the nitrogen comes out of the solution as expanding bubbles that stretch the membranes of each cell until they rupture and release the contents of the cell.

Nitrogen decompression is more protective of enzymes and organelles than ultrasonic and mechanical homogenizing methods and compares favorably to the controlled disruptive action obtained in a PTFE and glass mortar and pestle homogenizer. While other disruptive methods depend upon friction or a mechanical shearing action that generate heat, the nitrogen decompression procedure is accompanied by an adiabatic expansion that cools the sample instead of heating it.

The blanket of inert nitrogen gas that saturates the cell suspension and the homogenate offers protection against oxidation of cell components. Although other gases: carbon dioxide, nitrous oxide, carbon monoxide and compressed air have been used in this technique, nitrogen is preferred because of its non-reactive nature and because it does not alter the pH of the suspending medium. In addition, nitrogen is preferred because it is generally available at low cost and at pressures suitable for this procedure.

Once released, subcellular substances are not exposed to continued attrition that might denature the sample or produce unwanted damage. There is no need to watch for a peak between enzyme activity and percent disruption. Since nitrogen bubbles are generated within each cell, the same disruptive force is applied uniformly throughout the sample, thus ensuring unusual uniformity in the product. Cell-free homogenates can be produced.

The technique is used to homogenize cells and tissues, release intact organelles, prepare cell membranes, release labile biochemicals, and produce uniform and repeatable homogenates without subjecting the sample to extreme chemical or physical stress.

The method is particularly well suited for treating mammalian and other membrane-bound cells. It has also been used successfully for treating plant cells, for releasing virus from fertilized eggs and for treating fragile bacteria. It is not recommended for untreated bacterial cells. Yeast, fungus, spores and other materials with tough cell walls do not respond well to this method.

Cell Theory

In biology, cell theory is a scientific theory which describes the properties of cells. These cells are the basic unit of structure in all organisms and also the basic unit of reproduction. With continual improvements made to microscopes over time, magnification technology advanced enough to discover cells in the 17th century. This discovery is largely attributed to Robert Hooke, and began the scientific study of cells, also known as cell biology. Over a century later, many debates about cells began amongst scientists. Most of these debates involved the nature of cellular regeneration, and the idea of cells as a fundamental unit of life. Cell theory was eventually formulated in 1838. This is usually credited to Matthias Schleiden and Theodor Schwann. However, many other scientists like Rudolf Virchow contributed to the theory. Cell theory has become the foundation of biology and is the most widely accepted explanation of the function of cells.

The three tenets to the cell theory are as described below:

1. All living organisms are composed of one or more cells. (However, this is controversial because non-cellular life such as viruses are disputed as a life form).

2. The cell is the basic unit of structure and organization in organisms.
3. Cells come from pre-existing cells.

Microscopes

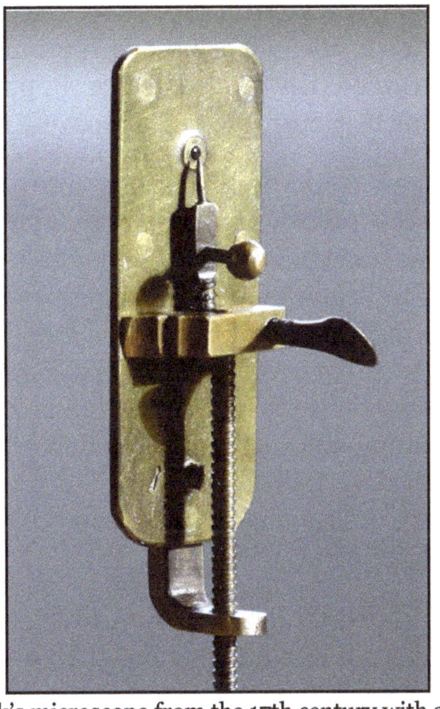

Anton van Leeuwenhoek's microscope from the 17th century with a magnification of 270x.

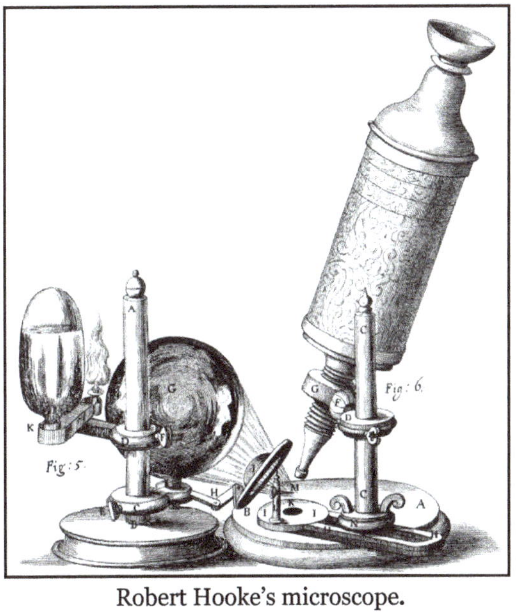

Robert Hooke's microscope.

The discovery of the cell was made possible through the invention of the microscope. In the first century BC, Romans were able to make glass, discovering that objects appeared to be larger under the glass. In Italy during the 12th century, Salvino D'Armate made a piece of glass fit over one eye, allowing for a magnification effect to that eye. It was not until the 1590s when a Dutch spectacle maker Zacharias Jansen began to test lenses that progress had been made to microscopes. Jansen was able to obtain about 9x magnification, but the objects appeared to be blurry. In 1595, Jansen and his father built the first compound microscope. While simple glasses were able to magnify objects, they were not considered to be a microscope. A compound microscope was defined by having two or more lenses in a hollow tube. In 1665, Robert Hooke used a microscope about six inches long with two convex lenses inside and examined specimens under reflected light for the observations in his book *Micrographia*. Hooke also used a simpler microscope with a single lens for examining specimens with directly transmitted light, because this allowed for a clearer image.

However, the first real invention and use of the microscope were done by Anton van Leeuwenhoek. He was a Dutch draper that took the interest in microscopes after seeing one while on an apprenticeship in Amsterdam in 1648. At some point in his life before 1668, he was able to learn how to grind lenses. This eventually led to Leeuwenhoek making his own unique microscope. His were instead simple powerful magnifying glasses, rather than a compound microscope. This was because he was able to use a single lens that was a small glass sphere but allowed for a magnification of 270x. This was a large progression since the magnification before was only a maximum of 50x. After Leeuwenhoek, there was not much progress for the microscopes until the 1850s, two hundred years later. Carl Zeiss, a German engineer who manufactured microscopes, began to make changes to the lenses used. But the optical quality did not improve until the 1880s when he hired Otto Schott and eventually Ernst Abbe.

These microscopes could focus on objects the size of a wavelength or larger, giving restrictions still to advancement in discoveries with objects smaller than a wavelength. Later in the 1920s, the electron microscope was developed, making it possible to view objects that are smaller than a wavelength, once again, changing the possibilities in science.

Discovery of Cells

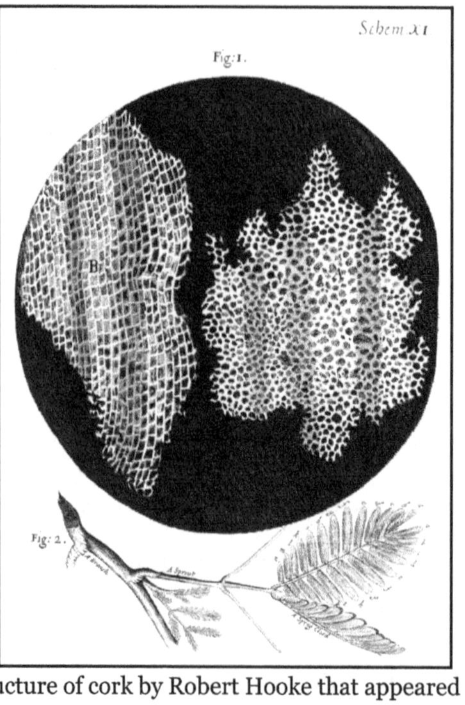

Drawing of the structure of cork by Robert Hooke that appeared in *Micrographia*.

The cell was first discovered by Robert Hooke in 1665, which can be found to be described in his book Micrographia. In this book, he gave 60 'observations' in detail of various objects under a coarse, compound microscope. One observation was from very thin slices of bottle cork. Hooke discovered a multitude of tiny pores that he named "cells". This came from the Latin word Cella, meaning 'a small room' like monks lived in and also Cellulae, which meant the six sided cell of a honeycomb. However, Hooke did not know their real structure or function. What Hooke had thought were cells, were actually empty cell walls of plant tissues. With microscopes during this time having a low magnification, Hooke was unable to see that there were other internal components to the cells he was observing. Therefore, he did not think the "cellulae" were alive. His cell observations gave no indication of the nucleus and other organelles found in most living cells. In Micrographia, Hooke also observed mould, bluish in color, found on leather. After studying it under his microscope, he was unable to observe "seeds" that would have indicated how the mould was multiplying in quantity. This led to Hooke suggesting that spontaneous generation, from either natural or artificial heat, was the cause. Since this was an old Aristotelian theory still accepted at the time, others did not reject it and was not disproved until Leeuwenhoek later discovers generation is achieved otherwise.

Anton van Leeuwenhoek is another scientist who saw these cells soon after Hooke did. He made use of a microscope containing improved lenses that could magnify objects almost 300-fold, or 270x. Under these microscopes, Leeuwenhoek found motile objects. In a letter to The Royal Society on October 9, 1676, he states that motility is a quality of life therefore these were living organisms. Over time, he wrote many more papers in which described many specific forms of microorganisms. Leeuwenhoek named these "animalcules," which included protozoa and other unicellular organisms, like bacteria. Though he did not have much formal education, he was able to identify the first accurate description of red blood cells and discovered bacteria after gaining interest in the sense of taste that resulted in Leeuwenhoek to observe the tongue of an ox, then leading him to study "pepper water" in 1676. He also found for the first time the sperm cells of animals and humans. Once discovering these types of cells, Leeuwenhoek saw that the fertilization process requires the sperm cell to enter the egg cell. This put an end to the previous theory of spontaneous generation. After reading letters by Leeuwenhoek, Hooke was the first to confirm his observations that were thought to be unlikely by other contemporaries.

The cells in animal tissues were observed after plants were because the tissues were so fragile and susceptible to tearing, it was difficult for such thin slices to be prepared for studying. Biologists believed that there was a fundamental unit to life, but were unsure what this was. It would not be until over a hundred years later that this fundamental unit was connected to cellular structure and existence of cells in animals or plants. This conclusion was not made until Henri Dutrochet. Besides stating "the cell is the fundamental element of organization", Dutrochet also claimed that cells were not just a structural unit, but also a physiological unit.

In 1804, Karl Rudolphi and J.H.F. Link were awarded the prize for "solving the problem of the nature of cells", meaning they were the first to prove that cells had independent cell walls by the Königliche Societät der Wissenschaft (Royal Society of Science), Göttingen. Before, it had been thought that cells shared walls and the fluid passed between them this way.

Cell Theory

Matthias Jakob Schleiden (1804–1881).

Theodor Schwann (1810–1882).

Credit for developing cell theory is usually given to two scientists: Theodor Schwann and Matthias Jakob Schleiden. While Rudolf Virchow contributed to the theory, he is not as credited for his attributions toward it. In 1838, Schleiden suggested that every structural part of a plant was made up of cells or the result of cells. He also suggested that cells were made by a crystallization process either within other cells or from the outside. However, this was not an original idea of Schlieden. He claimed this theory as his own, though Barthelemy Dumortier had stated it years before him. This crystallization process is no longer accepted with modern cell theory. In 1839, Theodor Schwann states that along with plants, animals are composed of cells or the product of cells in their structures. This was a major advancement in the field of biology since little was known about animal structure up to this point compared to plants. From these conclusions about plants and animals, two of the three tenets of cell theory were postulated.

1. All living organisms are composed of one or more cells.

2. The cell is the most basic unit of life.

Schleiden's theory of free cell formation through crystallization was refuted in the 1850s by Robert Remak, Rudolf Virchow, and Albert Kolliker. In 1855, Rudolf Virchow added the third tenet to cell theory. In Latin, this tenet states *Omnis cellula e cellula*. This translated to:

3. All cells arise only from pre-existing cells.

However, the idea that all cells come from pre-existing cells had in fact already been proposed by Robert Remak; it has been suggested that Virchow plagiarized Remak and did not give him credit. Remak published observations in 1852 on cell division, claiming Schleiden and Schawnn were incorrect about generation schemes. He instead said that binary fission, which was first introduced by Dumortier, was how reproduction of new animal cells were made. Once this tenet was added, the classical cell theory was complete.

Modern Interpretation

The generally accepted parts of modern cell theory include:

1. All known living things are made up of one or more cells.
2. All living cells arise from pre-existing cells by division.
3. The cell is the fundamental unit of structure and function in all living organisms.
4. The activity of an organism depends on the total activity of independent cells.
5. Energy flow (metabolism and biochemistry) occurs within cells.
6. Cells contain DNA which is found specifically in the chromosome and RNA found in the cell nucleus and cytoplasm.
7. All cells are basically the same in chemical composition in organisms of similar species.

The Modern Version of the Cell Theory

The modern version of the cell theory includes the ideas that:

- Energy flow occurs within cells.
- Heredity information (DNA) is passed on from cell to cell.
- All cells have the same basic chemical composition.

Opposing Concepts in Cell Theory: History and Background

The cell was first discovered by Robert Hooke in 1665 using a microscope. The first cell theory is credited to the work of Theodor Schwann and Matthias Jakob Schleiden in the 1830s. In this theory the internal contents of cells were called protoplasm and described as a jelly-like substance, sometimes called living jelly. At about the same time, colloidal chemistry began its development, and the concepts of bound water emerged. A colloid being something between a solution and a suspension, where Brownian motion is sufficient to prevent sedimentation. The idea of a semi-permeable membrane, a barrier that is permeable to solvent but impermeable to solute molecules was developed at about the same time. The term osmosis originated in 1827 and its importance to physiological phenomena realized, but it wasn't until 1877, when the botanist Pfeffer proposed the membrane theory of cell physiology. In this view, the cell was seen to be enclosed by a thin surface, the plasma membrane, and cell water and solutes such as a potassium ion existed in a physical state like that of a dilute solution. In 1889 Hamburger used hemolysis of erythrocytes to determine the permeability of various solutes. By measuring the time required for the cells to swell past their elastic limit, the rate at which solutes entered the cells could be estimated by the accompanying change in cell volume. He also found that there was an apparent nonsolvent volume of about 50% in red blood cells and later showed that this includes water of hydration in addition to the protein and other nonsolvent components of the cells.

Evolution of the Membrane and Bulk Phase Theories

Two opposing concepts developed within the context of studies on osmosis, permeability, and electrical properties of cells. The first held that these properties all belonged to the plasma membrane whereas the other predominant view was that the protoplasm was responsible for these properties. The membrane theory developed as a succession of ad-hoc additions and changes to the theory to overcome experimental hurdles. Overton (a distant cousin of Charles Darwin) first proposed the concept of a lipid (oil) plasma membrane in 1899. The major weakness of the lipid membrane was the lack of an explanation of the high permeability to water, so Nathansohn (1904) proposed the mosaic theory. In this view, the membrane is not a pure lipid layer, but a mosaic of areas with lipid and areas with semipermeable gel. Ruhland refined the mosaic theory to include pores to allow additional passage of small molecules. Since membranes are generally less permeable to anions, Leonor Michaelis concluded that ions are adsorbed to the walls of the pores, changing the permeability of the pores to ions by electrostatic repulsion. Michaelis demonstrated the membrane potential (1926) and proposed that it was related to the distribution of ions across the membrane. Harvey and Danielli (1939) proposed a lipid bilayer membrane covered on each side with a layer of protein to account for measurements of surface tension. In 1941 Boyle & Conway showed that the membrane of frog muscle was permeable to both K+ and Cl-, but apparently not to Na+, so the idea of electrical charges in the pores was unnecessary since a single critical pore size would explain the permeability to K+, H+, and Cl- as well as the impermeability to Na+, Ca+, and Mg++. Over the same time period, it was shown (Procter & Wilson, 1916) that gels, which do not have a semipermeable membrane, would swell in dilute solutions. Loeb (1920) also studied gelatin extensively, with and without a membrane, showing that more of the properties attributed to the plasma membrane could be duplicated in gels without a membrane. In particular, he found that an electrical potential difference between the gelatin and the outside medium could be developed, based on the H+ concentration. Some criticisms of the membrane theory developed in the 1930s, based on observations such as the ability of some cells to swell and increase their surface area by a factor of 1000. A lipid layer cannot stretch to that extent without becoming a patchwork (thereby losing its barrier properties. Such criticisms stimulated continued studies on protoplasm as the principal agent determining cell permeability properties. In 1938, Fischer and Suer proposed that water in the protoplasm is not free but in a chemically combined form—the protoplasm represents a combination of protein, salt and water—and demonstrated the basic similarity between swelling in living tissues and the swelling of gelatin and fibrin gels. Dimitri Nasonov (1944) viewed proteins as the central components responsible for many properties of the cell, including electrical properties. By the 1940s, the bulk phase theories were not as well developed as the membrane theories. In 1941, Brooks & Brooks published a monograph The Permeability of Living Cells, which rejects the bulk phase theories.

The Emergence of the steady-state Membrane Pump Concept

With the development of radioactive tracers, it was shown that cells are not impermeable to Na^+. This was difficult to explain with the membrane barrier theory, so the sodium pump was proposed to continually remove Na+ as it permeates cells. This drove the concept that cells are in a state of dynamic equilibrium, constantly using energy to maintain ion gradients. In 1935, Karl Lohmann discovered ATP and its role as a source of energy for cells, so the concept of a metabolically-driven sodium pump was proposed. The tremendous success of Hodgkin, Huxley, and Katz in the development of the membrane theory of cellular membrane potentials, with differential equations that modeled the phenomena correctly, provided even more support for the membrane pump hypothesis.

The modern view of the plasma membrane is of a fluid lipid bilayer that has protein components embedded within it. The structure of the membrane is now known in great detail, including 3D models of many of the hundreds of different proteins that are bound to the membrane. These major developments in cell physiology placed the membrane theory in a position of dominance and stimulated the imagination of most physiologists, who now apparently accept the theory as fact—there are, however, a few dissenters.

The Reemergence of the Bulk Phase Theories

In 1956, Afanasy S. Troshin published a book, *The Problems of Cell Permeability*, in Russian (1958 in German, 1961 in Chinese, 1966 in English) in which he found that permeability was of secondary importance in determination of the patterns of equilibrium between the cell and its environment. Troshin showed that cell water decreased in solutions of galactose or urea although these compounds did slowly permeate cells. Since the membrane theory requires an impermanent solute to sustain cell shrinkage, these experiments cast doubt on the theory. Others questioned whether the cell has enough energy to sustain the sodium/potassium pump. Such questions became even more urgent as dozens of new metabolic pumps were added as new chemical gradients were discovered.

In 1962, Gilbert Ling became the champion of the bulk phase theories and proposed his association-induction hypothesis of living cells.

Types of Cells

Cells can be subdivided into the following subcategories:

1. *Prokaryotes*: Prokaryotes are relatively small cells surrounded by the plasma membrane, with a characteristic cell wall that may differ in composition depending on the particular organism. Prokaryotes lack a nucleus (although they do have circular or linear DNA) and

other membrane-bound organelles (though they do contain ribosomes). The protoplasm of a prokaryote contains the chromosomal region that appears as fibrous deposits under the microscope, and the cytoplasm. Bacteria and Archaea are the two domains of prokaryotes.

2. *Eukaryotes*: Eukaryotic cells are also surrounded by the plasma membrane, but on the other hand, they have distinct nuclei bound by a nuclear membrane or envelope. Eukaryotic cells also contain membrane-bound organelles, such as (mitochondria, chloroplasts, lysosomes, rough and smooth endoplasmic reticulum, vacuoles). In addition, they possess organized chromosomes which store genetic material.

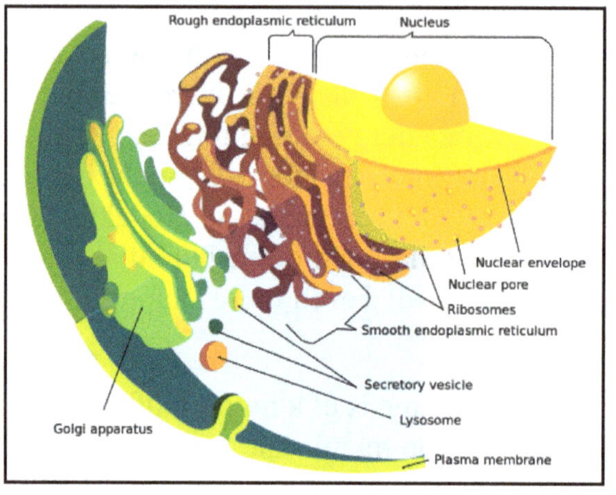

Eukaryote cell.

Animals have evolved a greater diversity of cell types in a multicellular body (100–150 different cell types), compared with 10–20 in plants, fungi, and protoctists.

Cell Culture

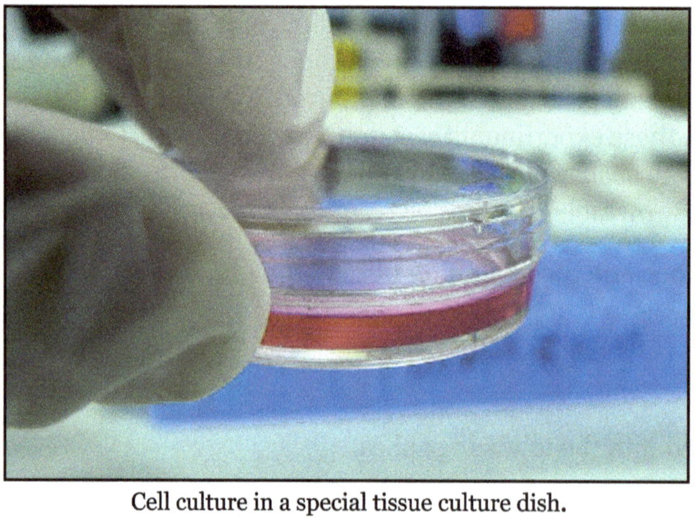

Cell culture in a special tissue culture dish.

Cell culture is the process by which cells are grown under controlled conditions, generally outside of their natural environment. In practice, the term "cell culture" now refers to the culturing of cells derived from multicellular eukaryotes, especially animal cells, in contrast with other types of culture that also grow cells, such as plant tissue culture, fungal culture, and microbiological culture (of microbes). The historical development and methods of cell culture are closely interrelated to those of tissue culture and organ culture. Viral culture is also related, with cells as hosts for the viruses.

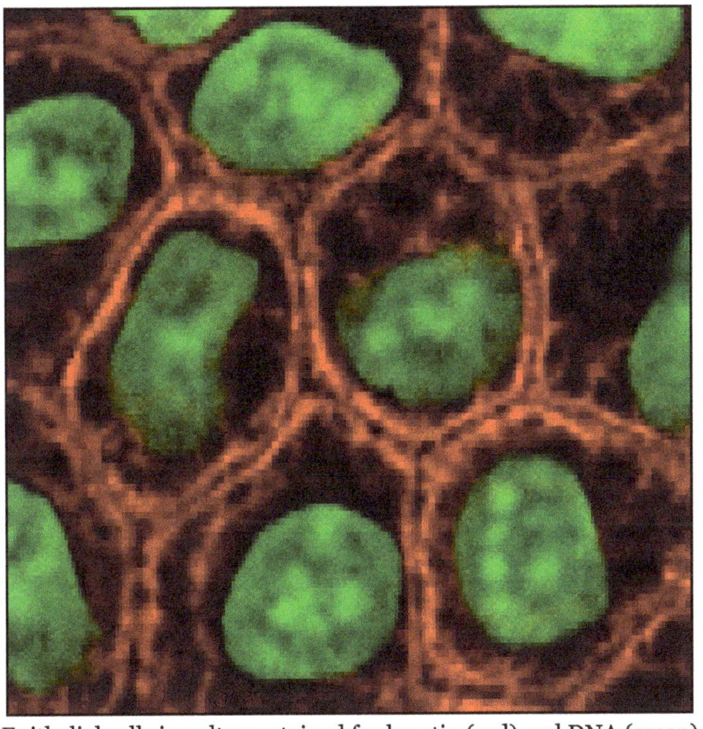

Epithelial cells in culture, stained for keratin (red) and DNA (green).

The laboratory technique of maintaining live cell lines (a population of cells descended from a single cell and containing the same genetic makeup) separated from their original tissue source became more robust in the middle 20th century.

History

The 19th-century English physiologist Sydney Ringer developed salt solutions containing the chlorides of sodium, potassium, calcium and magnesium suitable for maintaining the beating of an isolated animal heart outside of the body. In 1885, Wilhelm Roux removed a portion of the medullary plate of an embryonic chicken and maintained it in a warm saline solution for several days, establishing the principle of tissue culture. Ross Granville Harrison, working at Johns Hopkins Medical School and then at Yale University, published results of his experiments from 1907 to 1910, establishing the methodology of tissue culture.

Cell culture techniques were advanced significantly in the 1940s and 1950s to support research in virology. Growing viruses in cell cultures allowed preparation of purified viruses for the manufacture of vaccines. The injectable polio vaccine developed by Jonas Salk was one of the first products mass-produced using cell culture techniques. This vaccine was made possible by the cell culture research of John Franklin Enders, Thomas Huckle Weller, and Frederick Chapman Robbins, who were awarded a Nobel Prize for their discovery of a method of growing the virus in monkey kidney cell cultures.

Concepts in Mammalian Cell Culture

Isolation of Cells

Cells can be isolated from tissues for *ex vivo* culture in several ways. Cells can be easily purified from blood; however, only the white cells are capable of growth in culture. Mononuclear cells can be released from soft tissues by enzymatic digestion with enzymes such as collagenase, trypsin, or pronase, which break down the extracellular matrix. Alternatively, pieces of tissue can be placed in growth media, and the cells that grow out are available for culture. This method is known as explant culture.

Cells that are cultured directly from a subject are known as primary cells. With the exception of some derived from tumors, most primary cell cultures have limited lifespan.

An established or immortalized cell line has acquired the ability to proliferate indefinitely either through random mutation or deliberate modification, such as artificial expression of the telomerase gene. Numerous cell lines are well established as representative of particular cell types.

Maintaining Cells in Culture

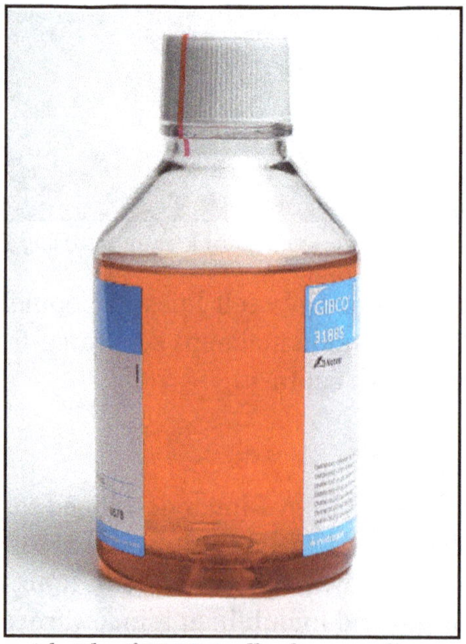

A bottle of DMEM cell culture medium.

For the majority of isolated primary cells, they undergo the process of senescence and stop dividing after a certain number of population doublings while generally retaining their viability (described as the Hayflick limit).

Cells are grown and maintained at an appropriate temperature and gas mixture (typically, 37 °C, 5% CO_2 for mammalian cells) in a cell incubator. Culture conditions vary widely for each cell type, and variation of conditions for a particular cell type can result in different phenotypes.

Aside from temperature and gas mixture, the most commonly varied factor in culture systems is the cell growth medium. Recipes for growth media can vary in pH, glucose concentration, growth factors, and the presence of other nutrients. The growth factors used to supplement media are often derived from the serum of animal blood, such as fetal bovine serum (FBS), bovine calf serum, equine serum, and porcine serum. One complication of these blood-derived ingredients is the potential for contamination of the culture with viruses or prions, particularly in medical biotechnology applications. Current practice is to minimize or eliminate the use of these ingredients wherever possible and use human platelet lysate (hPL). This eliminates the worry of cross-species contamination when using FBS with human cells. hPL has emerged as a safe and reliable alternative as a direct replacement for FBS or other animal serum. In addition, chemically defined media can be used to eliminate any serum trace (human or animal), but this cannot always be accomplished with different cell types. Alternative strategies involve sourcing the animal blood from countries with minimum BSE/TSE risk, such as The United States, Australia and New Zealand, and using purified nutrient concentrates derived from serum in place of whole animal serum for cell culture.

Plating density (number of cells per volume of culture medium) plays a critical role for some cell types. For example, a lower plating density makes granulosa cells exhibit estrogen production, while a higher plating density makes them appear as progesterone-producing theca lutein cells.

Cells can be grown either in suspension or adherent cultures. Some cells naturally live in suspension, without being attached to a surface, such as cells that exist in the bloodstream. There are also cell lines that have been modified to be able to survive in suspension cultures so they can be grown to a higher density than adherent conditions would allow. Adherent cells require a surface, such as tissue culture plastic or microcarrier, which may be coated with extracellular matrix (such as collagen and laminin) components to increase adhesion properties and provide other signals needed for growth and differentiation. Most cells derived from solid tissues are adherent. Another type of adherent culture is organotypic culture, which involves growing cells in a three-dimensional (3-D) environment as opposed to two-dimensional culture dishes. This 3D culture system is biochemically and physiologically more similar to *in vivo* tissue, but is technically challenging to maintain because of many factors (e.g. diffusion).

Cel l Line Cross- contamination

Cell line cross-contamination can be a problem for scientists working with cultured cells. Studies suggest anywhere from 15–20% of the time, cells used in experiments have been misidentified or contaminated with another cell line. Problems with cell line cross-contamination have even been detected in lines from the NCI-60 panel, which are used routinely for drug-screening studies. Major cell line repositories, including the American Type Culture Collection (ATCC), the European Collection of Cell Cultures (ECACC) and the German Collection of Microorganisms and Cell Cultures (DSMZ), have received cell line submissions from researchers that were misidentified by them. Such contamination poses a problem for the quality of research produced using cell culture lines, and the major repositories are now authenticating all cell line submissions. ATCC uses short tandem repeat (STR) DNA fingerprinting to authenticate its cell lines.

To address this problem of cell line cross-contamination, researchers are encouraged to authenticate their cell lines at an early passage to establish the identity of the cell line. Authentication should be repeated before freezing cell line stocks, every two months during active culturing and before any publication of research data generated using the cell lines. Many methods are used to identify cell lines, including isoenzyme analysis, human lymphocyte antigen (HLA) typing, chromosomal analysis, karyotyping, morphology and STR analysis.

One significant cell-line cross contaminant is the immortal HeLa cell line.

Other Technical Issues

As cells generally continue to divide in culture, they generally grow to fill the available area or volume. This can generate several issues:

- Nutrient depletion in the growth media.
- Changes in pH of the growth media.
- Accumulation of apoptotic/necrotic (dead) cells.
- Cell-to-cell contact can stimulate cell cycle arrest, causing cells to stop dividing, known as contact inhibition.
- Cell-to-cell contact can stimulate cellular differentiation.
- Genetic and epigenetic alterations, with a natural selection of the altered cells potentially leading to overgrowth of abnormal, culture-adapted cells with decreased differentiation and increased proliferative capacity.

Manipulation of Cultured Cells

Among the common manipulations carried out on culture cells are media changes, passaging cells, and transfecting cells. These are generally performed using tissue culture methods that rely on aseptic technique. Aseptic technique aims to avoid contamination with bacteria, yeast, or other cell lines. Manipulations are typically carried out in a biosafety hood or laminar flow cabinet to exclude contaminating micro-organisms. Antibiotics (e.g. penicillin and streptomycin) and antifungals (e.g.amphotericin B) can also be added to the growth media.

As cells undergo metabolic processes, acid is produced and the pH decreases. Often, a pH indicator is added to the medium to measure nutrient depletion.

Media Changes

In the case of adherent cultures, the media can be removed directly by aspiration, and then is replaced. Media changes in non-adherent cultures involve centrifuging the culture and resuspending the cells in fresh media.

Passaging Cells

Passaging (also known as subculture or splitting cells) involves transferring a small number of cells into a new vessel. Cells can be cultured for a longer time if they are split regularly, as it avoids the senescence associated with prolonged high cell density. Suspension cultures are easily passaged with a small amount of culture containing a few cells diluted in a larger volume of fresh media. For adherent cultures, cells first need to be detached; this is commonly done with a mixture of trypsin-EDTA; however, other enzyme mixes are now available for this purpose. A small number of detached cells can then be used to seed a new culture. Some cell cultures, such as RAW cells are mechanically scraped from the surface of their vessel with rubber scrapers.

Transfection and Transduction

Another common method for manipulating cells involves the introduction of foreign DNA by transfection. This is often performed to cause cells to express a gene of interest. More recently, the transfection of RNAi constructs have been realized as a convenient mechanism for suppressing the expression of a particular gene/protein. DNA can also be inserted into cells using viruses, in methods referred to as transduction, infection or transformation. Viruses, as parasitic agents, are well suited to introducing DNA into cells, as this is a part of their normal course of reproduction.

Established Human Cell Lines

Cell lines that originate with humans have been somewhat controversial in bioethics, as they may outlive their parent organism and later be used in the discovery of lucrative medical treatments. In the pioneering decision in this area, the Supreme Court of California held in *Moore v. Regents of the University of California* that human patients have no property rights in cell lines derived from organs removed with their consent.

It is possible to fuse normal cells with an immortalised cell line. This method is used to produce monoclonal antibodies. In brief, lymphocytes isolated from the spleen (or possibly blood) of an immunised animal are combined with an immortal myeloma cell line (B cell lineage) to produce a hybridoma which has the antibody specificity of the primary lymphocyte and the immortality of the myeloma. Selective growth medium (HA or HAT) is used to select against unfused myeloma cells; primary lymphoctyes die quickly in culture and only the fused cells survive. These are screened for production of the required antibody, generally in pools to start with and then after single cloning.

Cell Strains

A cell strain is derived either from a primary culture or a cell line by the selection or cloning of cells having specific properties or characteristics which must be defined. Cell strains are cells that have been adapted to culture but, unlike cell lines, have a finite division potential. Non-immortalized cells stop dividing after 40 to 60 population doublings and, after this, they lose their ability to proliferate (a genetically determined event known as senescence).

Applications of Cell Culture

Mass culture of animal cell lines is fundamental to the manufacture of viral vaccines and other products of biotechnology. Culture of human stem cells is used to expand the number of cells and differentiate the cells into various somatic cell types for transplantation. Stem cell culture is also used to harvest the molecules and exosomes that the stem cells release for the purposes of therapeutic development.

Biological products produced by recombinant DNA (rDNA) technology in animal cell cultures include enzymes, synthetic hormones, immunobiologicals (monoclonal antibodies, interleukins, lymphokines), and anticancer agents. Although many simpler proteins can be produced using rDNA in bacterial cultures, more complex proteins that are glycosylated (carbohydrate-modified) currently must be made in animal cells. An important example of such a complex protein is the hormone erythropoietin. The cost of growing mammalian cell cultures is high, so research is underway to produce such complex proteins in insect cells or in higher plants, use of single embryonic cell and somatic embryos as a source for direct gene transfer via particle bombardment, transit gene expression and confocal microscopy observation is one of its applications. It also offers to confirm single cell origin of somatic embryos and the asymmetry of the first cell division, which starts the process.

Cell Culture in Two Dimensions

Research in tissue engineering, stem cells and molecular biology primarily involves cultures of cells on flat plastic dishes. This technique is known as two-dimensional (2D) cell culture, and was first developed by Wilhelm Roux who, in 1885, removed a portion of the medullary plate of an embryonic chicken and maintained it in warm saline for several days on a flat glass plate. From the advance of polymer technology arose today's standard plastic dish for 2D cell culture, commonly known as the Petri dish. Julius Richard Petri, a German bacteriologist, is generally credited with this invention while working as an assistant to Robert Koch. Various researchers today also utilize culturing laboratory flasks, conicals, and even disposable bags like those used in single-use bioreactors.

Aside from Petri dishes, scientists have long been growing cells within biologically derived matrices such as collagen or fibrin, and more recently, on synthetic hydrogels such as polyacrylamide or PEG. They do this in order to elicit phenotypes that are not expressed on conventionally rigid substrates. There is growing interest in controlling matrix stiffness, a concept that has led to discoveries in fields such as:

- Stem cell self-renewal.
- Lineage specification.
- Cancer cell phenotype.
- Fibrosis.
- Hepatocyte function.
- Mechanosensing.

Cell Culture in Three Dimensions

Cell culture in three dimensions has been touted as "Biology's New Dimension". Nevertheless, the practice of cell culture remains based on varying combinations of single or multiple cell structures in 2D. That being said, there is an increase in use of 3D cell cultures in research areas including drug discovery, cancer biology, regenerative medicine and basic life science research. There are a variety of platforms used to facilitate the growth of three-dimensional cellular structures including scaffold systems such as hydrogel matrices and solid scaffolds, and scaffold-free systems such as low-adhesion plates, nanoparticle facilitated magnetic levitation, and hanging drop plates.

3D cell Culture in Hydrogels

As the natural extracellular matrix (ECM) is important in the survival, proliferation, differentiation and migration of the cells, different hydrogel culture matrices mimicking natural ECM structure are considered as potential approaches towards in vivo –like cell culturing. Hydrogels are composed of interconnected pores with high water retention, which enables efficient transport of e.g. nutrients and gases. Several different types of hydrogels from natural and synthetic materials are available for 3D cell culture, including e.g. animal ECM extract hydrogels, protein hydrogels, peptide hydrogels, polymer hydrogels, and wood-based nanocellulose hydrogel.

3D Cell Culturing by Magnetic Levitation

3D Cell Culturing by Magnetic Levitation method (MLM) is the application of growing 3D tissue by inducing cells treated with magnetic nanoparticle assemblies in spatially varying magnetic fields using neodymium magnetic drivers and promoting cell to cell interactions by levitating the cells up to the air/liquid interface of a standard petri dish. The magnetic nanoparticle assemblies consist of magnetic iron oxide nanoparticles, gold nanoparticles, and the polymer polylysine. 3D cell culturing is scalable, with the capability for culturing 500 cells to millions of cells or from single dish to high-throughput low volume systems.

Tissue Culture and Engineering

Cell culture is a fundamental component of tissue culture and tissue engineering, as it establishes the basics of growing and maintaining cells *in vitro*. The major application of human cell culture is in stem cell industry, where mesenchymal stem cells can be cultured and cryopreserved for future use. Tissue engineering potentially offers dramatic improvements in low cost medical care for hundreds of thousands of patients annually.

Vaccines

Vaccines for polio, measles, mumps, rubella, and chickenpox are currently made in cell cultures. Due to the H5N1 pandemic threat, research into using cell culture for influenza vaccines is being funded by the United States government. Novel ideas in the field include recombinant DNA-based vaccines, such as one made using human adenovirus (a common cold virus) as a vector, and novel adjuvants.

Culture of Non-mammalian Cells

Plant Cell Culture Methods

Plant cell cultures are typically grown as cell suspension cultures in a liquid medium or as callus cultures on a solid medium. The culturing of undifferentiated plant cells and calli requires the proper balance of the plant growth hormones auxin and cytokinin.

Insect Cell Culture

Cells derived from Drosophila melanogaster (most prominently, Schneider 2 cells) can be used for experiments which may be hard to do on live flies or larvae, such as biochemical studies or studies using siRNA. Cell lines derived from the army worm *Spodoptera frugiperda*, including Sf9 and Sf21, and from the cabbage looper *Trichoplusia ni*, High Five cells, are commonly used for expression of recombinant proteins using baculovirus.

Bacterial and Yeast Culture Methods

For bacteria and yeasts, small quantities of cells are usually grown on a solid support that contains nutrients embedded in it, usually a gel such as agar, while large-scale cultures are grown with the cells suspended in a nutrient broth.

Viral Culture Methods

The culture of viruses requires the culture of cells of mammalian, plant, fungal or bacterial origin as hosts for the growth and replication of the virus. Whole wild type viruses, recombinant viruses or viral products may be generated in cell types other than their natural hosts under the right conditions. Depending on the species of the virus, infection and viral replication may result in host cell lysis and formation of a viral plaque.

Common Cell Lines

Human cell lines:

- DU145 (prostate cancer).
- H295R (adrenocortical cancer).
- HeLa (cervical cancer).
- KBM-7 (chronic myelogenous leukemia).
- LNCaP (prostate cancer).
- MCF-7 (breast cancer).
- MDA-MB-468 (breast cancer).
- PC3 (prostate cancer).
- SaOS-2 (bone cancer).
- SH-SY5Y (neuroblastoma, cloned from a myeloma).
- T47D (breast cancer).
- THP-1 (acute myeloid leukemia).
- U87 (glioblastoma).
- National Cancer Institute's 60 cancer cell line panel (NCI60).

Primate cell lines:

- Vero (African green monkey *Chlorocebus* kidney epithelial cell line).

Mouse cell lines:

- MC3T3 (embryonic calvarium).

Rat tumor cell lines:

- GH3 (pituitary tumor).
- PC12 (pheochromocytoma).

Plant cell lines:

- Tobacco BY-2 cells (kept as cell suspension culture, they are model system of plant cell).

Other species cell lines:

- Dog MDCK kidney epithelial.
- Xenopus A6 kidney epithelial.
- Zebrafish AB9.

List of Cell Lines

This list is incomplete; you can help by expanding it.

Cell line	Meaning	Organism	Origin tissue	Morphology	Links
3T3-L1	"3-day transfer, inoculum 3 x 10^5 cells"	Mouse	Embryo	Fibroblast	ECACC Cellosaurus
4T1		Mouse	Mammary gland		ATCC Cellosaurus
9L		Rat	Brain	Glioblastoma	ECACC Cellosaurus
A172		Human	Brain	Glioblastoma	ECACC Cellosaurus
A20		Mouse	B lymphoma	B lymphocyte	Cellosaurus
A253		Human	Submandibular duct	Head and neck carcinoma	ATCC Cellosaurus
A2780		Human	Ovary	Ovarian carcinoma	ECACC Cellosaurus
A2780ADR		Human	Ovary	Adriamycin-resistant derivative of A2780	ECACC Cellosaurus
A2780cis		Human	Ovary	Cisplatin-resistant derivative of A2780	ECACC Cellosaurus
A431		Human	Skin epithelium	Squamous cell carcinoma	ECACC Cellosaurus
A549		Human	Lung	Lung carcinoma	ECACC Cellosaurus
AB9		Zebrafish	Fin	Fibroblast	ATCC Cellosaurus

Cell line	Meaning	Organism	Origin tissue	Morphology	Links
AHL-1	Armenian Hamster Lung-1	Hamster	Lung		ECACC Cellosaurus
ALC		Mouse	Bone marrow	Stroma	PMID 2435412 Cellosaurus
B16		Mouse	Melanoma		ECACC Cellosaurus
B35		Rat	Neuroblastoma		ATCC Cellosaurus
BCP-1		Human	PBMC	HIV+ primary effusion lymphoma	ATCC Cellosaurus
BEAS-2B	Bronchial epithelium + Adenovirus 12-SV40 virus hybrid (Ad12SV40)	Human	Lung	Epithelial	ECACC Cellosaurus
bEnd.3	Brain Endothelial 3	Mouse	Brain/cerebral cortex	Endothelium	Cellosaurus
BHK-21	Baby Hamster Kidney-21	Hamster	Kidney	Fibroblast	ECACC Cellosaurus
BOSC23	Packaging cell line derived from HEK 293	Human	Kidney (embryonic)	Epithelium	Cellosaurus
BT-20	Breast Tumor-20	Human	Breast epithelium	Breast carcinoma	ATCC Cellosaurus
BxPC3	Biopsy xenograft of Pancreatic Carcinoma line 3	Human	Pancreatic adenocarcinoma	Epithelial	ECACC Cellosaurus
C2C12		Mouse	Myoblast		ECACC Cellosaurus
C3H-10T1/2		Mouse	Embryonic mesenchymal cell line		ECACC Cellosaurus
C6		Rat	Brain astrocyte	Glioma	ECACC Cellosaurus
C6/36		Insect - Asian tiger mosquito	Larval tissue		ECACC Cellosaurus
Caco-2		Human	Colon	Colorectal carcinoma	ECACC Cellosaurus
Cal-27		Human	Tongue	Squamous cell carcinoma	ATCC Cellosaurus
CGR8		Mouse	Embryonic stem cells		ECACC Cellosaurus
CHO	Chinese Hamster Ovary	Hamster	Ovary	Epithelium	ECACC Cellosaurus
CML T1	Chronic myeloid leukemia T lymphocyte 1	Human	CML acute phase	T cell leukemia	DSMZ Cellosaurus

Cell line	Meaning	Organism	Origin tissue	Morphology	Links
CMT12	Canine Mammary Tumor 12	Dog	Mammary gland	Epithelium	Cellosaurus
COR-L23		Human	Lung	Lung carcinoma	ECACC Cellosaurus
COR-L23/5010		Human	Lung	Lung carcinoma	ECACC Cellosaurus
COR-L23/CPR		Human	Lung	Lung carcinoma	ECACC Cellosaurus
COR-L23/R23-		Human	Lung	Lung carcinoma	ECACC Cellosaurus
COS-7	*Cercopithecus aethiops*, origin-defective SV-40	Old World monkey - *Cercopithecus aethiops* (*Chlorocebus*)	Kidney	Fibroblast	ECACC Cellosaurus
COV-434		Human	Ovary	Ovarian granulosa cell carcinoma	PMID 8436435 ECACC Cellosaurus
CT26		Mouse	Colon	Colorectal carcinoma	Cellosaurus
D17		Dog	Lung metastasis	Osteosarcoma	ATCC Cellosaurus
DH82		Dog	Histiocytosis	Monocyte/macrophage	ECACC Cellosaurus
DU145		Human	Androgen insensitive prostate carcinoma		ATCC Cellosaurus
DuCaP	Dura mater cancer of the Prostate	Human	Metastatic prostate carcinoma	Epithelial	PMID 11317521 Cellosaurus
E14Tg2a		Mouse		Embryonic stem cells	ECACC Cellosaurus
EL4		Mouse		T cell leukemia	ECACC Cellosaurus
EM-2		Human	CML blast crisis	Ph+ CML line	DSMZ Cellosaurus
EM-3		Human	CML blast crisis	Ph+ CML line	DSMZ Cellosaurus
EMT6/AR1		Mouse	Mammary gland	Epithelial-like	ECACC Cellosaurus
EMT6/AR10.0		Mouse	Mammary gland	Epithelial-like	ECACC Cellosaurus
FM3		Human	Lymph node metastasis	Melanoma	ECACC Cellosaurus
GL261	Glioma 261	Mouse	Brain	Glioma	Cellosaurus
H1299		Human	Lung	Lung carcinoma	ATCC Cellosaurus

Cell line	Meaning	Organism	Origin tissue	Morphology	Links
HaCaT		Human	Skin	Keratinocyte	CLS Cellosaurus
HCA2		Human	Colon	Adenocarcinoma	ECACC Cellosaurus
HEK 293	Human Embryonic Kidney 293	Human	Kidney (embryonic)	Epithelium	ECACC Cellosaurus
HEK 293T	HEK 293 derivative	Human	Kidney (embryonic)	Epithelium	ECACC Cellosaurus
HeLa	"Henrietta Lacks"	Human	Cervix epithelium	Cervical carcinoma	ECACC Cellosaurus
Hepa1c1c7	Clone 7 of clone 1 hepatoma line 1	Mouse	Hepatoma	Epithelial	ECACC Cellosaurus
Hep G2		Human	Liver	Hepatoblastoma	ECACC Cellosaurus
High Five		Insect (moth) - *Trichoplusia ni*	Ovary		Cellosaurus
HL-60	Human Leukemia-60	Human	Blood	Myeloblast	ECACC Cellosaurus
HT-1080		Human		Fibrosarcoma	ECACC Cellosaurus
HT-29		Human	Colon epithelium	Adenocarcinoma	ECACC Cellosaurus
J558L		Mouse	Myeloma	B lymphocyte cell	ECACC Cellosaurus
Jurkat		Human	White blood cells	T cell leukemia	ECACC Cellosaurus
JY		Human	Lymphoblastoid	EBV-transformed B cell	ECACC Cellosaurus
K562		Human	Lymphoblastoid	CML blast crisis	ECACC Cellosaurus
KBM-7		Human	Lymphoblastoid	CML blast crisis	Cellosaurus
KCL-22		Human	Lymphoblastoid	CML	DSMZ Cellosaurus
KG1		Human	Lymphoblastoid	AML	ECACC Cellosaurus
Ku812		Human	Lymphoblastoid	Erythroleukemia	ECACC Cellosaurus
KYO-1	Kyoto-1	Human	Lymphoblastoid	CML	DSMZ Cellosaurus
L1210		Mouse	Lymphocytic leukemia	Ascitic fluid	ECACC Cellosaurus
L243		Mouse	Hybridoma	Secretes L243 mAb (against HLA-DR)	ATCC Cellosaurus
LNCaP	Lymph Node Cancer of the Prostate	Human	Prostatic adenocarcinoma	Epithelial	ECACC Cellosaurus

Cell line	Meaning	Organism	Origin tissue	Morphology	Links
MA2.1		Mouse	Hybridoma	Secretes MA2.1 mAb (against HLA-A2 and HLA-B17)	ATCC Cellosaurus
Ma-Mel 1, 2, 3....48		Human	Skin	A range of melanoma cell lines	ECACC Cellosaurus
MC-38	Mouse Colon-38	Mouse	Colon	Adenocarcinoma	Cellosaurus
MCF-7	Michigan Cancer Foundation-7	Human	Breast	Invasive breast ductal carcinoma ER+, PR+	ECACC Cellosaurus
MCF-10A	Michigan Cancer Foundation-10A	Human	Breast epithelium		ATCC Cellosaurus
MDA-MB-157	M.D. Anderson - Metastatic Breast-157	Human	Pleural effusion metastasis	Breast carcinoma	ECACC Cellosaurus
MDA-MB-231	M.D. Anderson - Metastatic Breast-231	Human	Pleural effusion metastasis	Breast carcinoma	ECACC Cellosaurus
MDA-MB-361	M.D. Anderson - Metastatic Breast-361	Human		Melanoma (contaminated by M14)	ECACC Cellosaurus
MDA-MB-468	M.D. Anderson - Metastatic Breast-468	Human	Pleural effusion metastasis	Breast carcinoma	ATCC Cellosaurus
MDCK II	Madin Darby Canine Kidney II	Dog	Kidney	Epithelium	ECACC Cellosaurus
MG63		Human	Bone	Osteosarcoma	ECACC Cellosaurus
MOR/0.2R		Human	Lung	Lung carcinoma	ECACC Cellosaurus
Mono-Mac-6		Human	White blood cells	Myeloid metaplasic AML	DSMZ Cellosaurus
MRC-5	Medical Research Council cell strain 5	Human	Lung (fetal)	Fibroblast	ECACC Cellosaurus
MTD-1A		Mouse		Epithelium	Cellosaurus
MyEnd	Myocardial Endothelial	Mouse		Endothelium	Cellosaurus
NCI-H69		Human	Lung	Lung carcinoma	ECACC Cellosaurus
NCI-H69/CPR		Human	Lung	Lung carcinoma	ECACC Cellosaurus
NCI-H69/LX10		Human	Lung	Lung carcinoma	ECACC Cellosaurus
NCI-H69/LX20		Human	Lung	Lung carcinoma	ECACC Cellosaurus
NCI-H69/LX4		Human	Lung	Lung carcinoma	ECACC Cellosaurus

Cell line	Meaning	Organism	Origin tissue	Morphology	Links
Neuro-2a		Mouse	Nerve/ neuroblastoma	Neuronal stem cells	ECACC Cellosaurus
NIH-3T3	NIH, 3-day transfer, inoculum 3 x 10^5 cells	Mouse	Embryo	Fibroblast	ECACC Cellosaurus
NALM-1		Human	Peripheral blood	Blast-crisis CML	ATCC Cellosaurus
Neuro2a		Mouse	Nerve/ neuroblastoma	Neuronal stem cells	Cellosaurus
NK-92		Human	Leukemia/ lymphoma		ATCC Cellosaurus
NTERA-2		Human	Lung metastasis	Embryonal carcinoma	ECACC Cellosaurus
NW-145		Human	Skin	Melanoma	ESTDAB Cellosaurus
OK	Opossum Kidney	Virginia opossum - *Didelphis virginiana*	Kidney		ECACC Cellosaurus
OPCN / OPCT cell lines		Human	Prostate	Range of prostate tumour lines	Cellosaurus
P3X63Ag8		Mouse	Myeloma		ECACC Cellosaurus
PC12		Rat	Adrenal medulla	Pheochromocytoma	ECACC Cellosaurus
PC-3	Prostate Cancer-3	Human	Bone metastasis	Prostate carcinoma	ECACC Cellosaurus
Peer		Human	T cell leukemia		DSMZ Cellosaurus
PNT1A		Human	Prostate	SV40-transformed tumour line	ECACC Cellosaurus
PNT2		Human	Prostate	SV40-transformed tumour line	ECACC Cellosaurus
Pt K2	The second cell line derived from *Potorous tridactylis*	Long-nosed potoroo - *Potorous tridactylus*	Kidney	Epithelial	ECACC Cellosaurus
Raji		Human	B lymphoma	Lymphoblast-like	ECACC Cellosaurus
RBL-1	Rat Basophilic Leukemia-1	Rat	Leukemia	Basophil cell	ECACC Cellosaurus
RenCa	Renal Carcinoma	Mouse	Kidney	Renal carcinoma	ATCC Cellosaurus
RIN-5F		Mouse	Pancreas		ECACC Cellosaurus
RMA-S		Mouse		T cell tumour	Cellosaurus

Cell line	Meaning	Organism	Origin tissue	Morphology	Links
S2	Schneider 2	Insect - *Drosophila melanogaster*	Late stage (20–24 hours old) embryos		ATCC Cellosaurus
SaOS-2	Sarcoma OSteogenic-2	Human	Bone	Osteosarcoma	ECACC Cellosaurus
Sf21	*Spodoptera frugiperda* 21	Insect (moth) - *Spodoptera frugiperda*	Ovary		ECACC Cellosaurus
Sf9	*Spodoptera frugiperda* 9	Insect (moth) - *Spodoptera frugiperda*	Ovary		ECACC Cellosaurus
SH-SY5Y		Human	Bone marrow metastasis	Neuroblastoma	ECACC Cellosaurus
SiHa		Human	Cervix epithelium	Cervical carcinoma	ATCC Cellosaurus
SK-BR-3	Sloan-Kettering Breast cancer 3	Human	Breast	Breast carcinoma	DSMZ Cellosaurus
SK-OV-3	Sloan-Kettering Ovarian cancer 3	Human	Ovary	Ovarian carcinoma	ECACC Cellosaurus
T2		Human		T cell leukemia/B cell line hybridoma	ATCC Cellosaurus
T-47D		Human	Breast	Breast ductal carcinoma	ECACC Cellosaurus
T84		Human	Lung metastasis	Colorectal carcinoma	ECACC Cellosaurus
T98G		Human	Glioblastoma-astrocytoma	Epithelium	ECACC Cellosaurus
THP-1		Human	Monocyte	Acute monocytic leukemia	ECACC Cellosaurus
U373		Human	Glioblastoma-astrocytoma	Epithelium	ECACC Cellosaurus
U87		Human	Glioblastoma-astrocytoma	Epithelial-like	ECACC Cellosaurus
U937		Human	Leukemic monocytic lymphoma		ECACC Cellosaurus
VCaP	Vertebral Cancer of the Prostate	Human	Vertebra metastasis	Prostate carcinoma	ECACC Cellosaurus
Vero	*Vero* (truth)	African green monkey - *Chlorocebus sabaeus*	Kidney epithelium		ECACC Cellosaurus
VG-1		Human		Primary effusion lymphoma	Cellosaurus
WM39		Human	Skin	Melanoma	ESTDAB Cellosaurus
WT-49		Human	Lymphoblastoid		ECACC Cellosaurus

Cell line	Meaning	Organism	Origin tissue	Morphology	Links
YAC-1		Mouse	Lymphoma		ECACC Cellosaurus
YAR		Human	Lymphoblastoid	EBV-transformed B cell	Human Immunology ECACC Cellosaurus

Autophagy

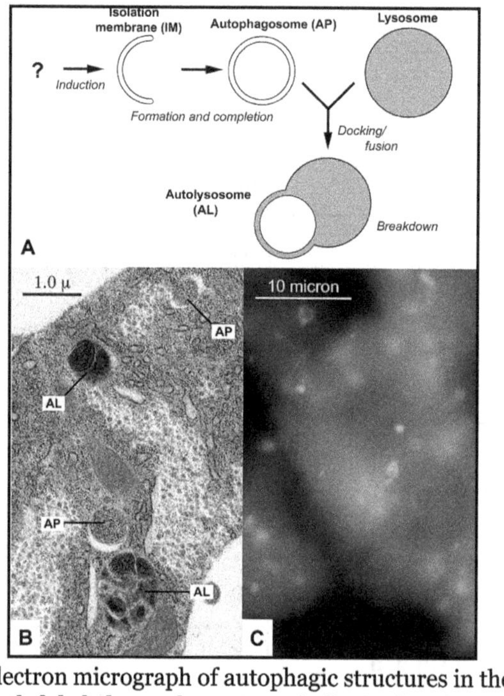

(A) Diagram of autophagy; (B) Electron micrograph of autophagic structures in the fatbody of a fruit fly larva; (C) Fluorescently labeled autophagosomes in liver cells of starved mice.

Autophagy (or autophagocytosis) (from the Greek *auto-*, "self" and *phagein*, "to eat"), is the natural, destructive mechanism that disassembles, through a regulated process, unnecessary or dysfunctional cellular components.

Autophagy allows the orderly degradation and recycling of cellular components. During this process, targeted cytoplasmic constituents are isolated from the rest of the cell within a double-membraned vesicle known as an autophagosome. The autophagosome then fuses with a lysosome and the contents are degraded and recycled. Three different forms of autophagy are commonly described: macroautophagy, microautophagy and chaperone-mediated autophagy. In the context of disease, autophagy has been seen as an adaptive response to stress, which promotes survival, whereas in other cases it appears to promote cell death and morbidity. In the extreme case of starvation, the breakdown of cellular components promotes cellular survival by maintaining cellular energy levels.

The name "autophagy" was coined by Belgian biochemist Christian de Duve in 1963.

History

Autophagy was first observed by Keith R. Porter and his student Thomas Ashford at the Rockefeller Institute. In January 1962 they reported an increased number of lysosomes in rat liver cells after addition of glucagon, and that some displaced lysosomes towards the centre of the cell contained other cell organelles such as mitochondria. They called this autolysis after Christian de Duve. However Porter and Ashford wrongly interpreted their data as lysosome formation (ignoring the pre-existing organelles). Lysosomes could not be cell organelles, but part of cytoplasm such as mitochondria, and that hydrolytic enzymes were produced by microbodies.

In 1963 researchers published a detailed ultrastructural description of "focal cytoplasmic degradation," which referenced a 1955 German study of injury-induced sequestration. The study recognized three continuous stages of maturation of the sequestered cytoplasm to lysosomes, and that the process was not limited to injury states that functioned under physiological conditions for "reutilization of cellular materials," and the "disposal of organelles" during differentiation.

Inspired by this discovery, de Duve christened the phenomena "autophagy". Unlike Porter and Ashford, de Duve conceived the term as a part of lysosomal function while describing the role of glucagon as a major inducer of cell degradation in the liver. With his student Peter, he established that lysosomes are responsible for glucagon-induced autophagy. This was the first time the fact that lysosomes were established as the sites of intracellular autophagy.

Process and Pathways

There are three pathways of autophagy and these are mediated by the autophagy-related genes and their associated enzymes.

Macroautophagy is the main pathway, used primarily to eradicate damaged cell organelles or unused proteins. This involves the formation of a double membrane known as an autophagosome, about the organelle marked for destruction. The autophagosome then travels through the cytoplasm of the cell to a lysosome, and the two organelles fuse. Within the lysosome, the contents of the autophagosome are degraded via acidic lysosomal hydrolases.

Microautophagy, on the other hand, involves the direct engulfment of cytoplasmic material into the lysosome. This occurs by invagination, meaning the inward folding of the lysosomal membrane, or cellular protrusion.

Chaperone-mediated autophagy, or CMA, is a very complex and specific pathway, which involves the recognition by the hsc70-containing complex. This means that a protein must contain the recognition site for this hsc70 complex which will allow it to bind to this chaperone, forming the CMA- substrate/chaperone complex. This complex then moves to the lysosomal membrane-bound protein that will recognise and bind with the CMA receptor, allowing it to enter the cell. Upon recognition, the substrate protein gets unfolded and it is translocated across the lysosome membrane with the assistance of the lysosomal hsc70 chaperone. CMA is significantly different from other types of autophagy because it translocates protein material in a one by one manner, and it is extremely selective about what material crosses the lysosomal barrier.

Molecular Biology

Autophagy is executed by autophagy-related (Atg) genes. The first autophagy genes were identified by genetic screens conducted in the budding yeast Saccharomyces cerevisiae. Following their identification those genes were functionally characterized and their orthologs in a variety of different organisms were identified and studied. In mammals, amino acid sensing and additional signals such as growth factors and reactive oxygen species regulate the activity of the protein kinases mTOR and AMPK. These two kinases regulate autophagy through inhibitory phosphorylation of the Unc-51-like kinases ULK1 and ULK2 (mammalian homologues of Atg1). Induction of autophagy results in the dephosphorylation and activation of the ULK kinases. ULK is part of a protein complex containing Atg13, Atg101 and FIP200. ULK phosphorylates and activates Beclin-1 (mammalian homologue of Atg6), which is also part of a protein complex. The autophagy-inducible Beclin-1 complex contains the proteins p150, Atg14L and the class III phosphatidylinositol 3-phosphate kinase (PI(3)K) VPS34. The active ULK and Beclin-1 complexes re-localize to the site of autophagosome initiation, the phagophore, where they both contribute to the activation of downstream autophagy components. Once active, VPS34 phosphorylates the lipid phosphatidylinositol to generate phosphatidylinositol 3-phosphate (PtdIns(3)P) on the surface of the phagophore. The generated PtdIns(3)P is used as a docking point for proteins harboring a PtdIns(3)P binding motif. WIPI2, a PtdIns(3)P binding protein of the WIPI (WD-repeat protein interacting with phosphoinositides) protein family, was recently shown to physically bind Atg16L1. Atg16L1 is a member of an E3-like protein complex involved in one of two ubiquitin-like conjugation systems essential for autophagosome formation. Its binding by WIPI2 recruits it to the phagophore and mediates its activity. The first of the two ubiquitin-like conjugation systems involved in autophagy covalently binds the ubiquitin-like protein Atg12 to Atg5. The resulting conjugate protein then binds Atg16L1 to form an E3-like complex which functions as part of the second ubiquitin-like conjugation system. This complex binds and activates Atg3, which covalently attaches mammalian homologues of the ubiquitin-like yeast protein Atg8 (LC3A-C, GATE16, and GABARAPL1-3), the most studied being LC3 proteins, to the lipid phosphatidylethanolamine (PE) on the surface of autophagosomes. Lipidated LC3 contributes to the closure of autophagosomes, and enables the docking of specific cargos and adaptor proteins such as p62. The completed autophagosome then fuses with a lysosome through the actions of multiple proteins, including SNAREs and UVRAG. Following the fusion LC3 is retained on the vesicle's inner side and degraded along with the cargo, while the LC3 molecules attached to the outer side are cleaved off by Atg4 and recycled. The contents of the autolysosome are subsequently degraded and their building blocks are released from the vesicle through the action of permeases.

Functions

Nutrient Starvation

Autophagy has roles in various cellular functions. One particular example is in yeasts, where the nutrient starvation induces a high level of autophagy. This allows unneeded proteins to be degraded and the amino acids recycled for the synthesis of proteins that are essential for survival. In higher eukaryotes, autophagy is induced in response to the nutrient depletion that occurs in animals at birth after severing of the trans-placental food supply, as well as that of nutrient starved cultured cells and tissues. Mutant yeast cells that have a reduced autophagic capability rapidly perish in nutrition-deficient conditions. Studies on the *apg* mutants suggest that autophagy via autophagic bodies is indispensable for protein degradation in the vacuoles under starvation conditions, and that at least 15 APG genes are involved in autophagy in yeast. A gene known as ATG7 has been implicated in nutrient-mediated autophagy, as mice studies have shown that starvation-induced autophagy was impaired in *atg7*-deficient mice.

Xenophagy

In microbiology, xenophagy is the autophagic degradation of infectious particles. Cellular autophagic machinery also play an important role in innate immunity. Intracellular pathogens, such as *Mycobacterium tuberculosis* (the bacterium which is responsible for tuberculosis) are targeted for degradation by the same cellular machinery and regulatory mechanisms that target host mitochondria for degradation. Incidentally, this is further evidence for the endosymbiotic hypothesis. This process generally leads to the destruction of the invasive organism, although some bacteria can block the maturation of phagosomes into degradative organelles called phagolysosomes. Stimulation of autophagy in infected cells can help overcome this phenomenon, restoring pathogen degradation.

Autophagy and Infection

Vesicular stomatitis virus is believed to be taken up by the autophagosome from the cytosol and translocated to the endosomes where detection takes place by a member of the PRRs called toll-like receptor 7, detecting single stranded RNA. Following activation of the toll-like receptor, intracellular signaling cascades are initiated, leading to induction of interferon and other antiviral cytokines. A subset of viruses and bacteria subvert the autophagic pathway to promote their own replication. Galectin-8 has recently been identified as an intracellular "danger receptor", able to initiate autophagy against intracellular pathogens. When galectin-8 binds to a damaged vacuole, it recruits autophagy adaptor such as NDP52 leading to the formation of an autophagosome and bacterial degradation.

Repair Mechanism

Autophagy degrades damaged organelles, cell membranes and proteins, and the failure of autophagy is thought to be one of the main reasons for the accumulation of cell damage and aging.

Programmed Cell Death

One of the mechanisms of programmed cell death (PCD) is associated with the appearance of autophagosomes and depends on autophagy proteins. This form of cell death most likely corresponds to a process that has been morphologically defined as autophagic PCD. One question that constantly arises, however, is whether autophagic activity in dying cells is the cause of death or is actually an attempt to prevent it. Morphological and histochemical studies so far did not prove a causative relationship between the autophagic process and cell death. In fact, there have recently been strong arguments that autophagic activity in dying cells might actually be a survival mechanism. Studies of the metamorphosis of insects have shown cells undergoing a form of PCD that appears distinct from other forms; these have been proposed as examples of autophagic cell death.

Recent pharmacological and biochemical studies have proposed that survival and lethal autophagy can be distinguished by the type and degree of regulatory signaling during stress particularly after viral infection. Although promising, these findings have not been examined in non-viral systems.

Autophagy and Caloric Restriction

Research suggests that autophagy is required for the lifespan-prolonging effects of caloric restriction. A 2010 French study of nematodes, mice and flies showed that inhibition of autophagy exposed cells to metabolic stress. Resveratrol and the dietary restriction prolonged the lifespan of normal, autophagy-proficient nematodes, but not of nematodes in which autophagy had been inhibited by knocking out Beclin 1 (a known autophagic modulator).

Autophagy and Exercise

Autophagy is essential for basal homeostasis; it is also extremely important in maintaining muscle homeostasis during physical exercise. Autophagy at the molecular level is only partially understood. A study of mice shows that autophagy is important for the ever changing demands of their nutritional and energy needs, particularly through the metabolic pathways of protein catabolism. In a 2012 study conducted by the University of Texas Southwestern Medical Center in Dallas, mutant mice (with a knock-in mutation of BCL2 phosphorylation sites to produce progeny that showed normal levels of basal autophagy yet were deficient in stress-induced autophagy) were tested to challenge this theory. Results showed that when compared to a control group, these mice illustrated a decrease in endurance and an altered glucose metabolism during acute exercise.

Another study demonstrated that skeletal muscle fibres of collagen VI knockout mice showed signs of degeneration due to an insufficiency of autophagy which led to an accumulation of damaged mitochondria and excessive cell death. Exercise-induced autophagy was unsuccessful however; but when autophagy was induced artificially post-exercise, the accumulation of damaged organelles in collagen VI deficient muscle fibres was prevented and cellular homeostasis was maintained. Both studies demonstrate that autophagy induction may contribute to the beneficial metabolic effects of exercise and that it is essential in the maintaining of muscle homeostasis during exercise, particularly in collagen VI fibres.

Work at the Institute for Cell Biology, University of Bonn, showed that a certain type of autophagy, i.e., chaperone-assisted selective autophagy (CASA), is induced in contracting muscles and is required for maintaining the muscle sarcomere under mechanical tension. The CASA chaperone complex recognizes mechanically damaged cytoskeleton components and directs these components through a ubiquitin-dependent autophagic sorting pathway to lysosomes for disposal. This is necessary for maintaining muscle activity.

Autophagy and Osteoarthritis

Because autophagy decreases with age and age is a major risk factor for osteoarthritis, the role of autophagy in the development of this disease is suggested. Proteins involved in autophagy are reduced with age in both human and mouse articular cartilage. Mechanical injury to cartilage explants in culture also reduced autophagy proteins. Autophagy is constantly activated in normal cartilage but it is compromised with age and precedes cartilage cell death and structural damage.

These results suggest autophagy is a normal protective process (chondroprotection) in the joint.

Autophagy and Cancer

Oftentimes, cancer occurs when several different pathways that regulate cell differentiation are disturbed. Autophagy plays an important role in cancer – both in protecting against cancer as well as potentially contributing to the growth of cancer. Autophagy can contribute to cancer by promoting survival of tumor cells that have been starved, or that degrade apoptotic mediators through autophagy: in such cases, use of inhibitors of the late stages of autophagy (such as chloroquine), on the cells that use autophagy to survive, increases the number of cancer cells killed by antineoplastic drugs.

The role of autophagy in cancer is one that has been highly researched and reviewed. There is evidence that emphasizes the role of autophagy both as a tumor suppressor as well as a factor in tumor cell survival. However, recent research has been able to show that autophagy is more likely to be used as a tumor suppressor according to several models.

Tumor Suppressor

Several experiments have been done with mice and varying Beclin1, a protein that regulates autophagy. When the Beclin1 gene was altered to be heterozygous (Beclin 1+/-), the mice were found to be tumor prone. However, when Beclin1 was overexpressed, tumor development was inhibited.

Necrosis and chronic inflammation also has been shown to be limited through autophagy which helps protect against the formation of tumor cells. Thus these experiments show autophagy's role as a tumor suppressor.

Tumor Cell Survival

Alternatively, autophagy has also been shown to play a huge role in tumor cell survival. In cancerous cells, autophagy is used as a way to deal with stress on the cell. Once these autophagy related genes were inhibited, cell death was potentiated. The increase in metabolic energy is offset by autophagy functions. These metabolic stresses include hypoxia, nutrient deprivation, and an increase in proliferation. These stresses activate autophagy in order to recycle ATP and maintain survival of the cancerous cells. Autophagy has been shown to enable continued growth of tumor cells by maintaining cellular energy production. By inhibiting autophagy genes in these tumors cells, regression of the tumor and extended survival of the organs affected by the tumors were found. Furthermore, inhibition of autophagy has also been shown to enhance the effectiveness of anticancer therapies.

Mechanism of Cell Death

Cells that undergo an extreme amount of stress experience cell death either through apoptosis or necrosis. Prolonged autophagy activation leads to a high turnover rate of proteins and organelles. A high rate above the survival threshold may kill cancer cells with a high apoptotic threshold. This technique can be utilized as a therapeutic cancer treatment.

Therapeutic Target

New developments in research have found that targeted autophagy may be a viable therapeutic solution in fighting cancer. As discussed above, autophagy plays both a role in tumor suppression and tumor cell survival. Thus, the qualities of autophagy can be used as a strategy for cancer prevention. The first strategy is to induce autophagy and enhance its tumor suppression attributes. The second strategy is to inhibit autophagy and thus induce apoptosis.

The first strategy has been tested by looking at dose-response anti-tumor effects during autophagy-induced therapies. These therapies have shown that autophagy increases in a dose-dependent manner. This is directly related to the growth of cancer cells in a dose-dependent manner as well. This data supports the development of therapies that will encourage autophagy. Secondly, inhibiting the protein pathways directly known to induce autophagy may also serve as an anticancer therapy.

The second strategy is based on the idea that autophagy is a protein degradation system used to maintain homeostasis and the findings that inhibition of autophagy often leads to apoptosis. Inhibition of autophagy is riskier as it may lead to cell survival instead of the desired cell death.

Mitophagy

Mitophagy is the selective degradation of mitochondria by autophagy. It often occurs to defective mitochondria following damage or stress. This process was first mentioned by J.J. Lemasters in 2005, although lysosomes in the liver that contained mitochondrial fragments had been seen as early as 1962, "As part of almost every lysosome in these glucagon-treated cells it is possible to recognize a mitochondrion or a remnant of one. It was also mentioned in 1977 by scientists studying metamorphosis in silkworms, "...mitochondria develop functional alterations which would activate autophagy." Mitophagy is key in keeping the cell healthy. It promotes turnover of mitochondria and prevents accumulation of dysfunctional mitochondria which can lead to cellular degeneration. It is mediated by Atg32 (in yeast) and NIP3-like protein X (NIX). Mitophagy is regulated by PINK1 and parkin protein. The occurrence of mitophagy is not limited to the damaged mitochondria but also involves undamaged ones.

Role

Organelles and bits of cytoplasm are sequestered and targeted for degradation by the lysosome for hydrolytic digestion by a process known as autophagy. Mitochondria metabolism leads to the creation of by-products that lead to DNA damage and mutations. Therefore, a healthy population of mitochondria is critical for the well-being of cells. Previously it was thought that targeted degradation of mitochondria was a stochastic event, but accumulating evidence suggest that mitophagy is a selective process.

Generation of ATP by oxidative phosphorylation leads to the production of various reactive oxygen species (ROS) in the mitochondria, and submitochondrial particles. Formation of ROS as a mitochondrial waste product will eventually lead to cytotoxicity and cell death. Because of their role in metabolism, mitochondria are very susceptible to ROS damage. Damaged mitochondria causes a depletion in ATP and a release of cytochrome *c*, which leads to activation of caspases and onset of apoptosis. Mitochondrial damage is not caused solely by oxidative stress or disease processes; normal mitochondria will eventually accumulate oxidative damage hallmarks overtime, which can be deleterious to mitochondria as well as to the cell. These faulty mitochondria can further deplete the cell from ATP, increase production of ROS, and release proapoptopic proteins such as caspases.

Because of the danger of having damaged mitochondria in the cell, the timely elimination of damaged and aged mitochondria is essential for maintaining the integrity of the cell. This turnover process consists of the sequestration and hydrolytic degradation by the lysosome, a process also known as mitophagy.

Mitochondrial depletion reduces a spectrum of senescence effectors and phenotypes while preserving ATP production via enhanced glycolysis.

Pathways

Mitophagy in Mammals

There are several ways that mitophagy is induced in mammalian cells. The PINK1 and Parkin pathway is, so far, the best characterized. This pathway starts in by deciphering the difference between healthy mitochondria and damaged mitochondria. A 64-kDa protein, PTEN-induced kinase 1 (PINK1), has been implicated to detect mitochondrial quality. PINK1 contains a mitochondrial targeting sequence (MTS) and is recruited to the mitochondria. In healthy mitochondria, PINK1 is imported through the outer membrane via the TOM complex, and partially through the inner mitochondrial membrane via the TIM complex, so it then spans the inner mitochondrial membrane. The process of import into the inner membrane is associated -----with the cleavage of PINK1 from 64-kDa into 60-kDa. Pink1 is then cleaved by PARL into 52-kDa. This new form of PINK1 is degraded by proteases within the mitochondria. This keeps the concentration of PINK1 in check in healthy mitochondria.

In unhealthy mitochondria, the interior mitochondrial membrane becomes depolarized. This membrane potential is necessary for the TIM-mediated protein import. In depolarized mitochondria, PINK1 is no longer imported into the inner membrane, is not cleaved by PARL and PINK1 concentration increases in the outer mitochondrial membrane. PINK1 can then recruit Parkin. It is thought that PINK1 phosphorylates Parkin ubiquitin at S65 which initiates Parkin recruitment at the mitochondria. Parkin is a cystolic E3 ubiquitin ligase 8. Once localized at the mitochondria, PINK1 phosphorylates Parkin at S65, homologous to the site where ubiquitin was phosphorylated, which activates Parkin by inducing dimerization and an active state. This allows for Parkin mediated ubiquitination on other proteins.

Because of the PINK1 mediated recruitment to the mitochondrial surface, Parkin can ubiquitylate proteins in the outer mitochondrial membrane. Some of these proteins include Mfn1/Mfn2 and mitoNEET. The ubiquitylation of mitochondrial surface proteins brings in mitophagy initiating factors. Parkin promotes ubiquitin chain linkages on both K63 and K48. K48 Ubiquitination initiates degradation of the proteins, and could allow for passive mitochondrial degradation. K63 ubiquitination is thought to recruit autophagy adaptors LC3/GABARAP which will then lead to mitophagy. It is still unclear which proteins are necessary and sufficient for mitophagy, and how these proteins, once ubiquitylated, initiate mitophagy.

Other pathways that can induce mitophagy includes mitophagy receptors on the outer mitochondrial membrane surface. These receptors include NIX1, BNIP3 and FUNDC1. All of these receptors contain LIR consensus sequences that bind LC3/GABARAP which can lead to the degradation of the mitochondria. In hypoxic conditions BNIP3 is upregulated by HIF1α. BNIP3 is then phosphorylated at its serine residues near the LIR sequence which promotes LC3 binding. FUNDCI is also hypoxia sensitive, although it is constitutively present at the outer mitochondrial membrane during normal conditions.

Mitophagy in Yeast

Mitophagy in yeast was first presumed after the discovery of Yeast Mitochondrial Escape genes (yme), specifically yme1. Yme1 like other genes in the family showed increase escapes of mtDNA, but was the only one that showed in increase in mitochondrial degradation. Through work on this gene that mediates the escape of mtDNA, researchers discovered that mitochondrial turnover is triggered by proteins.

More was discovered about genetic control of mitophagy after studies done of UTH1. After doing a screen for genes that regulate longevity. It was found in ΔUTH1 strains there was an inhibition of mitophagy, which occurred without affecting autophagy mechanisms. It also showed that Uth1p protein is necessary to move mitochondria to the vacuole. This suggested there is a specialized system for mitophagy. Other studies looked at AUP1, a mitochondrial phosphatase, and found Aup1 marks mitochondria for elimination.

Another yeast protein associated with mitophagy is a mitochondrial inner membrane protein, Mdm38p/Mkh1p. This protein is part of the complex that exchanges K+/H+ ions across the inner membrane. Deletions to this protein causes swelling, a loss of membrane potential, and mitochondrial fragmentation.

Recently, it has been shown that ATG32 (autophagy related gene 32) plays a crucial role in yeast mitophagy. It is localized to the mitochondria. Once mitophagy is initiated, Atg32 binds to Atg11 and the Atg32-associated mitochondria is transported to the vacuole. Atg32 silencing stops recruitment of autophagy machinery and mitochondrial degradation. Atg32 is not necessary for other forms of autophagy.

All of these proteins likely play a role in maintaining a healthy mitochondria, but mutations have shown that dysregulation can lead to a selective degradation of mitochondria. Whether these proteins work in concert, are main players in mitophagy, or the networks controlled still remain to be elucidated.

Relation to Disease

Mitophagy and Cancer

In 1920 Otto Warburg observed that certain cancerous tumors display a metabolic shift towards glycolysis. This hypothesis is referred to as the "Warburg Effect", in which cancer cells produce energy via the conversion of glucose into lactate, even in the presence of oxygen (aerobic glycolysis). Despite nearly a century since it was first described, a lot of questions remained unanswered regarding the Warburg effect. Initially, Warburg attributed this metabolic shift to mitochondrial dysfunction in cancer cells. Further studies in tumor biology have shown that the increased growth rate in cancer cells is due to an overdrive in glycolysis (glygolytic shift), which leads to a decrease in oxidative phosphorylation and mitochondrial density. As a consequence of the Warburg effect, cancer cells would produce large amounts of lactate. The excess lactate is then released to the extracellular environment which results in a decrease in extracellular pH. This micro-environment acidification can lead to cellular stress, which would lead to autophagy. Autophagy is activated in response to a range of stimuli, including nutrient depletion, hypoxia, and activated oncogenes. However, it appears that autophagy can help in cancer cell survival under conditions of metabolic stress and it may confer resistance to anti-cancer therapies such as radiation and chemotherapy. Additionally, in the microenvironment of cancer cells, there is an increase in hypoxia-inducible transcription factor 1-alpha (HIF1A), which promotes expression of BNIP3, an essential factor for mitophagy.

Mitophagy and Parkinson's Disease

Parkinson's disease is a neurodegenerative disorder partially caused by the cell death of dopamine creating cells in the substantia nigra. There are several genetic mutations implicated in Parkinson's disease, including loss of function PINK1 and Parkin. Loss of function in these genes can lead to damaged mitochondrial accumulation and protein aggregates than can lead to cellular degeneration.

Mitochondria is thought to be involved in Parkinson's disease. In spontaneous, usually aging related Parkinson's disease (non-genetically linked), the disease is commonly caused by dysfunctional mitochondria, cellular oxidative stress, autophagic alterations and the aggregation of proteins. These can lead to mitochondrial swelling and depolarization. It is important to keep the dysfunctional mitochondria regulated, because all of these traits could be induced by mitochondrial dysfunction and can induce cell death. Disorders in energy creation by mitochondria can cause cellular degeneration, like those seen in the substantia nigra.

References

- Kobayashi S (2015). "Choose Delicately and Reuse Adequately: The Newly Revealed Process of Autophagy". Biological & Pharmaceutical Bulletin 38 (8): 1098–103. doi:10.1248/bpb.b15-00096. PMID 26235572
- Cuervo AM, Bergamini E, Brunk UT, Dröge W, Ffrench M, Terman A (2005). "Autophagy and aging: the importance of maintaining "clean" cells". Autophagy 1 (3): 131–40. doi:10.4161/auto.1.3.2017. PMID 16874025
- Tavassoly, Iman (2015). Dynamics of Cell Fate Decision Mediated by the Interplay of Autophagy and Apoptosis in Cancer Cells. Springer International Publishing. ISBN 978-3-319-14962-2
- Derosa, Maria C.; Monreal, Carlos; Schnitzer, Morris; Walsh, Ryan; Sultan, Yasir (2010). "Nanotechnology in fertilizers". Nature Nanotechnology. 5 (2): 91. doi:10.1038/nnano.2010.2. PMID 20130583
- Baxevanis CN, Chrousos GP (2014). "New use for old drugs? Prospective targets of chloroquines in cancer therapy". Curr Drug Targets. 15 (9): 843–51. doi:10.2174/1389450115666140714121514. PMID 25023646
- Inwood, Stephen (2003). The man who knew too much: the strange and inventive life of Robert Hooke, 1635–1703. London: Pan. p. 72. ISBN 0-330-48829-5

- William A. Dunn Jr.; Laura A. Schroder; John P. Aris (2013). "Historical overview of autophagy". In Hong-Gang Wang. Autophagy and Cancer. Springer. pp. 3–4. ISBN 9781461465614
- Carames, B; Olmer, M; Kiosses, WB; Lotz, MK (2015). "The relationship of autophagy defects to Cartilage Damage During joint aging in a mouse model". Arthritis Rheumatol 67 (6): 1568–1576. doi:10.1002/art.39073. PMID 25708836
- Tavassoly, Iman (2015). Dynamics of Cell Fate Decision Mediated by the Interplay of Autophagy and Apoptosis in Cancer Cells. Springer International Publishing. ISBN 978-3-319-14962-2
- T. Proikas-Cezanne, Z. Takacs, P. Donnes, and O. Kohlbacher, 'Wipi Proteins: Essential Ptdins3p Effectors at the Nascent Autophagosome', J Cell Sci, 128 (2015), 207-17
- H. C. Dooley, M. Razi, H. E. Polson, S. E. Girardin, M. I. Wilson, and S. A. Tooze, 'Wipi2 Links Lc3 Conjugation with Pi3p, Autophagosome Formation, and Pathogen Clearance by Recruiting Atg12-5-16l1', Mol Cell, 55 (2014), 238-52
- Kalenderblatt Dezember 2013 – Mathematisch-Naturwissenschaftliche Fakultät – Universität Rostock. Math-nat.uni-rostock.de (2013-11-28). Retrieved 18, June 2020
- S. Park, S. G. Choi, S. M. Yoo, J. H. Son, and Y. K. Jung, 'Choline Dehydrogenase Interacts with Sqstm1/P62 to Recruit Lc3 and Stimulate Mitophagy', Autophagy, 10 (2014), 1906-20
- Y. M. Kim, C. H. Jung, M. Seo, E. K. Kim, J. M. Park, S. S. Bae, and D. H. Kim, 'Mtorc1 Phosphorylates Uvrag to Negatively Regulate Autophagosome and Endosome Maturation', Mol Cell (2014)
- C. A. Lamb, T. Yoshimori, and S. A. Tooze, 'The Autophagosome: Origins Unknown, Biogenesis Complex', Nat Rev Mol Cell Biol, 14 (2013), 759-74

Chapter 5

Cell Cycle

The events through which a cell grows and divides into two new cells is known as the cell cycle. It is a series of ordered steps during which the cell grows, replicates its DNA and finally separates to form two new cells through division. The various phases of the cell cycle are explained in a detailed manner in this chapter.

The cell cycle or cell-division cycle is the series of events that take place in a cell leading to its division and duplication of its DNA (DNA replication) to produce two daughter cells. In bacteria, which lack a cell nucleus, the cell cycle is divided into the B, C, and D periods. The B period extends from the end of cell division to the beginning of DNA replication. DNA replication occurs during the C period. The D period refers to the stage between the end of DNA replication and the splitting of the bacterial cell into two daughter cells. In cells with a nucleus, as in eukaryotes, the cell cycle is also divided into three periods: interphase, the mitotic (M) phase, and cytokinesis. During interphase, the cell grows, accumulating nutrients needed for mitosis, preparing it for cell division and duplicating its DNA. During the mitotic phase, the chromosomes separate. During the final stage, cytokinesis, the chromosomes and cytoplasm separate into two new daughter cells. To ensure the proper division of the cell, there are control mechanisms known as cell cycle checkpoints.

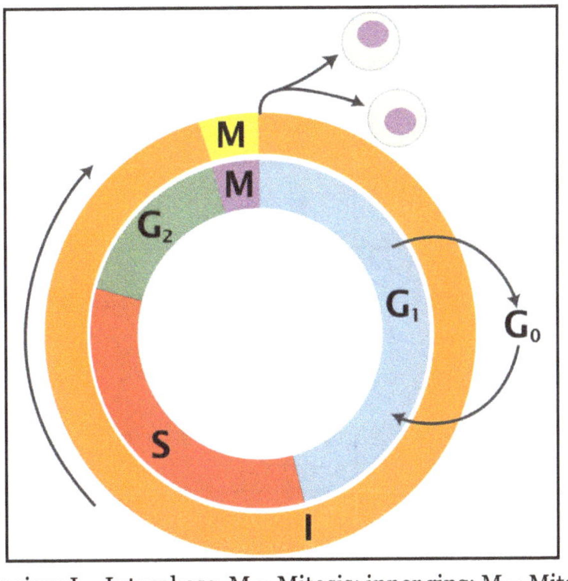

Schematic of the cell cycle. outer ring: I = Interphase, M = Mitosis; inner ring: M = Mitosis, G_1 = Gap 1, G_2 = Gap 2, S = Synthesis; not in ring: G_0 = Gap 0/Resting.

The cell-division cycle is a vital process by which a single-celled fertilized egg develops into a mature organism, as well as the process by which hair, skin, blood cells, and some internal organs are renewed. After cell division, each of the daughter cells begin the interphase of a new cycle. Although the various stages of interphase are not usually morphologically distinguishable, each phase of the cell cycle has a distinct set of specialized biochemical processes that prepare the cell for initiation of cell division.

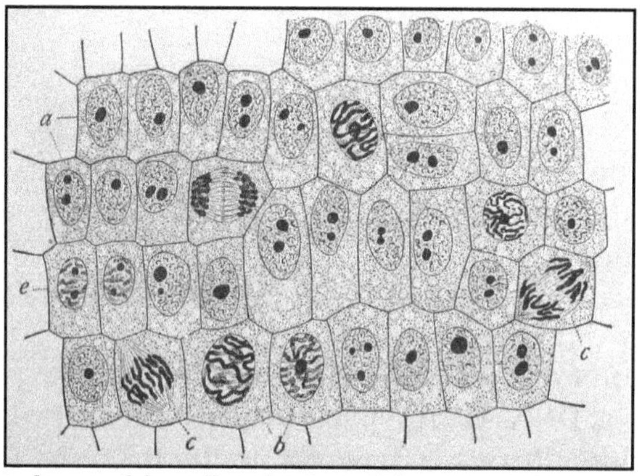

Onion (Allium) cells in different phases of the cell cycle. Growth in an organism is carefully controlled by regulating the cell cycle.

Cell Cycle Phases

State	Description	Abbreviation	:	
Quiescent senescent/ resting phase	Gap 0	**G_0**	A resting phase where the cell has left the cycle and has stopped dividing.Cell cycle starts with this phase.	
	Interphase	Gap 1	**G_1**	Cells increase in size in Gap 1. The *G_1 checkpoint* control mechanism ensures that everything is ready for DNA synthesis.
		Synthesis	**S**	DNA replication occurs during this phase.
Gap 2		**G_2**	During the gap between DNA synthesis and mitosis, the cell will continue to grow. The *G_2 checkpoint* control mechanism ensures that everything is ready to enter the M (mitosis) phase and divide.	
Cell division	Mitosis	**M**	Cell growth stops at this stage and cellular energy is focused on the orderly division into two daughter cells. A checkpoint in the middle of mitosis (*Metaphase Checkpoint*) ensures that the cell is ready to complete cell division.	

G_0 Phase (Quiescence)

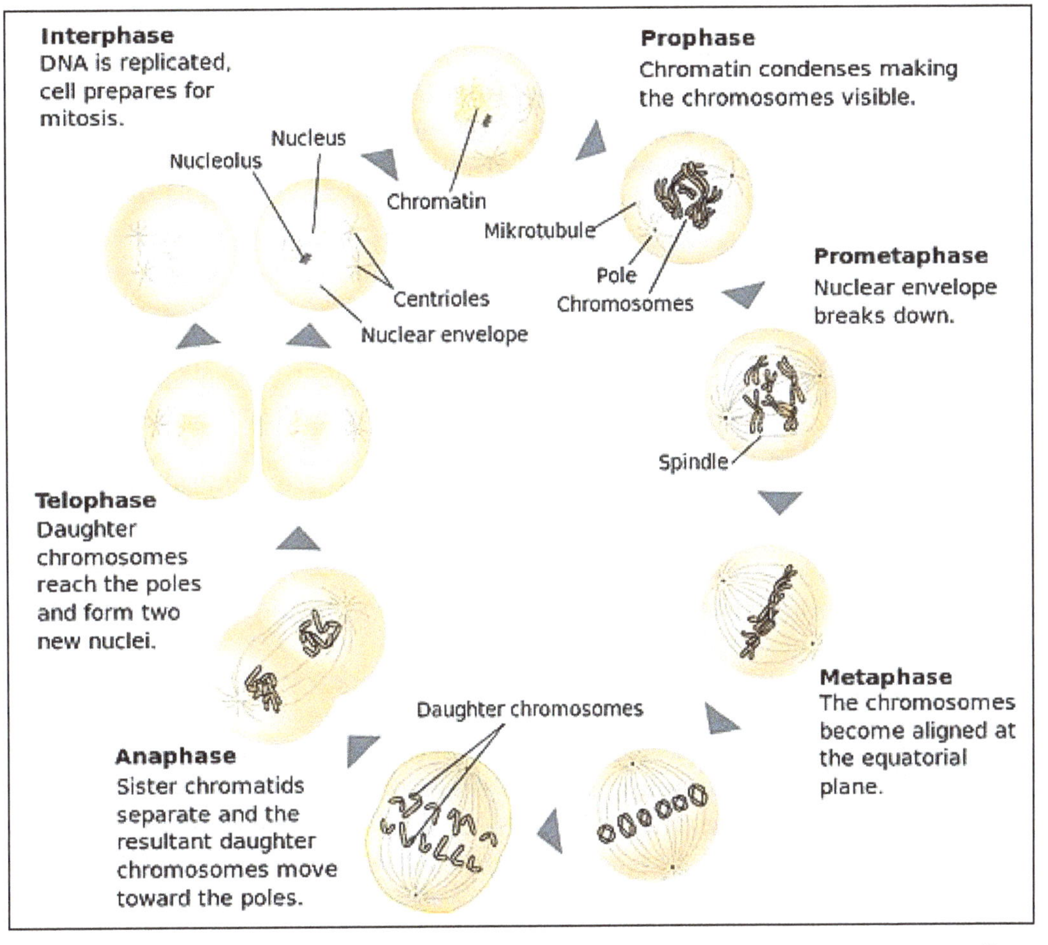

The word "post-mitotic" is sometimes used to refer to both quiescent and senescent cells. Nonproliferative cells in multicellular eukaryotes generally enter the quiescent G_0 state from G_1 and may remain quiescent for long periods of time, possibly indefinitely (as is often the case for neurons). This is very common for cells that are fully differentiated. Cellular senescence occurs in response to DNA damage or degradation that would make a cell's progeny nonviable; for example, become cancerous. Some cells enter the G_0 phase semi-permanently e.g., some liver, kidney, stomach cells. Many cells do not enter G_0 and continue to divide throughout an organism's life, e.g. epithelial cells.

Interphase (Intermitosis)

Before a cell can enter cell division, it needs to take in nutrients. All of the preparations are done during interphase. Interphase is a series of changes that takes place in a newly formed cell and its nucleus, before it becomes capable of division again. It is also called preparatory phase or intermitosis. Previously it was called resting stage because there is no apparent activity related to cell division.Typically interphase lasts for at least 90% of the total time required for the cell cycle.

Interphase proceeds in three stages, G_1, S, and G_2, followed by the cycle of mitosis and cytokinesis. The cell's nuclear chromosomes are duplicated during S phase.

G_1 Phase (Growth)

The first phase within interphase, from the end of the previous M phase until the beginning of DNA synthesis, is called G_1 (G indicating *gap*). It is also called the growth phase. During this phase, the biosynthetic activities of the cell, which are considerably slowed down during M phase, resume at a high rate. The duration of G_1 is highly variable, even among different cells of the same species. In this phase, the cell increases its supply of proteins, increases the number of organelles (such as mitochondria, ribosomes), and grows in size.

S Phase (DNA Replication)

The ensuing S phase starts when DNA replication commences; when it is completed, all of the chromosomes have been replicated, i.e., each chromosome has two (sister) chromatids. Thus, during this phase, the amount of DNA in the cell has effectively doubled, though the ploidy of the cell remains the same. During this phase, synthesis is completed as quickly as possible due to the exposed base pairs being sensitive to harmful external factors such as mutagens.

G_2 Phase (Growth)

G2 phase is a period of protein synthesis and rapid cell growth to prepare the cell for mitosis.

Mitotic Phase (Chromosome Separation)

The relatively brief *M phase* consists of nuclear division (karyokinesis). It is a relatively short period of the cell cycle. M phase is complex and highly regulated. The sequence of events is divided into phases, corresponding to the completion of one set of activities and the start of the next. These phases are sequentially known as:

- prophase,
- metaphase,
- anaphase,
- telophase

Mitosis is the process by which a eukaryotic cell separates the chromosomes in its cell nucleus into two identical sets in two nuclei. During the process of mitosis the pairs of chromosomes condense and attach to fibers that pull the sister chromatids to opposite sides of the cell.

Mitosis occurs exclusively in eukaryotic cells, but occurs in different ways in different species. For example, animals undergo an "open" mitosis, where the nuclear envelope breaks down before the chromosomes separate, while fungi such as *Aspergillus nidulans* and *Saccharomyces cerevisiae* (yeast) undergo a "closed" mitosis, where chromosomes divide within an intact cell nucleus. Prokaryotic cells, which lack a nucleus, divide by a process called binary fission.

Cytokinesis Phase (Separation of All Cell Components)

Mitosis is immediately followed by cytokinesis, which divides the nuclei, cytoplasm, organelles and cell membrane into two cells containing roughly equal shares of these cellular components.

Mitosis and cytokinesis together define the division of the mother cell into two daughter cells, genetically identical to each other and to their parent cell. This accounts for approximately 10% of the cell cycle.

Because cytokinesis usually occurs in conjunction with mitosis, "mitosis" is often used interchangeably with "M phase". However, there are many cells where mitosis and cytokinesis occur separately, forming single cells with multiple nuclei in a process called endoreplication. This occurs most notably among the fungi and slime moulds, but is found in various groups. Even in animals, cytokinesis and mitosis may occur independently, for instance during certain stages of fruit fly embryonic development. Errors in mitosis can either kill a cell through apoptosis or cause mutations that may lead to cancer.

Regulation of Eukaryotic Cell Cycle

Regulation of the cell cycle involves processes crucial to the survival of a cell, including the detection and repair of genetic damage as well as the prevention of uncontrolled cell division. The molecular events that control the cell cycle are ordered and directional; that is, each process occurs in a sequential fashion and it is impossible to "reverse" the cycle.

Role of Cyclins and CDKs

Nobel Laureate Paul Nurse.

Two key classes of regulatory molecules, cyclins and cyclin-dependent kinases (CDKs), determine a cell's progress through the cell cycle. Leland H. Hartwell, R. Timothy Hunt, and Paul M. Nurse won the 2001 Nobel Prize in Physiology or Medicine for their discovery of these central molecules. Many of the genes encoding cyclins and CDKs are conserved among all eukaryotes, but in general more complex organisms have more elaborate cell cycle control systems that incorporate more individual components. Many of the relevant genes were first identified by studying yeast, especially *Saccharomyces cerevisiae*; genetic nomenclature in yeast dubs many of these genes *cdc* (for "cell division cycle") followed by an identifying number, e.g. *cdc25* or *cdc20*.

Cyclins form the regulatory subunits and CDKs the catalytic subunits of an activated heterodimer; cyclins have no catalytic activity and CDKs are inactive in the absence of a partner cyclin. When activated by a bound cyclin, CDKs perform a common biochemical reaction called phosphorylation that activates or inactivates target proteins to orchestrate coordinated entry into the next phase of the cell cycle. Different cyclin-CDK combinations determine the downstream proteins targeted. CDKs are constitutively expressed in cells whereas cyclins are synthesised at specific stages of the cell cycle, in response to various molecular signals.

General Mechanism of Cyclin-CDK Interaction

Upon receiving a pro-mitotic extracellular signal, G_1 cyclin-CDK complexes become active to prepare the cell for S phase, promoting the expression of transcription factors that in turn promote the expression of S cyclins and of enzymes required for DNA replication. The G_1 cyclin-CDK complexes also promote the degradation of molecules that function as S phase inhibitors by targeting them for ubiquitination. Once a protein has been ubiquitinated, it is targeted for proteolytic degradation by the proteasome. However, results from a recent study of E2F transcriptional dynamics at the single-cell level argue that the role of G1 cyclin-CDK activities, in particular cyclin D-CDK4/6, is to tune the timing rather than the commitment of cell cycle entry.

Active S cyclin-CDK complexes phosphorylate proteins that make up the pre-replication complexes assembled during G_1 phase on DNA replication origins. The phosphorylation serves two purposes: to activate each already-assembled pre-replication complex, and to prevent new complexes from forming. This ensures that every portion of the cell's genome will be replicated once and only once. The reason for prevention of gaps in replication is fairly clear, because daughter cells that are missing all or part of crucial genes will die. However, for reasons related to gene copy number effects, possession of extra copies of certain genes is also deleterious to the daughter cells.

Mitotic cyclin-CDK complexes, which are synthesized but inactivated during S and G_2 phases, promote the initiation of mitosis by stimulating downstream proteins involved in chromosome condensation and mitotic spindle assembly. A critical complex activated during this process is a ubiquitin ligase known as the anaphase-promoting complex (APC), which promotes degradation of structural proteins associated with the chromosomal kinetochore. APC also targets the mitotic cyclins for degradation, ensuring that telophase and cytokinesis can proceed.

Specific Action of Cyclin-CDK Complexes

Cyclin D is the first cyclin produced in the cell cycle, in response to extracellular signals (e.g. growth factors). Cyclin D binds to existing CDK4, forming the active cyclin D-CDK4 complex. Cyclin D-CDK4 complex in turn phosphorylates the retinoblastoma susceptibility protein (Rb). The hyperphosphorylated Rb dissociates from the E2F/DP1/Rb complex (which was bound to the E2F responsive genes, effectively "blocking" them from transcription), activating E2F. Activation of E2F results in transcription of various genes like cyclin E, cyclin A, DNA polymerase, thymidine kinase, etc. Cyclin E thus produced binds to CDK2, forming the cyclin E-CDK2 complex, which pushes the cell from G_1 to S phase (G_1/S, which initiates the G_2/M transition). Cyclin B-cdk1 complex activation causes breakdown of nuclear envelope and initiation of prophase, and subsequently, its deactivation causes the cell to exit mitosis.

A quantitative study of E2F transcriptional dynamics at the single-cell level by using engineered fluorescent reporter cells provided a quantitative framework for understanding the control logic of cell cycle entry, challenging the canonical textbook model. Genes that regulate the amplitude of

E2F accumulation, such as Myc, determine the commitment into cell cycle and S phase entry. G1 cyclin-CDK activities are not the driver of cell cycle entry. Instead, they primarily tune the timing of E2F increase, thereby modulating the pace of cell cycle progression.

Inhibitors

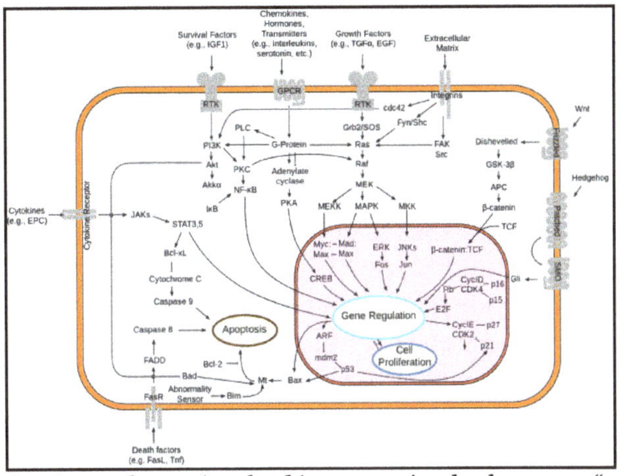

Overview of signal transduction pathways involved in apoptosis, also known as "programmed cell death".

Two families of genes, the *cip/kip* (*CDK interacting protein/Kinase inhibitory protein*) family and the INK4a/ARF (*In*hibitor of *K*inase 4/*A*lternative *R*eading *F*rame) family, prevent the progression of the cell cycle. Because these genes are instrumental in prevention of tumor formation, they are known as tumor suppressors.

The cip/kip family includes the genes p21, p27 and p57. They halt cell cycle in G_1 phase, by binding to, and inactivating, cyclin-CDK complexes. p21 is activated by p53 (which, in turn, is triggered by DNA damage e.g. due to radiation). p27 is activated by Transforming Growth Factor of β (TGF β), a growth inhibitor.

The INK4a/ARF family includes $p16^{INK4a}$, which binds to CDK4 and arrests the cell cycle in G_1 phase, and $p14^{ARF}$ which prevents p53 degradation.

Synthetic inhibitors of Cdc25 could also be useful for the arrest of cell cycle and therefore be useful as antineoplastic and anticancer agents.

Transcriptional Regulatory Network

Current evidence suggests that a semi-autonomous transcriptional network acts in concert with the CDK-cyclin machinery to regulate the cell cycle. Several gene expression studies in *Saccharomyces cerevisiae* have identified 800-1200 genes that change expression over the course of the cell cycle. They are transcribed at high levels at specific points in the cell cycle, and remain at lower levels throughout the rest of the cycle. While the set of identified genes differs between studies due to the computational methods and criteria used to identify them, each study indicates that a large portion of yeast genes are temporally regulated.

Many periodically expressed genes are driven by transcription factors that are also periodically expressed. One screen of single-gene knockouts identified 48 transcription factors (about 20%

of all non-essential transcription factors) that show cell cycle progression defects. Genome-wide studies using high throughput technologies have identified the transcription factors that bind to the promoters of yeast genes, and correlating these findings with temporal expression patterns have allowed the identification of transcription factors that drive phase-specific gene expression. The expression profiles of these transcription factors are driven by the transcription factors that peak in the prior phase, and computational models have shown that a CDK-autonomous network of these transcription factors is sufficient to produce steady-state oscillations in gene expression).

Experimental evidence also suggests that gene expression can oscillate with the period seen in dividing wild-type cells independently of the CDK machinery. Orlando *et al.* used microarrays to measure the expression of a set of 1,271 genes that they identified as periodic in both wild type cells and cells lacking all S-phase and mitotic cyclins (*clb1,2,3,4,5,6*). Of the 1,271 genes assayed, 882 continued to be expressed in the cyclin-deficient cells at the same time as in the wild type cells, despite the fact that the cyclin-deficient cells arrest at the border between G_1 and S phase. However, 833 of the genes assayed changed behavior between the wild type and mutant cells, indicating that these genes are likely directly or indirectly regulated by the CDK-cyclin machinery. Some genes that continued to be expressed on time in the mutant cells were also expressed at different levels in the mutant and wild type cells. These findings suggest that while the transcriptional network may oscillate independently of the CDK-cyclin oscillator, they are coupled in a manner that requires both to ensure the proper timing of cell cycle events. Other work indicates that phosphorylation, a post-translational modification, of cell cycle transcription factors by Cdk1 may alter the localization or activity of the transcription factors in order to tightly control timing of target genes.

While oscillatory transcription plays a key role in the progression of the yeast cell cycle, the CDK-cyclin machinery operates independently in the early embryonic cell cycle. Before the midblastula transition, zygotic transcription does not occur and all needed proteins, such as the B-type cyclins, are translated from maternally loaded mRNA.

DNA Replication and DNA Replication Origin Activity

Analyses of synchronized cultures of *Saccharomyces cerevisiae* under conditions that prevent DNA replication initiation without delaying cell cycle progression showed that origin licensing decreases the expression of genes with origins near their 3' ends, revealing that downstream origins can regulate the expression of upstream genes. This confirms previous predictions from mathematical modeling of a global causal coordination between DNA replication origin activity and mRNA expression, and shows that mathematical modeling of DNA microarray data can be used to correctly predict previously unknown biological modes of regulation.

Checkpoints

Cell cycle checkpoints are used by the cell to monitor and regulate the progress of the cell cycle. Checkpoints prevent cell cycle progression at specific points, allowing verification of necessary phase processes and repair of DNA damage. The cell cannot proceed to the next phase until checkpoint requirements have been met. Checkpoints typically consist of a network of regulatory proteins that monitor and dictate the progression of the cell through the different stages of the cell cycle.

There are several checkpoints to ensure that damaged or incomplete DNA is not passed on to daughter cells. Three main checkpoints exist: the G_1/S checkpoint, the G_2/M checkpoint and the metaphase (mitotic) checkpoint.

G_1/S transition is a rate-limiting step in the cell cycle and is also known as restriction point. This is where the cell checks whether it has enough raw materials to fully replicate its DNA (nucleotide bases, DNA synthase, chromatin, etc.). An unhealthy or malnourished cell will get stuck at this checkpoint.

The G_2/M checkpoint is where the cell ensures that it has enough cytoplasm and phospholipids for two daughter cells. But sometimes more importantly, it checks to see if it is the right time to replicate. There are some situations where many cells need to all replicate simultaneously (for example, a growing embryo should have a symmetric cell distribution until it reaches the mid-blastula transition). This is done by controlling the G_2/M checkpoint.

The metaphase checkpoint is a fairly minor checkpoint, in that once a cell is in metaphase, it has committed to undergoing mitosis. However that's not to say it isn't important. In this checkpoint, the cell checks to ensure that the spindle has formed and that all of the chromosomes are aligned at the spindle equator before anaphase begins.

While these are the three "main" checkpoints, not all cells have to pass through each of these checkpoints in this order to replicate. Many types of cancer are caused by mutations that allow the cells to speed through the various checkpoints or even skip them altogether. Going from S to M to S phase almost consecutively. Because these cells have lost their checkpoints, any DNA mutations that may have occurred are disregarded and passed on to the daughter cells. This is one reason why cancer cells have a tendency to exponentially accrue mutations. Aside from cancer cells, many fully differentiated cell types no longer replicate so they leave the cell cycle and stay in G_0 until their death. Thus removing the need for cellular checkpoints. An alternative model of the cell cycle response to DNA damage has also been proposed, known as the postreplication checkpoint.

Checkpoint regulation plays an important role in an organism's development. In sexual reproduction, when egg fertilization occurs, when the sperm binds to the egg, it releases signalling factors that notify the egg that it has been fertilized. Among other things, this induces the now fertilized oocyte to return from its previously dormant, G_0, state back into the cell cycle and on to mitotic replication and division.

p53 plays an important role in triggering the control mechanisms at both G_1/S and G_2/M checkpoints. In addition to p53, checkpoint regulators are being heavily researched for their roles in cancer growth and proliferation.

Role in Tumor Formation

A disregulation of the cell cycle components may lead to tumor formation. As mentioned above, when some genes like the cell cycle inhibitors, RB, p53 etc. mutate, they may cause the cell to multiply uncontrollably, forming a tumor. Although the duration of cell cycle in tumor cells is equal to or longer than that of normal cell cycle, the proportion of cells that are in active cell division (versus quiescent cells in G_0 phase) in tumors is much higher than that in normal tissue. Thus there is a net increase in cell number as the number of cells that die by apoptosis or senescence remains the same.

The cells which are actively undergoing cell cycle are targeted in cancer therapy as the DNA is relatively exposed during cell division and hence susceptible to damage by drugs or radiation. This fact is made use of in cancer treatment; by a process known as debulking, a significant mass of the tumor is removed which pushes a significant number of the remaining tumor cells from G_0 to G_1 phase (due to increased availability of nutrients, oxygen, growth factors etc.). Radiation or chemotherapy following the debulking procedure kills these cells which have newly entered the cell cycle.

The fastest cycling mammalian cells in culture, crypt cells in the intestinal epithelium, have a cycle time as short as 9 to 10 hours. Stem cells in resting mouse skin may have a cycle time of more than 200 hours. Most of this difference is due to the varying length of G_1, the most variable phase of the cycle. M and S do not vary much.

In general, cells are most radiosensitive in late M and G_2 phases and most resistant in late S phase.

For cells with a longer cell cycle time and a significantly long G_1 phase, there is a second peak of resistance late in G_1.

The pattern of resistance and sensitivity correlates with the level of sulfhydryl compounds in the cell. Sulfhydryls are natural substances that protect cells from radiation damage and tend to be at their highest levels in S and at their lowest near mitosis.

Fluorescence Imaging of the Cell Cycle

Fluorescent proteins visualize the cell cycle progression. IFP2.0-hGem(1/110) fluorescence is shown in green and highlights the S/G_2/M phases. smURFP-hCdtI(30/120) fluorescence is shown in red and highlights the G_0/G_1 phases.

Pioneering work by Atsushi Miyawaki and coworkers developed the fluorescent ubiquitination-based cell cycle indicator (FUCCI), which enables fluorescence imaging of the cell cycle. Originally, a green fluorescent protein, mAG, was fused to hGem(1/110) and an orange fluorescent protein (mKO_2) was fused to hCdt1(30/120). Note, these fusions are fragments that contain a nuclear localization signal and ubiquitination sites for degradation, but are not functional proteins. The green fluorescent protein is made during the S, G_2, or M phase and degraded during the G_0 or G_1 phase, while the orange fluorescent protein is made during the G_0 or G_1 phase and destroyed during the S, G_2, or M phase. A far-red and near-infrared FUCCI was developed using a cyanobacteria-derived fluorescent protein (smURFP) and a bacteriophytochrome-derived fluorescent protein (movie found at this link).

Mitosis

In cell biology, Mitosis is a part of the cell cycle in which chromosomes in a cell nucleus are separated into two identical sets of chromosomes, and each set ends up in its own nucleus. In general, mitosis (division of the nucleus) is preceded by the S stage of interphase (during which the DNA is replicated) and is often accompanied or followed by cytokinesis, which divides the cytoplasm, organelles and cell membrane into two new cells containing roughly equal shares of these cellular components. Mitosis and cytokinesis together define the mitotic (M) phase of an animal cell cycle—the division of the mother cell into two daughter cells, genetically identical to each other and to their parent cell.

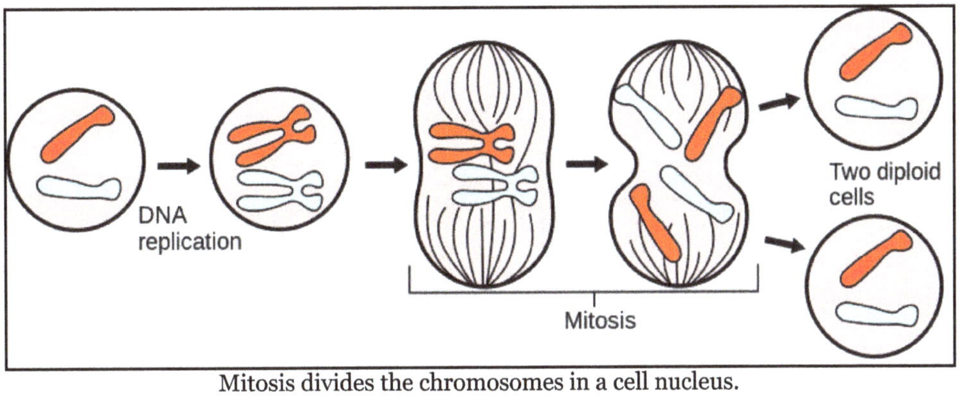

Mitosis divides the chromosomes in a cell nucleus.

The process of mitosis is divided into stages corresponding to the completion of one set of activities and the start of the next. These stages are prophase, prometaphase, metaphase, anaphase, and telophase. During mitosis, the chromosomes, which have already duplicated, condense and attach to spindle fibers that pull one copy of each chromosome to opposite sides of the cell. The result is two genetically identical daughter nuclei. The rest of the cell may then continue to divide by cytokinesis to produce two daughter cells. Producing three or more daughter cells instead of normal two is a mitotic error called tripolar mitosis or multipolar mitosis (direct cell triplication / multiplication). Other errors during mitosis can induce apoptosis (programmed cell death) or cause mutations. Certain types of cancer can arise from such mutations.

Mitosis occurs only in eukaryotic cells and the process varies in different organisms. For example, animals undergo an "open" mitosis, where the nuclear envelope breaks down before the chromosomes separate, while fungi undergo a "closed" mitosis, where chromosomes divide within an intact cell nucleus. Furthermore, most animal cells undergo a shape change, known as mitotic cell rounding, to adopt a near spherical morphology at the start of mitosis. Prokaryotic cells, which lack a nucleus, divide by a different process called binary fission.

Discovery

German zoologist Otto Bütschli might have claimed the discovery of the process presently known as "mitosis", a term coined by Walther Flemming in 1882.

Mitosis was discovered in frog, rabbit, and cat cornea cells in 1873 and described for the first time by the Polish histologist Wacław Mayzel in 1875. The term is derived from the Greek word μίτος *mitos* "warp thread".

Overview of Mitosis

The primary result of mitosis and cytokinesis is the transfer of a parent cell's genome into two daughter cells. The genome is composed of a number of chromosomes—complexes of tightly coiled DNA that contain genetic information vital for proper cell function. Because each resultant daughter cell should be genetically identical to the parent cell, the parent cell must make a copy of each chromosome before mitosis. This occurs during the S phase of interphase. Chromosome duplication results in two identical *sister chromatids* bound together by cohesin proteins at the *centromere.*

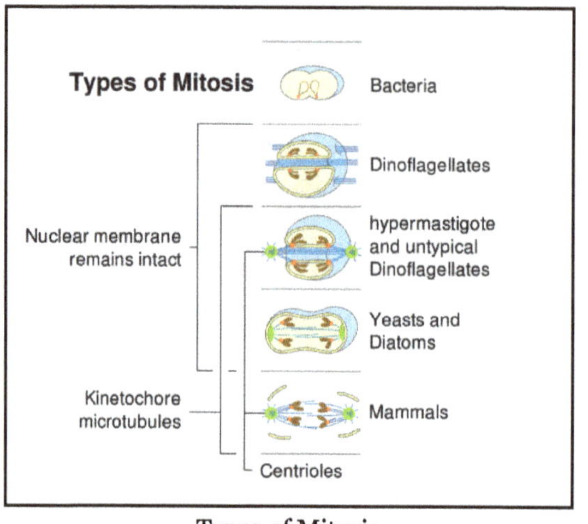

Types of Mitosis.

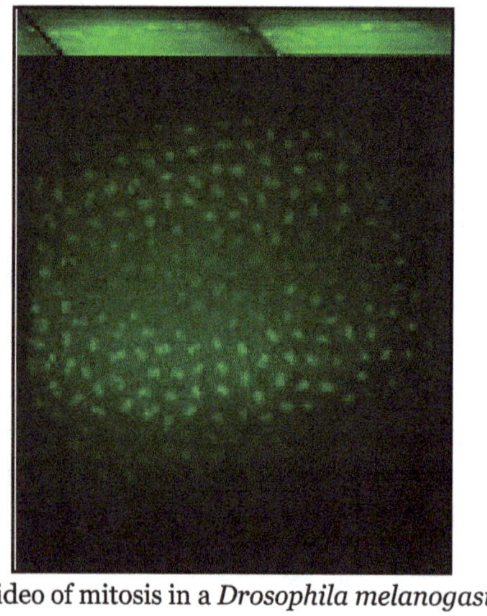

Time-lapse video of mitosis in a *Drosophila melanogaster* embryo.

When mitosis begins, the chromosomes condense and become visible. In some eukaryotes, for example animals, the nuclear envelope, which segregates the DNA from the cytoplasm, disintegrates into small vesicles. The nucleolus, which makes ribosomes in the cell, also disappears. Microtubules project from opposite ends of the cell, attach to the centromeres, and align the chromosomes centrally within the cell. The microtubules then contract to pull the sister chromatids of each chromosome apart. Sister chromatids at this point are called *daughter chromosomes*. As the cell elongates, corresponding daughter chromosomes are pulled toward opposite ends of the cell and condense maximally in late anaphase. A new nuclear envelope forms around the separated daughter chromosomes, which decondense to form interphase nuclei.

During mitotic progression, typically after anaphase onset, the cell may undergo cytokinesis. In animal cells, a cell membrane pinches inward between the two developing nuclei to produce two new cells. In plant cells, a cell plate forms between the two nuclei. Cytokinesis does not always occur; coenocytic (a type of multinucleate condition) cells undergo mitosis without cytokinesis.

Phases of Cell Cycle and Mitosis

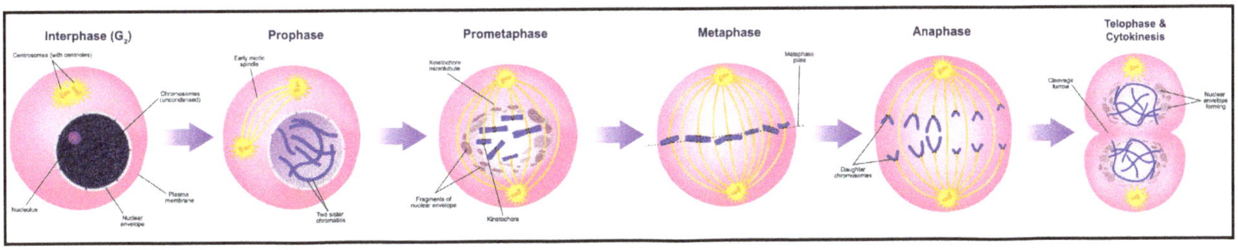

Interphase

The mitotic phase is a relatively short period of the cell cycle. It alternates with the much longer *interphase*, where the cell prepares itself for the process of cell division. Interphase is divided into three phases: G_1 (first gap), S (synthesis), and G_2 (second gap). During all three phases, the cell grows by producing proteins and cytoplasmic organelles. However, chromosomes are replicated only during the S phase. Thus, a cell grows (G_1), continues to grow as it duplicates its chromosomes (S), grows more and prepares for mitosis (G_2), and finally divides (M) before restarting the cycle. All these phases in the cell cycle are highly regulated by cyclins, cyclin-dependent kinases, and other cell cycle proteins. The phases follow one another in strict order and there are "checkpoints" that give the cell cues to proceed from one phase to another. Cells may also temporarily or permanently leave the cell cycle and enter G_0 phase to stop dividing. This can occur when cells become overcrowded (density-dependent inhibition) or when they differentiate to carry out specific functions for the organism, as is the case for human heart muscle cells and neurons. Some G_0 cells have the ability to re-enter the cell cycle.

Preprophase (Plant Cells)

In plant cells only, prophase is preceded by a pre-prophase stage. In highly vacuolated plant cells, the nucleus has to migrate into the center of the cell before mitosis can begin. This is achieved through the formation of a phragmosome, a transverse sheet of cytoplasm that bisects the cell along the future plane of cell division. In addition to phragmosome formation, preprophase is characterized by the formation of a ring of microtubules and actin filaments (called preprophase band) underneath the plasma membrane around the equatorial plane of the future mitotic spindle. This band marks the position where the cell will eventually divide. The cells of higher plants (such as the flowering plants) lack centrioles; instead, microtubules form a spindle on the surface of the nucleus and are then organized into a spindle by the chromosomes themselves, after the nuclear envelope breaks down. The preprophase band disappears during nuclear envelope breakdown and spindle formation in prometaphase.

Prophase

During prophase, which occurs after G_2 interphase, the cell prepares to divide by tightly condensing its chromosomes and initiating mitotic spindle formation, this process is called chromosome condensation. During interphase, the genetic material in the nucleus consists of loosely packed chromatin. At the onset of prophase, chromatin fibers condense into discrete chromosomes that are typically visible at high magnification through a light microscope. In this stage, chromosomes centrosomes begin pulling the chromosomes towards opposite ends of the cell. The resulting ten-are long, thin and thread-like. Each chromosome has two chromatids. The two chromatids are joined at a place called centromere.

Gene transcription ceases during prophase and does not resume until late anaphase to early G1 phase. The nucleolus also disappears during early prophase.

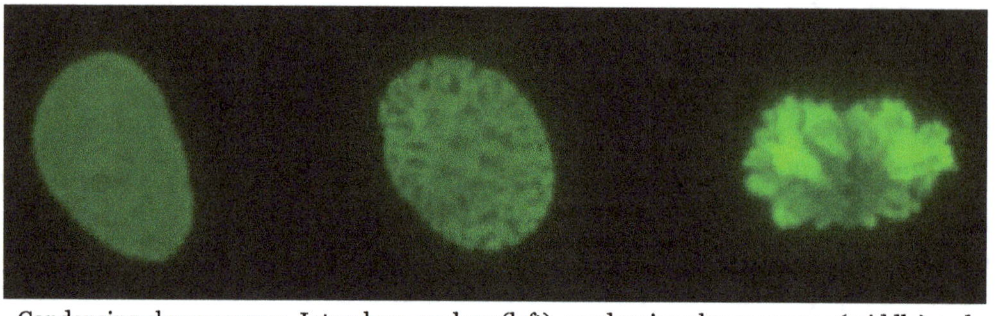

Condensing chromosomes. Interphase nucleus (left), condensing chromosomes (middle) and condensed chromosomes (right).

Close to the nucleus of animal cells are structures called centrosomes, consisting of a pair of centrioles surrounded by a loose collection of proteins. The centrosome is the coordinating center for the cell's microtubules. A cell inherits a single centrosome at cell division, which is duplicated by the cell before a new round of mitosis begins, giving a pair of centrosomes. The two centrosomes polymerize tubulin to help form a microtubule spindle apparatus. Motor proteins then push the centrosomes along these microtubules to opposite sides of the cell. Although centrosomes help organize microtubule assembly, they are not essential for the formation of the spindle apparatus, since they are absent from plants, and are not absolutely required for animal cell mitosis.

Prometaphase

At the beginning of prometaphase in animal cells, phosphorylation of nuclear lamins causes the nuclear envelope to disintegrate into small membrane vesicles. As this happens, microtubules invade the nuclear space. This is called *open mitosis*, and it occurs in some multicellular organisms. Fungi and some protists, such as algae or trichomonads, undergo a variation called *closed mitosis* where the spindle forms inside the nucleus, or the microtubules penetrate the intact nuclear envelope.

In late prometaphase, *kinetochore microtubules* begin to search for and attach to chromosomal kinetochores. A *kinetochore* is a proteinaceous microtubule-binding structure that forms on the chromosomal centromere during late prophase. A number of *polar microtubules* find and interact with corresponding polar microtubules from the opposite centrosome to form the mitotic spindle. Although the kinetochore structure and function are not fully understood, it is known that it contains some form of molecular motor. When a microtubule connects with the kinetochore, the motor activates, using energy from ATP to "crawl" up the tube toward the originating centrosome. This motor activity, coupled with polymerisation and depolymerisation of microtubules, provides the pulling force necessary to later separate the chromosome's two chromatids.

Metaphase

After the microtubules have located and attached to the kinetochores in prometaphase, the two

sion causes the chromosomes to align along the *metaphase plate* or *equatorial plane*, an imaginary line that is centrally located between the two centrosomes (at approximately the midline of the cell). To ensure equitable distribution of chromosomes at the end of mitosis, the *metaphase checkpoint* guarantees that kinetochores are properly attached to the mitotic spindle and that the chromosomes are aligned along the metaphase plate. If the cell successfully passes through the metaphase checkpoint, it proceeds to anaphase.

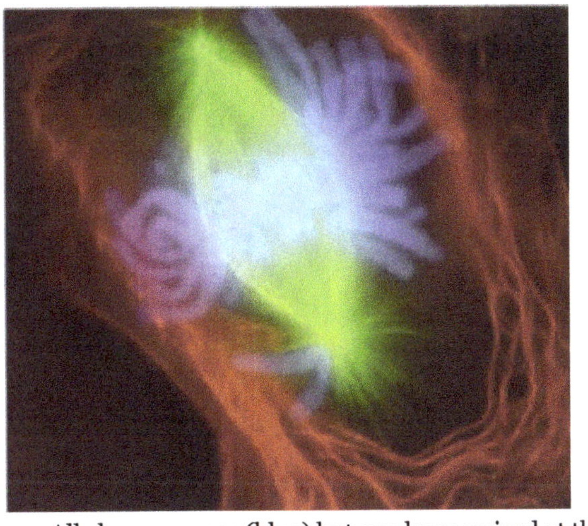

A cell in late metaphase. All chromosomes (blue) but one have arrived at the metaphase plate.

Anaphase

During *anaphase A*, the cohesins that bind sister chromatids together are cleaved, forming two identical daughter chromosomes. Shortening of the kinetochore microtubules pulls the newly formed daughter chromosomes to opposite ends of the cell. During *anaphase B*, polar microtubules push against each other, causing the cell to elongate. In late anaphase, chromosomes also reach their overall maximal condensation level, to help chromosome segregation and the re-formation of the nucleus. In most animal cells, anaphase A precedes anaphase B, but some vertebrate egg cells demonstrate the opposite order of events.

Telophase

Telophase is a reversal of prophase and prometaphase events. At telophase, the polar microtubules continue to lengthen, elongating the cell even more. If the nuclear envelope has broken down, a new nuclear envelope forms using the membrane vesicles of the parent cell's old nuclear envelope. The new envelope forms around each set of separated daughter chromosomes (though the membrane does not enclose the centrosomes) and the nucleolus reappears. Both sets of chromosomes, now surrounded by new nuclear membrane, begin to "relax" or decondense. Mitosis is complete. Each daughter nucleus has an identical set of chromosomes. Cell division may or may not occur at this time depending on the organism.

Cytokinesis

Cytokinesis is not a phase of mitosis but rather a separate process, necessary for completing cell division. In animal cells, a cleavage furrow (pinch) containing a contractile ring develops where the metaphase plate used to be, pinching off the separated nuclei. In both animal and plant cells, cell division is also driven by vesicles derived from the Golgi apparatus, which move along microtubules to the middle of the cell. In plants, this structure coalesces into a cell plate at the center of the phragmoplast and develops into a cell wall, separating the two nuclei. The phragmoplast is a microtubule structure typical for higher plants, whereas some green algae use a phycoplast microtubule array during cytokinesis. Each daughter cell has a complete copy of the genome of its parent cell. The end of cytokinesis marks the end of the M-phase.

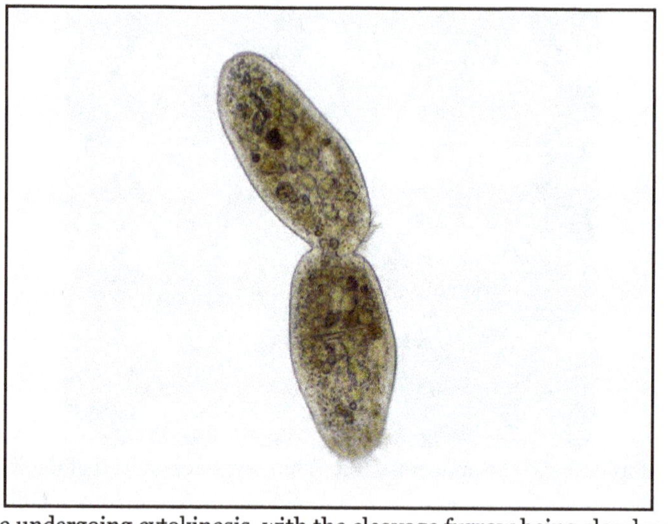

Cilliate undergoing cytokinesis, with the cleavage furrow being clearly visible.

There are many cells where mitosis and cytokinesis occur separately, forming single cells with multiple nuclei. The most notable occurrence of this is among the fungi, slime molds, and coenocytic algae, but the phenomenon is found in various other organisms. Even in animals, cytokinesis and mitosis may occur independently, for instance during certain stages of fruit fly embryonic development.

Significance

Mitosis is important for the maintenance of the chromosomal set; each cell formed receives chromosomes that are alike in composition and equal in number to the chromosomes of the parent cell.

Mitosis occurs in the following circumstances:

Development and growth

> The number of cells within an organism increases by mitosis. This is the basis of the development of a multicellular body from a single cell, i.e., zygote and also the basis of the growth of a multicellular body.

Cell replacement

> In some parts of body, e.g. skin and digestive tract, cells are constantly sloughed off and replaced by new ones. New cells are formed by mitosis and so are exact copies of the cells being replaced. In like manner, red blood cells have short lifespan (only about 4 months) and new RBCs are formed by mitosis.

Regeneration

Some organisms can regenerate body parts. The production of new cells in such instances is achieved by mitosis. For example, starfish regenerate lost arms through mitosis.

Asexual reproduction

Some organisms produce genetically similar offspring through asexual reproduction. For example, the hydra reproduces asexually by budding. The cells at the surface of hydra undergo mitosis and form a mass called a bud. Mitosis continues in the cells of the bud and this grows into a new individual. The same division happens during asexual reproduction or vegetative propagation in plants.

Cell Shape Changes During Mitosis

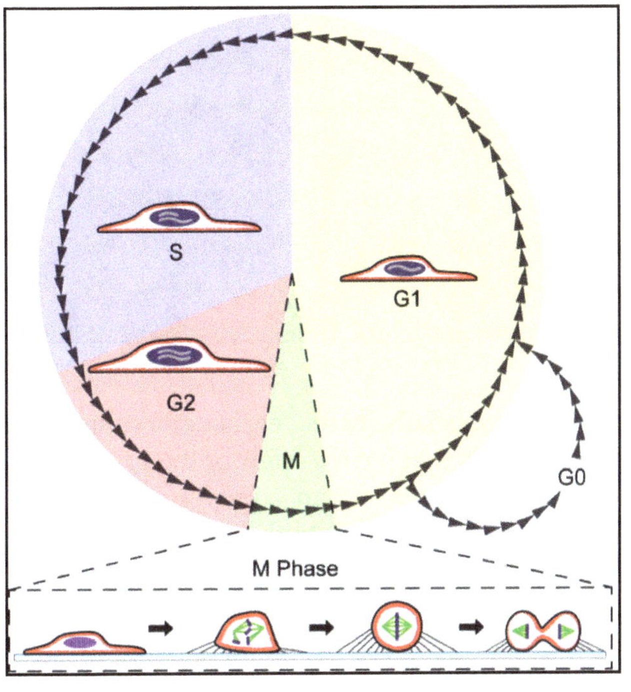

Cell shape changes through mitosis for a typical animal cell cultured on a flat surface. The cell undergoes mitotic cell rounding during spindle assembly and then divides via cytokinesis. The actomyosin cortex is depicted in red, DNA/chromosomes purple, microtubules green, and membrane and retraction fibers in black. Rounding also occurs in live tissue, as described in the text.

In animal tissue, most cells round up to a near-spherical shape during mitosis. In epithelia and epidermis, an efficient rounding process is correlated with proper mitotic spindle alignment and subsequent correct positioning of daughter cells. Moreover, researchers have found that if rounding is heavily suppressed it may result in spindle defects, primarily pole splitting and failure to efficiently capture chromosomes. Therefore, mitotic cell rounding is thought to play a protective role in ensuring accurate mitosis.

Rounding forces are driven by reorganization of F-actin and myosin (actomyosin) into a contractile homogeneous cell cortex that 1) rigidifies the cell periphery and 2) facilitates generation of intracellular hydrostatic pressure (up to 10 fold higher than interphase). The generation of intracellular pressure is particularly critical under confinement, such as would be important in a tissue scenario, where outward forces must be produced to round up against surrounding cells and/or the extracellular matrix. Generation of pressure is dependent on formin-mediated F-actin nucleation and Rho kinase (ROCK)-mediated myosin II contraction, both of which are governed upstream by signaling pathways RhoA and ECT2 through the activity of Cdk1. Due to its importance in mitosis, the molecular components and dynamics of the mitotic actomyosin cortex is an area of active research.

Errors and Variations of Mitosis

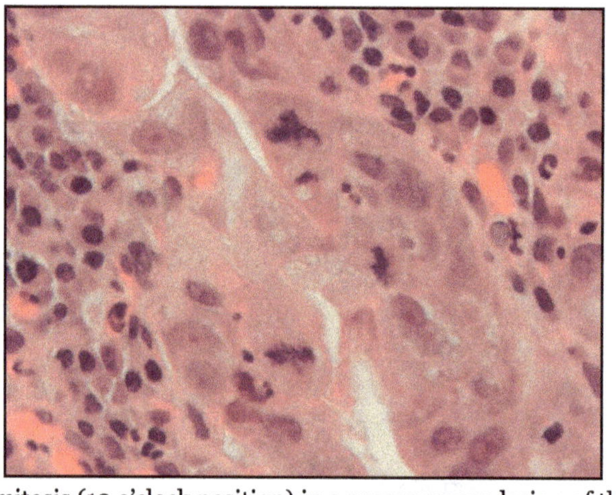

An abnormal (tripolar) mitosis (12 o'clock position) in a precancerous lesion of the stomach (H&E stain).

Errors can occur during mitosis, especially during early embryonic development in humans. Mitotic errors can create aneuploid cells that have too few or too many of one or more chromosomes, a condition associated with cancer. Early human embryos, cancer cells, infected or intoxicated cells can also suffer from pathological division into three or more daughter cells (tripolar or multipolar mitosis), resulting in severe errors in their chromosomal complements.

In *nondisjunction*, sister chromatids fail to separate during anaphase. One daughter cell receives both sister chromatids from the nondisjoining chromosome and the other cell receives none. As a result, the former cell gets three copies of the chromosome, a condition known as *trisomy*, and the latter will have only one copy, a condition known as *monosomy*. On occasion, when cells experience nondisjunction, they fail to complete cytokinesis and retain both nuclei in one cell, resulting in binucleated cells.

Anaphase lag occurs when the movement of one chromatid is impeded during anaphase. This may be caused by a failure of the mitotic spindle to properly attach to the chromosome. The lagging chromatid is excluded from both nuclei and is lost. Therefore, one of the daughter cells will be monosomic for that chromosome.

Endoreduplication (or endoreplication) occurs when chromosomes duplicate but the cell does not subsequently divide. This results in polyploid cells or, if the chromosomes duplicates repeatedly, polytene chromosomes. Endoreduplication is found in many species and appears to be a normal part of development. Endomitosis is a variant of endoreduplication in which cells replicate their chromosomes during S phase and enter, but prematurely terminate, mitosis. Instead of being divided into two new daughter nuclei, the replicated chromosomes are retained within the original nucleus. The cells then re-enter G_1 and S phase and replicate their chromosomes again. This may occur multiple times, increasing the chromosome number with each round of replication and endomitosis. Platelet-producing megakaryocytes go through endomitosis during cell differentiation.

Timeline in Pictures

Mitotic cells can be visualized microscopically by staining them with fluorescent antibodies and dyes.

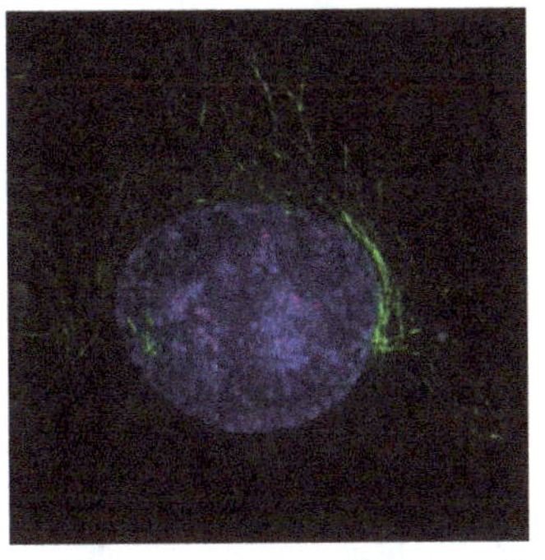

Early prophase: Nonkinetochore microtubules, shown as green strands, have established a matrix around the disassembling nucleus, with the condensing chromosomes in blue. The red nodules are the centromeres.

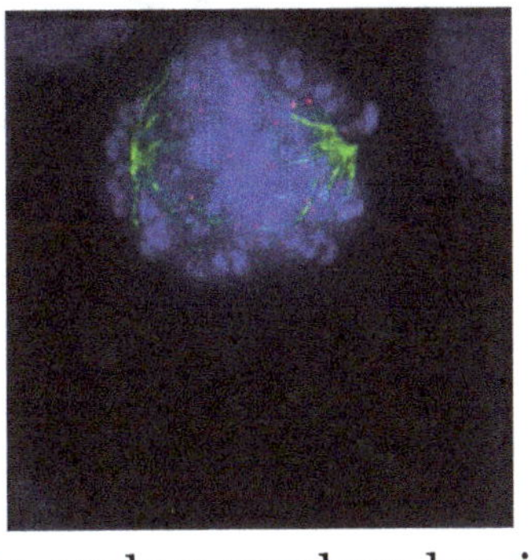

Early prometaphase: The nuclear envelope has just disassembled, allowing the microtubules to quickly interact with the kinetochores, which assemble on the centromeres of the condensing chromosomes.

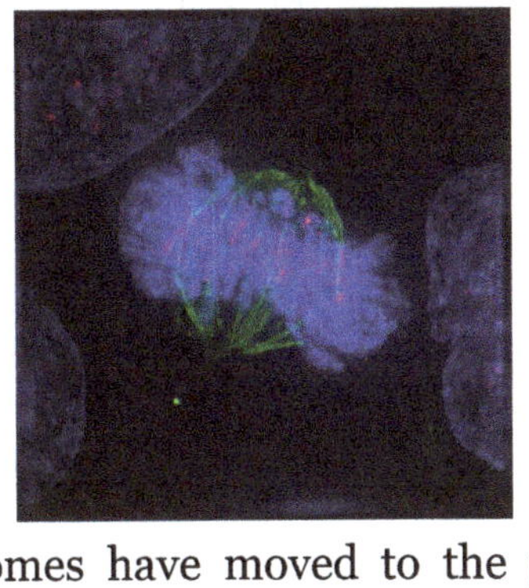

Metaphase: The centrosomes have moved to the poles of the cell and have established the mitotic spindle. The chromosomes have congressed at the metaphase plate.

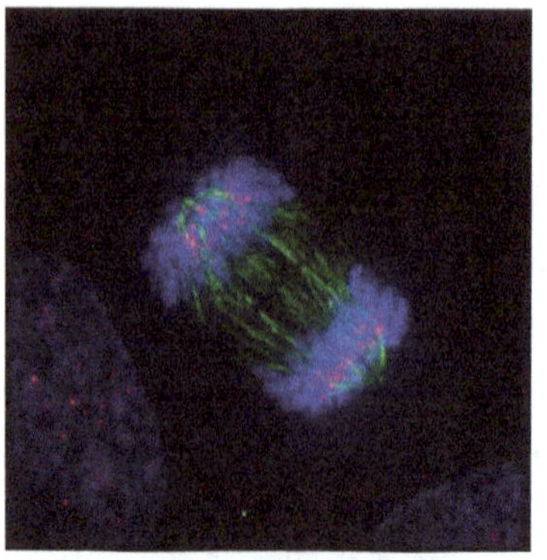

Anaphase: Kinetochore microtubules pull the two sets of chromosomes apart, and lengthening nonkinetochore microtubules push the halves of the dividing cell further apart, while chromosomes are condensed maximally.

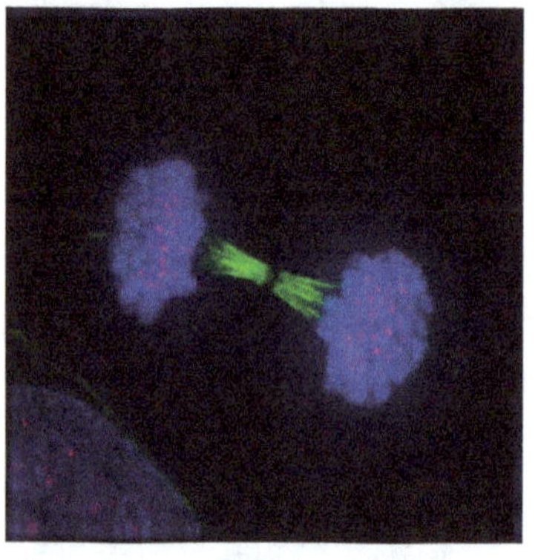

Telophase: Reversal of prophase and prometaphase events and thus completing the cell cycle.

Cell-cycle Analysis

Cell-cycle analysis is a method in cell biology that employs flow cytometry to distinguish cells in different phases of the cell cycle. Before analysis, the cells are permeabilised and treated with a fluorescent dye that stains DNA quantitatively, usually propidium iodide (PI). The fluorescence intensity of the stained cells at certain wavelengths will therefore correlate with the amount of DNA they contain. As the DNA content of cells duplicates during the S phase of the cell cycle, the relative amount of cells in the G_0 phase and G_1 phase (before S phase), in the S phase, and in the G_2 phase and M phase (after S phase) can be determined, as the fluorescence of cells in the G_2/M phase will be twice as high as that of cells in the G_0/G_1 phase.

Cell-cycle anomalies can be symptoms for various kinds of cell damage, for example DNA damage, which cause the cell to interrupt the cell cycle at certain checkpoints to prevent transformation into a cancer cell (carcinogenesis). Other possible reasons for anomalies include lack of nutrients, for example after serum deprivation.

Cell cycle analysis was first described in 1969 at Los Alamos Scientific Laboratory by a group from the University of California, using the Feulgen staining technique. The first protocol for cell-cycle analysis using propidium iodide staining was presented in 1975 by Awtar Krishan from Harvard Medical School and is still widely cited today.

Experimental Procedure

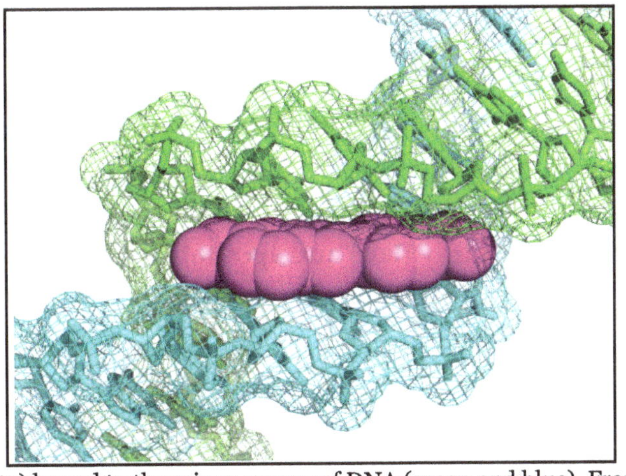

DAPI (magenta) bound to the minor groove of DNA (green and blue). From PDB: 1D30.

The first step in preparing cells for cell-cycle analysis is permeabilisation of the cells' plasma membranes. This is usually done by incubating them in a buffer solution containing a detergent such as Triton X-100 or NP-40, or by fixating them in ethanol. Most fluorescent DNA dyes are not membrane permeable, that is, unable to pass through an intact cell membrane. Permeabilisation is therefore crucial for the success of the next step, the staining of the cells.

Prior to or during the staining step, the cells will usually be treated with RNase A to remove RNAs from the cells. This is important because dyes that stain DNA will also stain RNA, thus creating artefacts that would distort the results.

Aside from propidium iodide, quantifiable dyes that are frequently used include (but are not limited to) DRAQ5, 7-Aminoactinomycin D, DAPI and Hoechst 33342.

When the cells pass through the flow cytometer's laser, a fluorescence pulse is generated that correlates with the amount of dye associated with the DNA and thus with the total amount of DNA in the cell.

Doublet Discrimination

Because cells and especially fixated cells tend to stick together, cell aggregates have to be excluded from analysis through a process called *doublet discrimination*. This is important because a doublet of two cells in the G_0/G_1 phase has the same total amount of DNA and thus the same fluorescence intensity as a single cell in the G_2/M phase. G_0/G_1 doublets would therefore create false positive results for G_2/M cells.

Related Methods

Nicoletti Assay

The Nicoletti assay, named after its inventor, the Italian physician Ildo Nicoletti, is a modified form of cell cycle analysis. It is used to detect and measure apoptosis, a form of programmed cell death, by analysing cells with a DNA content less than 2n ("sub-G_1 cells"). Such cells are usually the result of apoptotic DNA fragmentation: during apoptosis, the DNA is degraded by cellular endonucleases. Therefore, nuclei of apoptotic cells contain less DNA than nuclei of healthy G_0/G_1 cells, resulting in a sub-G_0/G_1 peak in the fluorescence histogram that can be used to determine the relative amount of apoptotic cells in a sample.

This method was developed and first described in 1991 by Nicoletti and co-workers at Perugia University School of Medicine. An optimised protocol developed by two of the authors of the original publication was published in 2006.

Nicoletti assay with healthy (left) and apoptotic (middle and right) cells

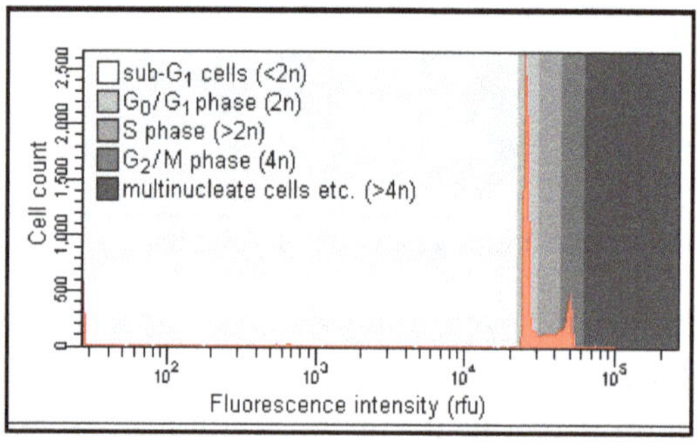

Healthy cells. Note the absence of a sub-G_0/G_1 peak.

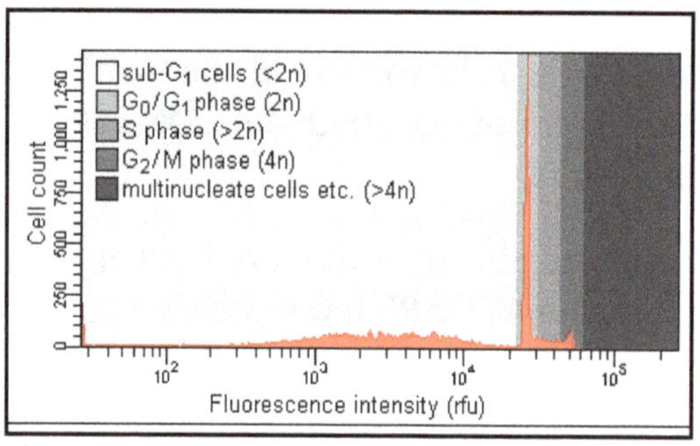

Apoptotic cells one day after apoptosis induction. Note the presence of a sub-G_0/G_1 peak.

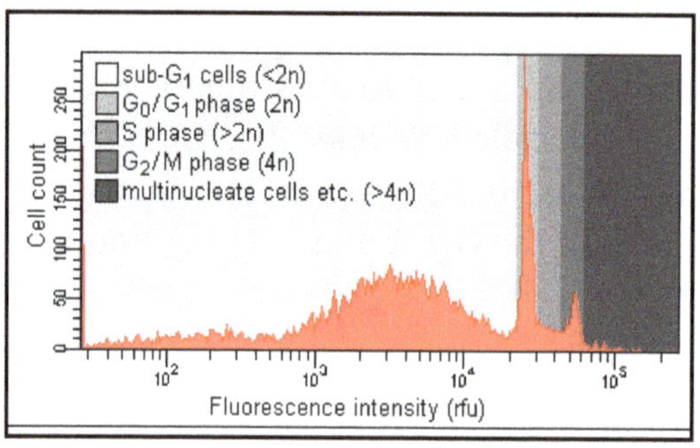

Apoptotic cells several days after apoptosis induction. Note the relative increase of the sub-G_0/G_1 peak.

Cell Division

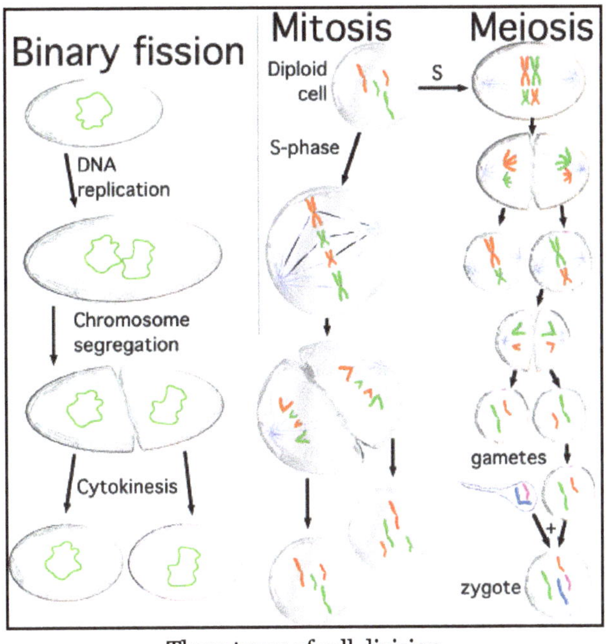

Three types of cell division.

Cell division is the process by which a parent cell divides into two or more daughter cells. Cell division usually occurs as part of a larger cell cycle. In eukaryotes, there are two distinct types of cell division: a vegetative division, whereby each daughter cell is genetically identical to the parent cell (mitosis), and a reproductive cell division, whereby the number of chromosomes in the daughter cells is reduced by half, to produce haploid gametes (meiosis). Meiosis results in four haploid daughter cells by undergoing one round of DNA replication followed by two divisions: homologous chromosomes are separated in the first division, and sister chromatids are separated in the second division. Both of these cell division cycles are used in sexually reproducing organisms at some point in their life cycle, and both are believed to be present in the last eukaryotic common ancestor. Prokaryotes also undergo a vegetative cell division known as binary fission, where their genetic material is segregated equally into two daughter cells. All cell divisions, regardless of organism, are preceded by a single round of DNA replication.

For simple unicellular organisms[Note 1] such as the amoeba, one cell division is equivalent to reproduction – an entire new organism is created. On a larger scale, mitotic cell division can create progeny from multicellular organisms, such as plants that grow from cuttings. Cell division also enables sexually reproducing organisms to develop from the one-celled zygote, which itself was produced by cell division from gametes. And after growth, cell division allows for continual construction and repair of the organism. A human being's body experiences about 10 quadrillion cell divisions in a lifetime.

The primary concern of cell division is the maintenance of the original cell's genome. Before division can occur, the genomic information that is stored in chromosomes must be replicated, and the duplicated genome must be separated cleanly between cells. A great deal of cellular infrastructure is involved in keeping genomic information consistent between "generations".

Variants

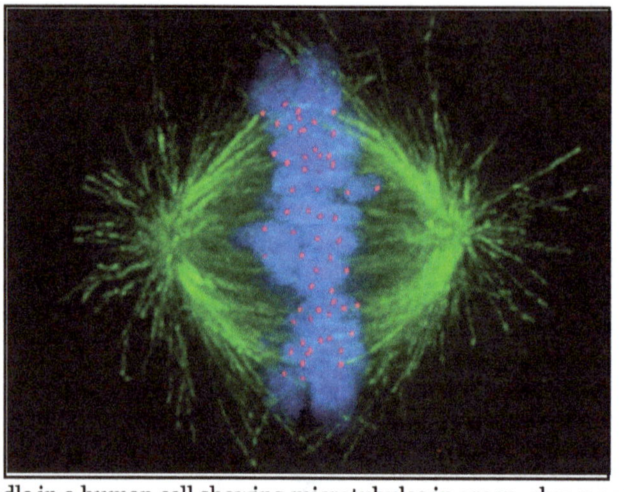

Image of the mitotic spindle in a human cell showing microtubules in green, chromosomes (DNA) in blue, and kinetochores in red.

Cells are classified into two main categories: simple, non-nucleated prokaryotic cells, and complex, nucleated eukaryotic cells. Owing to their structural differences, eukaryotic and prokaryotic cells do not divide in the same way. Also, the pattern of cell division that transforms eukaryotic stem cells into gametes (sperm cells in males or ova – egg cells – in females) is different from that of the somatic cell division in the cells of the body.

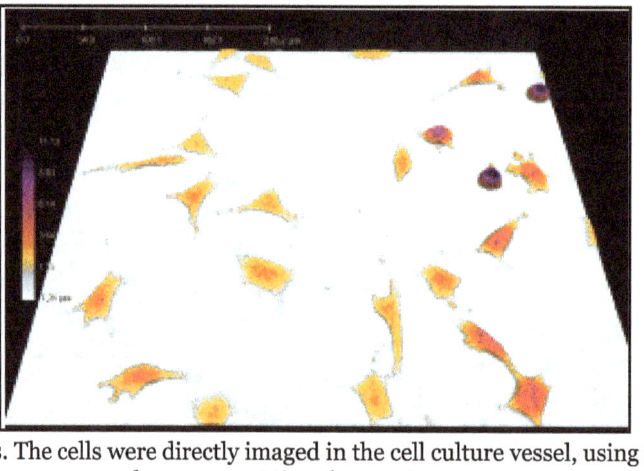

Cell division over 42 hours. The cells were directly imaged in the cell culture vessel, using non-invasive quantitative phase contrast time-lapse microscopy.

Degradation

Multicellular organisms replace worn-out cells through cell division. In some animals, however, cell division eventually halts. In humans this occurs on average, after 52 divisions, known as the Hayflick limit. The cell is then referred to as senescent. Cells stop dividing because the telomeres, protective bits of DNA on the end of a chromosome required for replication, shorten with each copy, eventually being consumed. Cancer cells, on the other hand, are not thought to degrade in this way, if at all. An enzyme called telomerase, present in large quantities in cancerous cells, rebuilds the telomeres, allowing division to continue indefinitely.

Cellular Differentiation

In developmental biology, cellular differentiation is the process where a cell changes from one cell type to another. Most commonly this is a less specialized type becoming a more specialized type, such as during cell growth. Differentiation occurs numerous times during the development of a multicellular organism as it changes from a simple zygote to a complex system of tissues and cell types. Differentiation continues in adulthood as adult stem cells divide and create fully differentiated daughter cells during tissue repair and during normal cell turnover. Some differentiation occurs in response to antigen exposure. Differentiation dramatically changes a cell's size, shape, membrane potential, metabolic activity, and responsiveness to signals. These changes are largely due to highly controlled modifications in gene expression and are the study of epigenetics. With a few exceptions, cellular differentiation almost never involves a change in the DNA sequence itself. Thus, different cells can have very different physical characteristics despite having the same genome.

A cell that can differentiate into all cell types of the adult organism is known as *pluripotent*. Such cells are called embryonic stem cells in animals and meristematic cells in higher plants. A cell that can differentiate into all cell types, including the placental tissue, is known as *totipotent*. In mammals, only the zygote and subsequent blastomeres are totipotent, while in plants many differentiated cells can become totipotent with simple laboratory techniques. In cytopathology, the level of cellular differentiation is used as a measure of cancer progression. "Grade" is a marker of how differentiated a cell in a tumor is.

Mammalian Cell Types

Three basic categories of cells make up the mammalian body: germ cells, somatic cells, and stem cells. Each of the approximately 100 trillion (10^{14}) cells in an adult human has its own copy or copies of the genome except certain cell types, such as red blood cells, that lack nuclei in their fully differentiated state. Most cells are diploid; they have two copies of each chromosome. Such cells, called somatic cells, make up most of the human body, such as skin and muscle cells. Cells differentiate to specialize for different functions.

Germ line cells are any line of cells that give rise to gametes—eggs and sperm—and thus are continuous through the generations. Stem cells, on the other hand, have the ability to divide for indefinite periods and to give rise to specialized cells. They are best described in the context of normal human development.

Development begins when a sperm fertilizes an egg and creates a single cell that has the potential to form an entire organism. In the first hours after fertilization, this cell divides into identical cells. In humans, approximately four days after fertilization and after several cycles of cell division, these cells begin to specialize, forming a hollow sphere of cells, called a blastocyst. The blastocyst has an outer layer of cells, and inside this hollow sphere, there is a cluster of cells called the inner cell mass. The cells of the inner cell mass go on to form virtually all of the tissues of the human body. Although the cells of the inner cell mass can form virtually every type of cell found in the human body, they cannot form an organism. These cells are referred to as pluripotent.

Pluripotent stem cells undergo further specialization into multipotent progenitor cells that then give rise to functional cells. Examples of stem and progenitor cells include:

- *Radial glial cells* (embryonic neural stem cells) that give rise to excitatory neurons in the fetal brain through the process of neurogenesis.
- Hematopoietic stem cells (adult stem cells) from the bone marrow that give rise to red blood cells, white blood cells, and platelets.
- *Mesenchymal stem cells* (adult stem cells) from the bone marrow that give rise to stromal cells, fat cells, and types of bone cells.
- *Epithelial stem cells* (progenitor cells) that give rise to the various types of skin cells.
- *Muscle satellite cells* (progenitor cells) that contribute to differentiated muscle tissue.

A pathway that is guided by the cell adhesion molecules consisting of four amino acids, arginine, glycine, asparagine, and serine, is created as the cellular blastomere differentiates from the single-layered blastula to the three primary layers of germ cells in mammals, namely the ectoderm, mesoderm and endoderm (listed from most distal (exterior) to proximal (interior)). The ectoderm ends up forming the skin and the nervous system, the mesoderm forms the bones and muscular tissue, and the endoderm forms the internal organ tissues.

Dedifferentiation

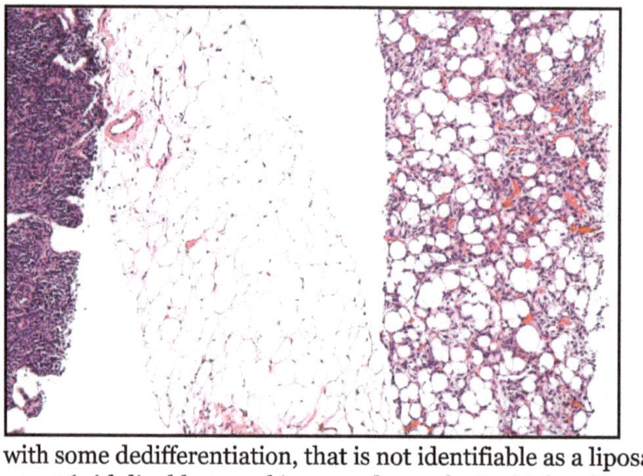

Micrograph of a liposarcoma with some dedifferentiation, that is not identifiable as a liposarcoma, (left edge of image) and a differentiated component (with lipoblasts and increased vascularity (right of image)). Fully differentiated (morphologically benign) adipose tissue (center of the image) has few blood vessels. H&E stain.

Dedifferentiation, or integration is a cellular process often seen in more basal life forms such as worms and amphibians in which a partially or terminally differentiated cell reverts to an earlier developmental stage, usually as part of a regenerative process. Dedifferentiation also occurs in plants. Cells in cell culture can lose properties they originally had, such as protein expression, or change shape. This process is also termed dedifferentiation.

Some believe dedifferentiation is an aberration of the normal development cycle that results in cancer, whereas others believe it to be a natural part of the immune response lost by humans at some point as a result of evolution.

A small molecule dubbed reversine, a purine analog, has been discovered that has proven to induce dedifferentiation in myotubes. These dedifferentiated cells could then redifferentiate into osteoblasts and adipocytes.

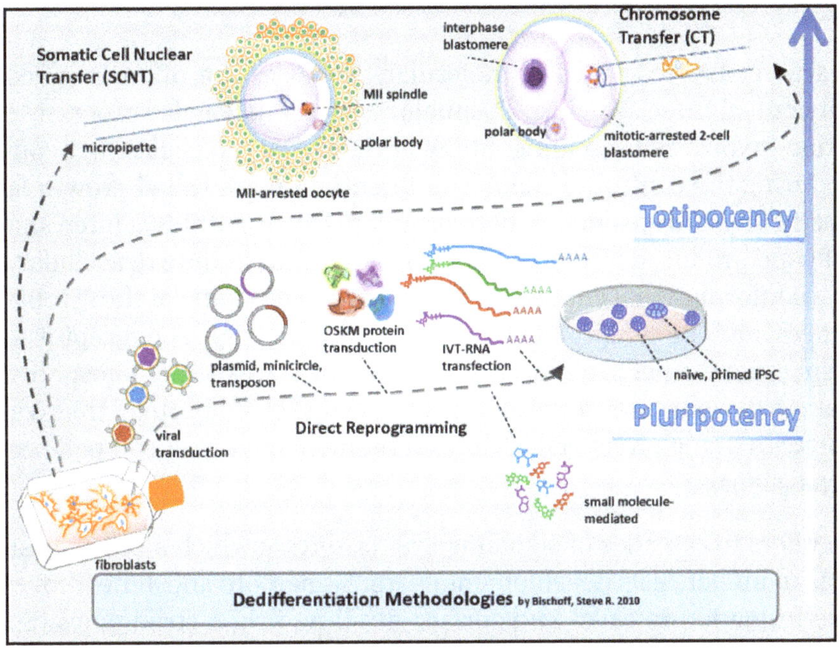

Diagram exposing several methods used to revert adult somatic cells to totipotency or pluripotency.

Mechanisms

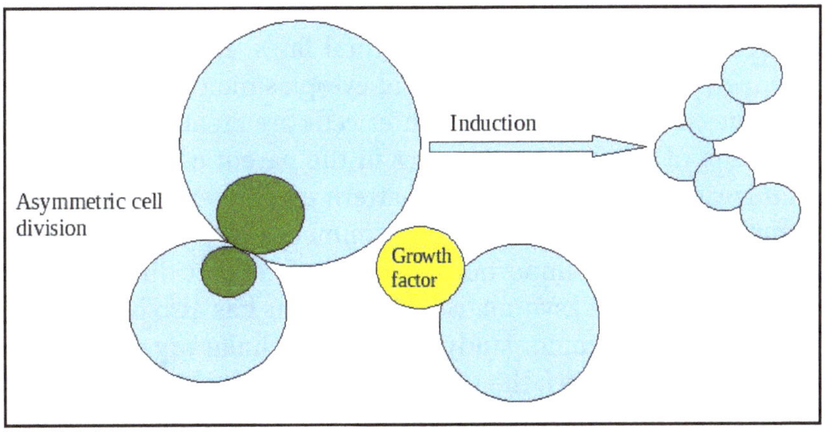

Mechanisms of cellular differentiation.

Each specialized cell type in an organism expresses a subset of all the genes that constitute the genome of that species. Each cell type is defined by its particular pattern of regulated gene expression. Cell differentiation is thus a transition of a cell from one cell type to another and it involves a switch from one pattern of gene expression to another. Cellular differentiation during development can be understood as the result of a gene regulatory network. A regulatory gene and its cis-regulatory modules are nodes in a gene regulatory network; they receive input and create output elsewhere in the network. The systems biology approach to developmental biology emphasizes the importance of investigating how developmental mechanisms interact to produce predictable patterns (morphogenesis). (However, an alternative view has been proposed recently. Based on stochastic gene expression, cellular differentiation is the result of a Darwinian selective process occurring among cells. In this frame, protein and gene networks are the result of cellular processes and not their cause.)

A few evolutionarily conserved types of molecular processes are often involved in the cellular mechanisms that control these switches. The major types of molecular processes that control cellular differentiation involve cell signaling. Many of the signal molecules that convey information from cell to cell during the control of cellular differentiation are called growth factors. Although the details of specific signal transduction pathways vary, these pathways often share the following general steps. A ligand produced by one cell binds to a receptor in the extracellular region of another cell, inducing a conformational change in the receptor. The shape of the cytoplasmic domain of the receptor changes, and the receptor acquires enzymatic activity. The receptor then catalyzes reactions that phosphorylate other proteins, activating them. A cascade of phosphorylation reactions eventually activates a dormant transcription factor or cytoskeletal protein, thus contributing to the differentiation process in the target cell. Cells and tissues can vary in competence, their ability to respond to external signals.

Signal induction refers to cascades of signaling events, during which a cell or tissue signals to another cell or tissue to influence its developmental fate. Yamamoto and Jeffery investigated the role of the lens in eye formation in cave- and surface-dwelling fish, a striking example of induction. Through reciprocal transplants, Yamamoto and Jeffery found that the lens vesicle of surface fish can induce other parts of the eye to develop in cave- and surface-dwelling fish, while the lens vesicle of the cave-dwelling fish cannot.

Other important mechanisms fall under the category of asymmetric cell divisions, divisions that give rise to daughter cells with distinct developmental fates. Asymmetric cell divisions can occur because of asymmetrically expressed maternal cytoplasmic determinants or because of signaling. In the former mechanism, distinct daughter cells are created during cytokinesis because of an uneven distribution of regulatory molecules in the parent cell; the distinct cytoplasm that each daughter cell inherits results in a distinct pattern of differentiation for each daughter cell. A well-studied example of pattern formation by asymmetric divisions is body axis patterning in Drosophila. RNA molecules are an important type of intracellular differentiation control signal. The molecular and genetic basis of asymmetric cell divisions has also been studied in green algae of the genus *Volvox*, a mo del system for studying how unicellular organisms can evolve into multicellular organisms. In *Volvox carteri*, the 16 cells in the anterior hemisphere of a 32-cell embryo divide asymmetrically, each producing one large and one small daughter cell. The size of the cell at the end of all cell divisions determines whether it becomes a specialized germ or somatic cell.

Epigenetic Control Of Cellular Differentiation

Since each cell, regardless of cell type, possesses the same genome, determination of cell type must occur at the level of gene expression. While the regulation of gene expression can occur through cis- and trans-regulatory elements including a gene's promoter and enhancers, the problem arises as to how this expression pattern is maintained over numerous generations of cell division. As it turns out, epigenetic processes play a crucial role in regulating the decision to adopt a stem, progenitor, or mature cell fate. This section will focus primarily on mammalian stem cells.

In systems biology and mathematical modeling of gene regulatory networks, cell-fate determination is predicted to exhibit certain dynamics, such as attractor-convergence (the attractor can be an equilibrium point, limit cycle or strange attractor) or oscillatory.

Importance of Epigenetic Control

The first question that can be asked is the extent and complexity of the role of epigenetic processes in the determination of cell fate. A clear answer to this question can be seen in the 2011 paper by Lister R, *et al.* on aberrant epigenomic programming in human induced pluripotent stem cells. As induced pluripotent stem cells (iPSCs) are thought to mimic embryonic stem cells in their pluripotent properties, few epigenetic differences should exist between them. To test this prediction, the authors conducted whole-genome profiling of DNA methylation patterns in several human embryonic stem cell (ESC), iPSC, and progenitor cell lines.

Female adipose cells, lung fibroblasts, and foreskin fibroblasts were reprogrammed into induced pluripotent state with the OCT4, SOX2, KLF4, and MYC genes. Patterns of DNA methylation in ESCs, iPSCs, somatic cells were compared. Lister R, *et al.* observed significant resemblance in methylation levels between embryonic and induced pluripotent cells. Around 80% of CG dinucleotides in ESCs and iPSCs were methylated, the same was true of only 60% of CG dinucleotides in somatic cells. In addition, somatic cells possessed minimal levels of cytosine methylation in non-CG dinucleotides, while induced pluripotent cells possessed similar levels of methylation as embryonic stem cells, between 0.5 and 1.5%. Thus, consistent with their respective transcriptional activities, DNA methylation patterns, at least on the genomic level, are similar between ESCs and iPSCs.

However, upon examining methylation patterns more closely, the authors discovered 1175 regions of differential CG dinucleotide methylation between at least one ES or iPS cell line. By comparing these regions of differential methylation with regions of cytosine methylation in the original somatic cells, 44-49% of differentially methylated regions reflected methylation patterns of the respective progenitor somatic cells, while 51-56% of these regions were dissimilar to both the progenitor and embryonic cell lines. In vitro-induced differentiation of iPSC lines saw transmission of 88% and 46% of hyper and hypo-methylated differentially methylated regions, respectively.

Two conclusions are readily apparent from this study. First, epigenetic processes are heavily involved in cell fate determination, as seen from the similar levels of cytosine methylation between induced pluripotent and embryonic stem cells, consistent with their respective patterns of transcription. Second, the mechanisms of de-differentiation (and by extension, differentiation) are very complex and cannot be easily duplicated, as seen by the significant number of differentially methylated regions between ES and iPS cell lines. Now that these two points have been established, we can examine some of the epigenetic mechanisms that are thought to regulate cellular differentiation.

Mechanisms of Epigenetic Regulation

Pioneering Factors (Oct4, Sox2, Nanog)

Three transcription factors, OCT4, SOX2, and NANOG – the first two of which are used in iPSC reprogramming – are highly expressed in undifferentiated embryonic stem cells and are necessary for the maintenance of their pluripotency. It is thought that they achieve this through alterations in chromatin structure, such as histone modification and DNA methylation, to restrict or permit the transcription of target genes.

Polycomb Repressive Complex (PRC2)

In the realm of gene silencing, Polycomb repressive complex 2, one of two classes of the Polycomb group (PcG) family of proteins, catalyzes the di- and tri-methylation of histone H3 lysine 27 (H3K27me2/me3). By binding to the H3K27me2/3-tagged nucleosome, PRC1 (also a complex of PcG family proteins) catalyzes the mono-ubiquitinylation of histone H2A at lysine 119 (H2AK119Ub1), blocking RNA polymerase II activity and resulting in transcriptional suppression. PcG knockout ES cells do not differentiate efficiently into the three germ layers, and deletion of the PRC1 and PRC2 genes leads to increased expression of lineage-affiliated genes and unscheduled differentiation. Presumably, PcG complexes are responsible for transcriptionally repressing differentiation and development-promoting genes.

Trithorax Group Proteins (TrxG)

Alternately, upon receiving differentiation signals, PcG proteins are recruited to promoters of pluripotency transcription factors. PcG-deficient ES cells can begin differentiation but cannot maintain the differentiated phenotype. Simultaneously, differentiation and development-promoting genes are activated by Trithorax group (TrxG) chromatin regulators and lose their repression. TrxG proteins are recruited at regions of high transcriptional activity, where they catalyze the trimethylation of histone H3 lysine 4 (H3K4me3) and promote gene activation through histone acetylation. PcG and TrxG complexes engage in direct competition and are thought to be functionally antagonistic, creating at differentiation and development-promoting loci what is termed a “bivalent domain” and rendering these genes sensitive to rapid induction or repression.

DNA Methylation

Regulation of gene expression is further achieved through DNA methylation, in which the DNA methyltransferase-mediated methylation of cytosine residues in CpG dinucleotides maintains heritable repression by controlling DNA accessibility. The majority of CpG sites in embryonic stem cells are unmethylated and appear to be associated with H3K4me3-carrying nucleosomes. Upon differentiation, a small number of genes, including OCT4 and NANOG, are methylated and their promoters repressed to prevent their further expression. Consistently, DNA methylation-deficient embryonic stem cells rapidly enter apoptosis upon in vitro differentiation.

Nucleosome Positioning

While the DNA sequence of most cells of an organism is the same, the binding patterns of transcription factors and the corresponding gene expression patterns are different. To a large extent, differences in transcription factor binding are determined by the chromatin accessibility of their binding sites through histone modification and/or pioneer factors. In particular, it is important to know whether a nucleosome is covering a given genomic binding site or not. Recent studies have elucidated the role of nucleosome positioning during stem cell development.

Role of Signaling in Epigenetic Control

A final question to ask concerns the role of cell signaling in influencing the epigenetic processes governing differentiation. Such a role should exist, as it would be reasonable to think that extrinsic signaling can lead to epigenetic remodeling, just as it can lead to changes in gene expression through the activation or repression of different transcription factors. Interestingly, little direct data is available concerning the specific signals that influence the epigenome, and the majority of current knowledge consist of speculations on plausible candidate regulators of epigenetic remodeling. We will first discuss several major candidates thought to be involved in the induction and maintenance of both embryonic stem cells and their differentiated progeny, and then turn to one example of specific signaling pathways in which more direct evidence exists for its role in epigenetic change.

The first major candidate is Wnt signaling pathway. The Wnt pathway is involved in all stages of differentiation, and the ligand Wnt3a can substitute for the overexpression of c-Myc in the generation of induced pluripotent stem cells. On the other hand, disruption of ß-catenin, a component of the Wnt signaling pathway, leads to decreased proliferation of neural progenitors.

Growth factors comprise the second major set of candidates of epigenetic regulators of cellular differentiation. These morphogens are crucial for development, and include bone morphogenetic proteins, transforming growth factors (TGFs), and fibroblast growth factors (FGFs). TGFs and FGFs have been shown to sustain expression of OCT4, SOX2, and NANOG by downstream signaling to Smad proteins. Depletion of growth factors promotes the differentiation of ESCs, while genes with bivalent chromatin can become either more restrictive or permissive in their transcription.

Several other signaling pathways are also considered to be primary candidates. Cytokine leukemia inhibitory factors are associated with the maintenance of mouse ESCs in an undifferentiated state. This is achieved through its activation of the Jak-STAT3 pathway, which has been shown to be necessary and sufficient towards maintaining mouse ESC pluripotency. Retinoic acid can induce differentiation of human and mouse ESCs, and Notch signaling is involved in the proliferation and self-renewal of stem cells. Finally, Sonic hedgehog, in addition to its role as a morphogen, promotes embryonic stem cell differentiation and the self-renewal of somatic stem cells.

The problem, of course, is that the candidacy of these signaling pathways was inferred primarily on the basis of their role in development and cellular differentiation. While epigenetic regulation is necessary for driving cellular differentiation, they are certainly not sufficient for this process. Direct modulation of gene expression through modification of transcription factors plays a key role that must be distinguished from heritable epigenetic changes that can persist even in the absence of the original environmental signals. Only a few examples of signaling pathways leading to epigenetic changes that alter cell fate currently exist, and we will focus on one of them.

Expression of Shh (Sonic hedgehog) upregulates the production of BMI1, a component of the PcG complex that recognizes H3K27me3. This occurs in a Gli-dependent manner, as Gli1 and Gli2 are downstream effectors of the Hedgehog signaling pathway. In culture, Bmi1 mediates the Hedgehog pathway's ability to promote human mammary stem cell self-renewal. In both humans and mice, researchers showed Bmi1 to be highly expressed in proliferating immature cerebellar granule cell precursors. When Bmi1 was knocked out in mice, impaired cerebellar development resulted, leading to significant reductions in postnatal brain mass along with abnormalities in motor control and behavior. A separate study showed a significant decrease in neural stem cell proliferation along with increased astrocyte proliferation in Bmi null mice.

In summary, the role of signaling in the epigenetic control of cell fate in mammals is largely unknown, but distinct examples exist that indicate the likely existence of further such mechanisms.

Effect of Matrix Elasticity on Differentiation

In order to fulfill the purpose of regenerating a variety of tissues, adult stems are known to migrate from their niches, adhere to new extracellular matrices (ECM) and differentiate. The ductility of these microenvironments are unique to different tissue types. The ECM surrounding brain, muscle and bone tissues range from soft to stiff. The transduction of the stem cells into these cells types is not directed solely by chemokine cues and cell to cell signaling. The elasticity of the microenvironment can also affect the differentiation of mesenchymal stem cells (MSCs which originate in bone marrow.) When MSCs are placed on substrates of the same stiffness as brain, muscle and bone ECM, the MSCs take on properties of those respective cell types. Matrix sensing requires the cell to pull against the matrix at focal adhesions, which triggers a cellular mechano-transducer to generate a signal to be informed what force is needed to deform the matrix. To determine the key players in matrix-elasticity-driven lineage specification in MSCs, different matrix microenvironments were mimicked. From these experiments, it was concluded that focal adhesions of the MSCs were the cellular mechano-transducer sensing the differences of the matrix elasticity. The non-muscle myosin IIa-c isoforms generates the forces in the cell that lead to signaling of early commitment markers. Nonmuslce myosin IIa generates the least force increasing to non-muscle myosin IIc. There are also factors in the cell that inhibit non-muscle myosin II, such as blebbistatin. This makes the cell effectively blind to the surrounding matrix. Researchers have obtained some success in inducing stem cell-like properties in HEK 239 cells by providing a soft matrix without the use of diffusing factors. The stem-cell properties appear to be linked to tension in the cells' actin network. One identified mechanism for matrix-induced differentiation is tension-induced proteins, which remodel chromatin in response to mechanical stretch. The RhoA pathway is also implicated in this process.

References

- Griffiths, Anthony J.F.; Wessler, Susan R.; Carroll, Sean B.; Doebley, John (2012). Introduction to Genetic Analysis (10 ed.). New York: W.H. Freeman and Company. p. 35. ISBN 978-1-4292-2943-2
- Alberts B, Johnson A, Lewis J, Raff M, Roberts K, Walter P (2002). Molecular Biology of the Cell. Garland Science. ISBN 0-8153-3218-1
- Iourov, I.Y.; Vorsanova, S.G.; Yurov, Y.B. (2006). "Chromosomal Variations in Mammalian Neuronal Cells: Known Facts and Attractive Hypotheses". In Jeon, Kwang J. International Review Of Cytology: A Survey of Cell Biology. 249. Waltham, MA: Academic Press. p. 146. ISBN 9780080463506
- Stocum DL (2004). "Amphibian regeneration and stem cells". Curr. Top. Microbiol. Immunol. Current Topics in Microbiology and Immunology. 280: 1–70. doi:10.1007/978-3-642-18846-6_1. ISBN 978-3-540-02238-1. PMID 14594207

- Alberts B, Johnson A, Lewis J, Raff M, Roberts K, Walter P (2002). Molecular Biology of the Cell. Garland Science. ISBN 0-8153-3218-1
- Robbins, Stanley L; Cotran, Ramzi S (2004). Vinay Kumar; Abul K Abbas; Nelson Fausto, eds. Pathological Basis of Disease. Elsevier. ISBN 81-8147-528-3
- Knisely, Karen; Gilbert, Scott F. (2009). Developmental Biology (8th ed.). Sunderland, Mass: Sinauer Associates. p. 147. ISBN 0-87893-371-9
- Olson, Mark O. J. (2011). The Nucleolus. Volume 15 of Protein Reviews. Berlin: Springer Science & Business Media. p. 15. ISBN 9781461405146
- Robert.S Hine, ed. (2008). Oxford Dictionary Biology (6th ed.). New York: Oxford University Press. p. 113. ISBN 978-0-19-920462-5

Permissions

All chapters in this book are published with permission under the Creative Commons Attribution Share Alike License or equivalent. Every chapter published in this book has been scrutinized by our experts. Their significance has been extensively debated. The topics covered herein carry significant information for a comprehensive understanding. They may even be implemented as practical applications or may be referred to as a beginning point for further studies.

We would like to thank the editorial team for lending their expertise to make the book truly unique. They have played a crucial role in the development of this book. Without their invaluable contributions this book wouldn't have been possible. They have made vital efforts to compile up to date information on the varied aspects of this subject to make this book a valuable addition to the collection of many professionals and students.

This book was conceptualized with the vision of imparting up-to-date and integrated information in this field. To ensure the same, a matchless editorial board was set up. Every individual on the board went through rigorous rounds of assessment to prove their worth. After which they invested a large part of their time researching and compiling the most relevant data for our readers.

The editorial board has been involved in producing this book since its inception. They have spent rigorous hours researching and exploring the diverse topics which have resulted in the successful publishing of this book. They have passed on their knowledge of decades through this book. To expedite this challenging task, the publisher supported the team at every step. A small team of assistant editors was also appointed to further simplify the editing procedure and attain best results for the readers.

Apart from the editorial board, the designing team has also invested a significant amount of their time in understanding the subject and creating the most relevant covers. They scrutinized every image to scout for the most suitable representation of the subject and create an appropriate cover for the book.

The publishing team has been an ardent support to the editorial, designing and production team. Their endless efforts to recruit the best for this project, has resulted in the accomplishment of this book. They are a veteran in the field of academics and their pool of knowledge is as vast as their experience in printing. Their expertise and guidance has proved useful at every step. Their uncompromising quality standards have made this book an exceptional effort. Their encouragement from time to time has been an inspiration for everyone.

The publisher and the editorial board hope that this book will prove to be a valuable piece of knowledge for students, practitioners and scholars across the globe.

Index

A

Active Transport, 76, 91-92, 94-99, 139

Adenosine Triphosphate, 32, 115, 142, 144, 155

Alginic Acid, 106

Amino Acid, 50, 52, 54-58, 62, 72, 75, 81-82, 86-89, 96-99, 101-102, 127, 132, 138, 192

Apoptosis, 129, 133, 135, 148-149, 154, 195-197, 199-200, 205, 207, 209, 211, 222, 230

Archaebacteria, 10

Autophagy, 17-19, 190-200

B

Binary Fission, 157-158, 171, 204, 211, 223

C

Cell Culture, 1, 174-177, 179-181, 226

Cell Cycle, 23, 125, 129, 132-134, 143, 155, 157, 163, 178, 201-210, 213, 220, 222-223, 232

Cell Division, 14, 17-19, 24, 26, 35, 37-38, 45, 47, 63, 108, 122, 129-130, 132, 139, 142, 155, 157, 171, 180, 201-203, 205, 209-210, 213-216, 223-225, 227-228

Cell Organelles, 29-30, 108, 143, 191

Cell Polarity, 19-21, 71-72

Cell Theory, 166, 170-171

Cell Wall, 3, 11-13, 30, 34, 38, 80-81, 95, 105-107, 165, 173, 216

Cellular Process, 226

Cellular Respiration, 144, 151

Central Vacuole, 112

Centrioles, 17-20, 34-35, 212-213

Centrosome, 18, 20-21, 34, 129, 214

Chemiosmosis, 153

Chromatin, 120, 122, 125, 129, 132-134, 139-141, 209, 213, 230-232

Chromosomes, 13-14, 18, 30-31, 102, 118, 126, 128-129, 133-134, 139, 141-142, 156, 162, 174, 201, 203-204, 209-220, 223-224

Cilia, 15, 19, 25, 34-35, 101-102

Citric Acid Cycle, 6-7, 147, 149, 151-152

Cyanobacteria, 11, 210

Cytoplasm, 11-12, 26, 28, 30, 35-36, 47-55, 59-61, 66-67, 78, 88, 91, 96-104, 108-112, 125-129, 132, 141-142, 145, 153, 157, 161, 171, 174, 191, 196, 201, 204, 212-213, 228

Cytoplasmic Matrix, 111, 117, 137, 140

Cytoplasmic Streaming, 108

Cytosine, 229-230

Cytoskeleton, 12-13, 20-25, 30, 33-34, 37-38, 109, 114, 116-118, 151, 194

Cytosol, 12-13, 22-23, 47-50, 57, 66-67, 93-95, 98, 108-117, 119, 136-137, 139-140, 146, 149, 151, 193

D

Daughter Cells, 17-19, 129-130, 157, 201-202, 205-206, 209, 211, 217-218, 223, 225, 228

Deoxyribonucleic Acid, 26, 142

DNA Double Helix, 27

DNA Replication, 124, 132-134, 201-202, 204, 206, 208, 223, 233

E

Electron Microscopy, 2, 125, 143, 157, 165

Electron Transport Chain, 147, 152-153, 159

Endocrine Cells, 91, 154

Endocytosis, 63, 116, 145

Endoplasmic Reticulum, 15-16, 30-31, 33, 41-42, 95, 101-102, 110, 120, 132, 137, 139-140, 146, 151, 154, 174

Energy Metabolism, 13, 154

Enzymes, 4-7, 12, 27, 30, 32-33, 37-38, 47, 49-50, 61-62, 64-65, 71-73, 96-97, 111, 114-115, 126-127, 136, 138, 141, 144, 146-149, 154-156, 164-166, 176, 179, 191, 206

Epithelial Cell, 71-72, 102, 182

Escherichia Coli, 11

Eukaryotes, 10, 12-13, 28, 33-34, 37-38, 110-111, 114, 116-117, 129, 131, 144-145, 156-157, 174-175, 192, 201, 203, 205, 212, 223

Eukaryotic Cells, 9-10, 12-14, 28-31, 47-49, 117, 126, 131, 144-145, 174, 204, 211, 224

F

Fluorescence Microscopy, 2, 148, 157

G

Gene Expression, 24, 118, 120, 125-127, 139, 142, 180, 207-208, 225, 227-228, 230-231

Gene Transfer, 180

Genome, 14, 24, 26, 31, 50, 114, 118, 133, 139, 143-144, 147, 155-156, 161-162, 206, 208, 211, 216, 223, 225, 227-229

Glycolysis, 4-6, 8, 72-73, 108, 116, 126, 138, 145, 152-153, 197-199

Glycoproteins, 13, 40-41, 47-49, 107

Golgi Apparatus, 17-18, 20-21, 30, 33, 110, 137, 216

Guanine, 136

H

Haploid Gametes, 223

Host Cell, 102, 147, 182

I

Intestinal Epithelium, 210

Intracellular Fluid, 111-112, 117

K

Kinetochores, 133, 219, 224

L

Lipid Bilayer, 24, 30, 35-36, 39-42, 44-58, 61-72, 74, 76, 78-83, 89-91, 150, 172-173

Lipid Rafts, 64-65

Lipids, 13, 22, 26-27, 31, 39-55, 62-67, 107, 110, 116-117, 137, 143, 146, 155

Lysosomes, 13, 19, 96, 110, 137, 174, 191, 194, 196

M

Macromolecules, 6-7, 37-38, 110-111, 113-114, 120, 127-128, 135, 138

Meiosis, 133, 223

Membrane Proteins, 24-26, 40-42, 45-59, 61-73, 75, 80-81, 83, 91-92, 132-133, 135, 139, 146

Metabolic Pathway, 6, 8, 115

Microscopy, 1-2, 68-69, 108, 111, 124-125, 139, 143, 148, 157, 165, 180, 224

Mitochondria, 13, 17-18, 20, 22-23, 29-32, 37-38, 45-46, 58-59, 108, 110-111, 115, 122, 130-131, 137-138, 142-162, 174, 191, 193-194, 196-199, 204

Mitophagy, 196-200

Mitosis, 2, 18, 116, 119, 125, 129, 133-135, 201-206, 209-219, 223

Mitotic Spindle, 17-19, 128-129, 133, 206, 213-215, 217-219, 224

Molecular Biology, 1, 64, 180, 192, 233

Morphology, 3, 9, 22-25, 157, 178, 183-190, 211

N

Natural Selection, 158, 178

Nuclear Envelope, 20, 22, 118-120, 123, 126, 129, 132-134, 139-140, 204, 206, 211-213, 215, 219

Nuclear Lamina, 108, 118, 121, 129, 132-135, 139

Nuclear Transport, 108, 118, 120, 127-128, 135-136

Nucleic Acids, 1, 6, 26-28, 91, 93, 113, 120, 141, 165

Nucleolus, 30, 118, 120, 122-123, 125, 139-142, 212, 215, 233

Nucleotide Sequence, 56-59

Nucleus, 9, 11-13, 17-21, 27-28, 30-31, 33, 103, 108, 111, 114, 116-120, 123-133, 135-144, 147, 154, 156-157, 169, 171, 173, 201, 203-204, 210-211, 213-215, 219

O

Oxidative Stress, 153-154, 159-160, 197, 199

P

Peroxisomes, 13, 18-19

Plasma Membrane, 10-12, 19-21, 25-27, 30, 33, 35-42, 44-45, 47-50, 53-57, 61-62, 64-76, 78-84, 86-87, 91-93, 95-99, 101-104, 114, 116, 145, 172-174, 213

Polyethylene Glycol, 47-49, 66-67

Polysaccharide, 107

Prokaryotic Cells, 9, 12, 14, 29, 64, 144, 204, 211, 224

Prophase, 129, 133, 139, 204, 206, 211, 213, 215, 219-220

Protein Complex, 192

R

Ribonucleic Acid, 26, 142

Ribosomes, 2, 12, 17-18, 30, 32-33, 108, 110, 116, 118, 120, 126-127, 130, 140-142, 144-145, 147, 174, 204, 212

S

Single-celled Organisms, 11, 13

Sister Chromatids, 129, 204, 212, 218, 223

Somatic Cells, 140, 225, 227, 229

Stroma, 32, 184

T

Thylakoid Membranes, 37-38

Tissue Culture, 1, 174-175, 177-178, 181

Transmembrane Domain, 52, 54-55, 57-59, 93-94

V

Vacuoles, 13, 35, 96, 108, 110, 174, 193

Viscoelastic Fluid, 111

Z

Zygote, 157, 159, 216, 223, 225

www.ingramcontent.com/pod-product-compliance
Lightning Source LLC
LaVergne TN
LVHW080904210926
848485LV00004B/592

* 9 7 8 1 6 3 9 8 9 4 3 5 2 *